# Light Emitting Diodes for Agriculture

S. Dutta Gupta
Editor

# Light Emitting Diodes for Agriculture

## Smart Lighting

Springer

*Editor*
S. Dutta Gupta
Department of Agricultural and Food
Engineering
Indian Institute of Technology Kharagpur
Kharagpur
India

ISBN 978-981-10-5806-6 ISBN 978-981-10-5807-3 (eBook)
DOI 10.1007/978-981-10-5807-3

Library of Congress Control Number: 2017946641



Printed on acid-free paper

This Springer imprint is published by Springer Nature
The registered company is Springer Nature Singapore Pte Ltd.
The registered company address is: 152 Beach Road, #21-01/04 Gateway East, Singapore 189721, Singapore

Dedicated to the Memory of
My Beloved Wife 'Rina'

Let the light shine!

# Foreword

We are at the beginning of a technological revolution that will have immense long-term impact on all of our lives. The majority of all of the lighting in the world is transitioning from conventional lighting technologies: incandescent, fluorescent, metal-halide, and low- and high-pressure sodium to LED lighting. In the USA., LED lighting technology is projected to reduce the total energy budget, which includes all primary energy consumption, by 5% by 2035. This is a massive energy saving that equates to about $50B per year in energy savings in 2035, not to mention all of the benefits of $CO_2$ reduction associated with this savings.

While the initial driver for this shift was improved energy efficiency and resulting energy savings, the value proposition for LED lighting technology has moved well beyond this initial and important benefit. Not only is LED technology more efficient than conventional sources, it is longer lived and can provide improved lighting performance across the board. Due to their improved efficiency, LEDs run cooler reducing thermal load on heating, ventilation, and air-conditioning "HVAC" systems. They have smaller optical source size, enabling improved control of optical distribution. They can last 50,000 h or more. They can be turned on and off instantaneously, and they are fundamentally dimmable. Finally, the spectral power density of the emitted light can be finely engineered and even made to be actively tunable. At this time, early 2017, most products do not fully engage all of these advancements due to cost, form factor, or engineering trade-offs, but consumers are learning to expect more and developers of LED technology are rapidly improving the lighting value with fewer compromises out of their lighting products.

The same technology advancements that are improving general illumination are also being applied to other lighting applications, in particular the use of LED lighting for controlled environment agriculture. LED lighting technology enables a more highly controlled growth environment that can improve productivity and control of the horticultural product. LED lighting may even enable new crops to be effectively produced in controlled environments. New levels of control over spectral power distribution, optical intensity distribution, form factor, and active color tuning can be used to tailor the light to specific crops, improve productivity, and control aspects of the plant growth such as height, bushiness, and color or nutritional content. As these new levels of control are being explored for various plant growth and development applications, increasing the value of the light, the cost of LED lighting products continues to decrease.

Not only can the features of LED lighting be used to improve production but the new control can also be used as a highly configurable research tool to refine our knowledge of plant physiological responses to light at a rapid pace. This book serves to connect the latest research in plant and biological responses to light with developments in LED lighting technology. There is a vast range of plant physiological responses to light for a vast range of plant species and cultivars. And now we have a vast range of control over the light they experience in terms of color, intensity, optical distribution, and changes in these factors over time. Understanding and harnessing the impacts of LED lighting on agriculture requires a long-term research effort. This book provides a range of research results in terms of lighting attributes, plant and cellular physiological responses, and even economics of lighting for controlled environment agriculture. Configurable LED lighting is now relatively inexpensive, allowing for researchers across the globe to conduct meaningful experiments and add to the body of knowledge for this important topic. Academic, commercial, and neophyte researchers can use the research described in this book as a starting point for their own research efforts.

This book contains fourteen chapters, contributed by pioneers who are leading the emergence of LED technology for controlled environment agriculture across the globe. The chapters follow a sequence from fundamental features of LED, their use as supplemental lighting system, economics and various applications in controlled environment agriculture and their role in regulating plant morphogenesis both in vivo and in vitro. I am confident that the present book will motivate plant scientist and biotechnologists to enter into this fascinating field of application of semiconductor lighting technology for the improvement of plant growth and development.

The use of LED lighting for agricultural/horticultural applications has profound implications for our world. LED lighting is a key and enabling component of controlled environment agriculture, which allows for growth of crops in new regions of the world at any time of year. This changes how crops and growth locations are chosen with respect to targeted markets. Energy, water, chemical, and nutrient inputs for plant growth are also dramatically changed with controlled environment agriculture. The long-term impacts on our global food supply are likely to be more localized production, increased self sufficiency, more nutritious produce available year-round, and increased opportunity for consistent small-scale food production, just to name a few of the likely impacts. While the full global impact of LED-enabled controlled environment agriculture with the knowledge of role of light in plant morphogenesis is difficult to anticipate, LED-regulated plant growth and development are certainly poised to play an expanded role in how the world gets its food and understanding the concepts put forth in this book will be critical to making this vision a reality.

P. Morgan Pattison, Ph.D.
President and Founder of
Solid State Lighting Services, Inc.
Senior Technical Advisor
United States Department of Energy
Solid State Lighting R&D Program
Washington, USA

# Preface

Light plays a pivotal role in regulating plant growth and development. Both quality and intensity of light as well as the photoperiod are very critical for plant morphogenesis. The significance of plant photoreceptors as key regulatory proteins that govern metabolic events and developmental changes within plants has been well documented. Complex, multiple photoreceptor systems respond to light and thereby regulate plant morphogenetic changes, functioning of the photosynthetic apparatus, and the trend of metabolic reactions. Moreover, photooxidative changes evoked by lighting condition may lead to the altered action of antioxidant defense system. Thus in combination with other agro-technical means, light, creating the mild photo-stress, might be an effective tool for phytochemical rich plant cultivation.

Crop failure due to unpredictable climate change is a matter of global concern. Threats such as pest attacks and diseases further aggravate the uncertainty of crop yields. Geo-climatic limitations of traditional agriculture and its dependence on environmentally hazardous fertilizers and pesticides have impelled the advancements in controlled environment farming techniques. The concept of controlled environment agriculture in greenhouses and closed plant production system has emerged as a reliable and sustainable alternative means of crop production. These "plant factories" for vertical farming are now becoming an indispensable part of the global food security system. However, the feasibility and sustainability of such systems are largely dependent on the power requirements. The large power requirements mainly from the electric lamps that provide the actinic light which drives the light reactions of photosynthesis, accounting for 40% of the recurring cost of plant factories, are the major bottlenecks to make controlled environment agriculture profitable.

The light source generally used for controlled environment agriculture is fluorescent light, metal-halide, high-pressure sodium, and incandescent lamps. Among them, fluorescent lamp has been the most popular. However, these lighting systems have a wide range of wavelengths from 350 to 750 nm and are of low quality for promoting plant growth and development. They also emit light with low photosynthetic photon flux and had limited lifetime of operation which restricts their utilization in plant lighting systems when the goal is to sustain high crop productivity.

The steady development of the light-emitting diode (LED) technology with the emergence of new types of semiconductor materials has made it possible to apply it in an increasing number of new areas including plant growth and development. As an alternative to conventional lighting system, LED has been demonstrated to be an artificial smart lighting source for controlled environment agriculture and in vitro studies of plant morphogenesis. Various morphological, anatomical, and physiological attributes of plants grown both in vivo and in vitro have found to be regulated by spectral properties of LED. Apart from its regulatory role in plant growth and development, LED affects the amplification of functional components which contribute toward the selective control of antioxidative attributes. Since the LED emits over specific spectral regions, they can be used to regulate the levels of photosynthetically active and photomorphogenic radiation necessary for plant growth and development. This feature allows implementation of LED with specific spectral ranges that are involved in plant responses and also ensures the independent control of each spectral range and precise manipulation of spectral quality and light intensity. The flexibility of matching wavelengths of LED to plant photoreceptors may provide optimal production influencing plant morphology and metabolism. These solid-state light sources are therefore ideal for use in plant lighting designs for controlled environment agriculture as well as for studies on photomorphogenesis.

The present book aims to present a comprehensive treatise on the advancements made in the use of LEDs for sustainable crop production and to describe research achievements on photomorphogenesis. This book introduces readers to the fundamentals and design features of LEDs applicable for plant growth and development and illustrates their various advantages over the traditional lighting systems with cost analysis. It contains 14 chapters, and organizes the information in order to present a wide spectrum of applications of LEDs covering a diverse domain of plant sciences relevant to controlled environment agriculture and in vitro plant morphogenesis. The scope of this book has been expanded by including chapters that deal with the role of LEDs in regulating cellular redox balance, nutritional quality, and gene expression. The chapters are written by a team of international experts who are pioneers, and have made significant achievements in this emerging interdisciplinary enterprise. I am indebted to the chapter contributors for sharing their research outcomes and kind support. I am grateful to Dr. P. Morgan Pattison for sparing his valuable time to write the "Foreword." Thanks are also due to Mr. Arjun Karmakar and Ms. Nirlipta Saha for their help in checking the cited references.

It is the invisible inspiration and encouragement of my beloved wife Rina (Dr. Rina Dutta Gupta) that raise me up to take the task of compilation of this book on LED lighting and their impacts on plant growth and development. She holds the light from her heavenly abode throughout the path of my endeavor and no words can describe and acknowledge such bestowed strength which motivates me.

Kharagpur, India S. Dutta Gupta
February 2017

# Contents

# About the Editor

**Dr. S. Dutta Gupta** is currently a Professor in the Department of Agricultural and Food Engineering, Indian Institute of Technology Kharagpur, Kharagpur, India. He has been engaged in teaching and research on plant biotechnology for more than 30 years. He is a pioneer in the application of artificial intelligence and imaging techniques in plant tissue culture system and brings engineering–plant tissue culture link to a new dimension of understanding. He has also made a significant contribution to light-emitting diode (LED)-assisted modifications of oxidative status during shoot organogenesis. Dr. Dutta Gupta has received fellowships from various agencies and governments such as the United States Department of Agriculture (USDA), Lockheed Martin, the Ministry of Human Resource Development (MHRD), the Indian National Science Academy (INSA), the Council of Scientific and Industrial Research (CSIR), the Department of Science and Technology (DST), the Czech Academy of Sciences, and the Japan Society for the Promotion of Science (JSPS). He has published more than 100 scientific articles and edited four books.

# Chapter 1
# Artificial Lighting System for Plant Growth and Development: Chronological Advancement, Working Principles, and Comparative Assessment

**S. Dutta Gupta and A. Agarwal**

## 1.1 Introduction

Solar radiation is the primary source of energy that sustains life on earth. The spectral distribution of solar radiation has a broad waveband ranging from 300 to 1000 nm. However, only 50% of the radiant energy is available to plants as photosynthetically active radiation (PAR) and comprises the wavelength region from 400 to 700 nm (Boyle 2004). Specialized photoreceptors present in the plant leaves capture the photons and convert the sun's radiant energy to chemical energy following the process of photosynthesis. The process utilizes light absorbed by chlorophyll *a* and *b*, the most important photosynthetic pigments, at 662 and 642 nm, respectively. Plants have also developed intricate mechanisms for transducing the different wavebands of the incoming solar radiation into specific chemical signals for regulating various complex growth and developmental processes. Other than high-energy-dependent process of photosynthesis, photomorphogenesis, photoperiodism, and phototropism are also significantly influenced by the ambient light conditions. Photomorphogenesis is defined as light-mediated plant development that also includes differentiation of cells, tissues, and organs and depends on far-red radiation in the range of 730–735 nm, whereas photoperiodism refers to the ability of plants to sense and respond to the changes in the photoperiod: the relative lengths of day and night. The growth movement of the plants toward the direction of its light source is termed as phototropism. Light in the wavelengths range of 400–500 nm triggers the phototropic processes.

Unpredictable changes in the natural lighting conditions, insufficient daylight during the winter season, and climate change phenomenon lead to suboptimal yields and crop failures in many parts of the world. In order to mitigate this low

S. Dutta Gupta (✉) · A. Agarwal
Agricultural and Food Engineering Department, Indian Institute of Technology Kharagpur, Kharagpur 721302, India
e-mail: sdg@agfe.iitkgp.ernet.in



S. Dutta Gupta (ed.), *Light Emitting Diodes for Agriculture*,
DOI 10.1007/978-981-10-5807-3_1

crop productivity, the concept of protected cultivation in greenhouses and controlled environment crop production facilities with artificial lighting came into existence (Mpelkas 1980). Artificial light sources are used to augment insufficient sunlight in greenhouse-based open production system, whereas crop and/or transplant production in closed production system relies upon electrical lighting as the sole light source. Plant tissue cultures maintained under in vitro conditions depend entirely upon artificial light sources for illumination.

The earliest reports of plant growth under artificial lighting were published in the 1860s by H. Mangon, E. Prilleux, and others. However, commercial application of artificial lighting for crop production took place only after the development of more robust and long-lasting electrical lamps in the early twentieth century (Pinho and Halonen 2014). In the present scenario, electrical lamps have become an indispensable tool for controlled environment agriculture as a steady and reliable source of plant lighting. Technological advancement in artificial lighting over more than a century has made it possible to attain the present state of the art in electric lamp designs.

The most preliminary electrical lamps were designed in the first half of the nineteenth century. The model for an "electric arc" lamp was demonstrated by Sir Humphry Davy in 1809, whereas the first prototype for an "incandescent" lamp was revealed by Warren de la Rue in 1840. The era of artificial electric lighting actually started with the development of incandescent lamp designed by Thomas Edison in 1879. The proposed models were too costly for commercial application and had very short life spans. Various "carbon-filament"-based models were designed for the incandescent lamp in the mid- and late nineteenth century. However, it was only in the first part of the twentieth century that tungsten-based incandescent lamps were developed. Gas discharge lamps, the next state of electro-optical advancement, were first fabricated by Heinrich Geissler in 1857 by using various noble gases in an electric arc tube. The fluorescent lamp is the most widely used gas discharge lamp and utilized extensively in plant growth applications due to its reasonable energy efficiency and life span. Afterward, the introduction of metals such as mercury and sodium into the discharge tube improved the illumination as the electrical current was channelized through the vaporized metal. The first widely accepted design for the mercury vapor lamp was produced by P.C. Hewitt in 1901. This design was further improved by various others, and in 1936, the first modern high-pressure gas discharge lamp was launched by Philips. In the following decade gas discharge lamps having higher luminous efficacy and better spectral output such as metal-halide lamps and high-pressure sodium lamps were developed. High-pressure discharge lamps have been the preferred light source for crop production in controlled environment agriculture. The high PAR emission with relatively high percentage of blue radiation, long life span, and the electrical efficiency in the range of 25–40% make these lamps an option to replace daylight totally or partially supplementing it for year-round cultivation (Simpson 2003). However, conventional light sources suffer from the poor ability of efficient use of energy. Further, the spectral quality specific to photosynthesis as well as photomorphogenesis cannot be controlled during lighting treatment. Such limitations of conventional light sources accelerated the emergence of LEDs as potentially viable and promising artificial light source in controlled environment agriculture. The practical implementation of LEDs originated from the experiment of Henry Josef Round, a radio engineer in Marconi Labs, who observed the emission of light from a silicon carbide crystal when a current flowed through the material. This was the very first demonstration of a solid-state lighting, and the light produced is based on an electroluminescence effect (Round 1907). In spite of this breakthrough, technological advancement of LEDs was relatively slow until the 1960s (Schubert 2003). Since the invention of the first commercial LED in the late 1960s, there has been a gradual improvement in LED design with the advancement of semiconductor technology. The new-generation LEDs have also become a promising light source for plant growth research and cultivation, besides its popular applications as indicators and optoelectronic devices.

Impact of electrical lighting on plant growth and development was studied by many scientists using contemporary incandescent lamps and electric arc lamps. As reported by Siemens in 1880, plants illuminated by carbon arc lamps in addition to sunlight displayed improved growth when compared to the naturally growing plants. Studies on various food crops under tungsten-based incandescent lamps suggested that it could be possible to grow crops independent of sunlight and could be made to reach maturity and set seed even during the winters (Harvey 1922). In 1926, Pfeiffer reported that the duration of artificial lighting had a significant impact on the phyto-constituents of various plants. Over the years, various electrical lamps such as incandescent lamps (ILs), fluorescent lamps (FLs), high-pressure mercury vapor lamps (HPMLs), high-pressure sodium vapor lamps (HPSLs), and metal-halide lamps (MHLs) were employed for experimental plant growth applications and commercial plant cultivation. However, the potential of light-emitting diodes (LEDs) as a photosynthetic radiation source for plant growth was first explored in the early 1990s (Bula et al. 1991, 1992). The outcome of these studies unveiled some of the advantageous features of LEDs and clarified certain plant morphogenic responses related to the spectral quality of lighting source. A major breakthrough in the LED technology was attained with the development of first viable high-brightness blue LED by Shuji Nakamura in 1993 (Nakamura and Fasol 1997; Nakamura et al. 2000). This achievement paved the way for utilization of LEDs in plant growth and development.

The overall aim of this chapter is to present the reader a basic introduction to artificial lighting systems used in plant growth and development with their technological advancement over time, working principles and attributes with respect to spectral quality, luminous efficacy, power consumption, heat generation, and life span. Finally, a comparative assessment of the various performance parameters of the different light sources has been presented to highlight the advantageous features of LEDs and its potential as a photosynthetic radiation source for growing plants in controlled environment. With a basic understanding of the electrical and optical properties of the artificial lighting system, readers with plant science background will be well placed to comprehend the specific function and applications of LEDs discussed in the rest of the book.

## 1.2 History of Development and Working Principles of Conventional Lamps

Conventional plant lighting sources include incandescent, fluorescent, high-pressure mercury, high-pressure sodium, and metal-halide lamps. Incandescent lamps emit light upon the heating up of a metal filament, the phenomenon being called incandescence. The fluorescent, high-pressure mercury, high-pressure sodium, and metal-halide lamps are gas discharge lamps (GDLs) since they emit light generated by an electrical discharge through an ionized gas. Emission from incandescent lamps consists of thermal radiations, whereas GDLs emit photons by the release of energy from thermionically excited electrons. Among the GDLs, the FLs are low-pressure lamps, whereas the HPMLs, HPSLs, and MHLs are termed as high-intensity discharge lamps (HIDLs) due to the high pressure of gases in the arc tube. This section outlines the technological advancement toward the present-day conventional lamps from the preliminary models and illustrates their working principles.

### *1.2.1 Incandescent Lamps (ILs)*

Incandescence, the working principle behind ILs, is the phenomenon by which a solid starts emanating electromagnetic radiations in the visible range upon being heated (Kitsinelis 2011). The oldest design for the incandescent lamp proposed by Rue in 1840 consisted of a platinum coil enclosed in an evacuated glass tube. The design was not cost-effective due to the platinum coil. Experimental designs replacing the platinum with charcoal, carbonized paper, and carbonized bamboo filaments were proposed by others. The challenges faced by the inventors included the short life span of the filament and the blackening of the bulb caused by the burning of the filament. The lamp developed by Woodward and Evans in 1874 comprised of a glass bulb filled with nitrogen housing a carbon rod connected to two electrodes. In spite of this improvement in the design feature, it was not accepted for commercialization. In 1879, Edison purchased the patent for this design and using the model developed an incandescent bulb that not only performed better but was also more convenient to use in a cost-effective manner. He received the patent for this bulb in 1880 and commercialized it. However, it was only in 1904 that Hanaman and Just developed the first tungsten filament-based incandescent lamp which was further improved by General Electric (commonly known as GE) in the following years. Further refinements in the tungsten filament lamp involved the production of improved filament and the use of noble gases instead of evacuating the bulb.

Modern ILs are composed of an airtight glass bulb with a tungsten filament connected to lead-in wires (Fig. 1.1). The bulb is essentially made devoid of oxygen by evacuation or by filling up with an inert gas to prevent the burning up of the filament. The filament is made of a metal having high melting point and low

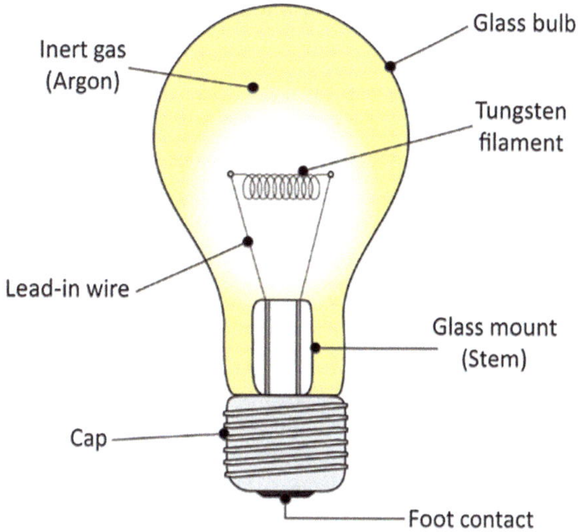


**Fig. 1.1** Structure of an incandescent lamp

coefficient of thermal expansion. Tungsten, possessing both these properties, has practically been the only metal used for producing the filament for ILs since the early twentieth century. The two lead-in wires connected to either ends of the filament are connected to the external circuit. The lamp operates when the electrical current flows in from one lead wire, through the filament and out of the second lead wire. As the filament has higher resistivity than the lead-in wires, it impedes the flow of electrons. The inelastic collisions between the moving electrons and the electrons within the filament lead to the conversion of the kinetic energy of the moving electrons into atomic vibrational energy. This causes the filament to gradually heat up and start dissipating energy as electromagnetic radiations. As the temperature of the filament rises up to almost 2800 K, it emits radiations in the entire visible range, with the intensity of radiations increasing from 400 to 700 nm. A significant portion of the energy is also dissipated as far-red emission which can reach up to 60% of the total PAR.

Early trials with ILs exhibited their potential for usage in indoor cultivation during winters as they not only produced a broad-spectrum emission but also provided warmth to the plants. However, the operations were not deemed economically feasible owing to the low luminous output in exchange of the high electricity input. Heat losses and poor electrical efficiency outweighed the gain in plant growth and yield. The energy conversion efficiency for the various modern ILs ranged between 1 and 5%, with the luminous efficacy never exceeding 20 lm/W (lumens/watt). Availability of power-efficient and long-lasting GDLs gradually replaced the ILs as a light source for indoor cultivation. Moreover, high power consumption along with low luminous efficacy has led to the phasing out of ILs, banning their manufacture, import, and sales in many countries.

### *1.2.2 Gas Discharge Lamps (GDLs)*

The carbon arc lamp demonstrated by Sir Humphry Davy in 1809 was the predecessor of all modern GDLs (Zissis and Kitsinelis 2009). The arc lamp worked on the principle of sustaining an electric arc or flow of electricity between two electrodes via an intervening gaseous medium. Davy's arc lamp involved the electrical breakdown of air or ionization of air molecules which maintained an electrical discharge between two carbon electrodes resulting in thermionic excitation of electrons leading to the emission of faint light. Lightning, a common natural phenomenon, is an example of an electric arc formed by the breakdown of molecules present in the air. In 1857, Heinrich Geissler demonstrated the world's first low-pressure mercury vapor discharge lamp. The mercury vapor discharge lamp produced a strong greenish-blue glow but had a short operating life. Peter Cooper Hewitt patented the mercury vapor lamp in 1901 after making certain improvements in Geissler's design. However, the application of this lamp was limited owing to the characteristic color of light it gave off. During that period, many scientists including Edison and Tesla tried to improve gas discharge lamps but success was limited. In 1906, a high-pressure mercury vapor lamp having a quartz arc tube was developed by Küch and Retschinsky. The next major step was the successful application of fluorescent coatings on the inside of the glass arc tube of mercury lamps by Compton in 1934. Application of halophosphate phosphor coatings resulted in the emission of white light from the low-pressure mercury vapor lamps. Philips launched the first high-pressure mercury vapor lamps in 1936, whereas General Electric became the first to commercially produce fluorescent lamps in 1938. Several experiments revealed that vaporized metals had a better emission spectrum at high pressures than at low pressures. However, glass arc tubes that could withstand such high pressures along with the high operating temperature without reacting with the vaporized metal were not available at that period. In 1955, R.L. Coble developed an aluminum oxide ceramic that could be used for making the arc tube for high-pressure sodium lamps. In 1962, metal-halide lamps were developed by Robert Reiling who introduced halides of metals in the high-pressure mercury lamp, resulting in a better emission spectrum than the mercury vapor ones. High-pressure sodium lamps emitting bright white light developed by Homonnay, Louden, and Schmidt were launched commercially in 1964.

#### 1.2.2.1 Fluorescent Lamps (FLs)

As mentioned earlier, FLs are low-pressure mercury vapor discharge lamps that produce visible light due to the fluorescence of a phosphor coating. FLs may be divided into two classes on the basis of their shape and size—tubular and compact (Fig. 1.2). Although the luminous efficacies of the two designs differ significantly, the working principle for both types of FLs is essentially the same. Both of them consist of an airtight hollow glass tube filled with a mixture of mercury and argon

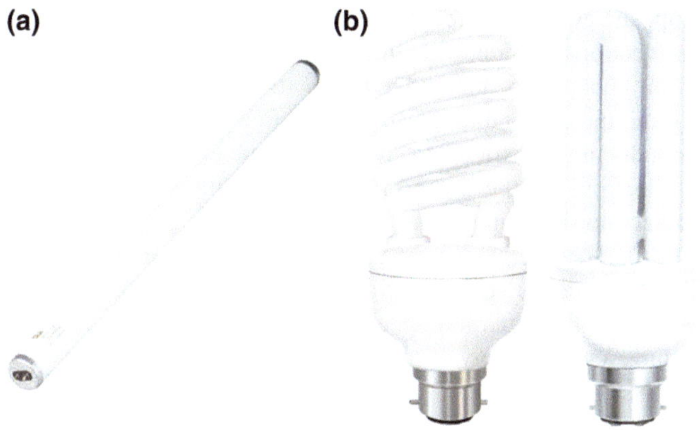


**Fig. 1.2** Tubular (**a**) and compact (**b**) fluorescent lamps

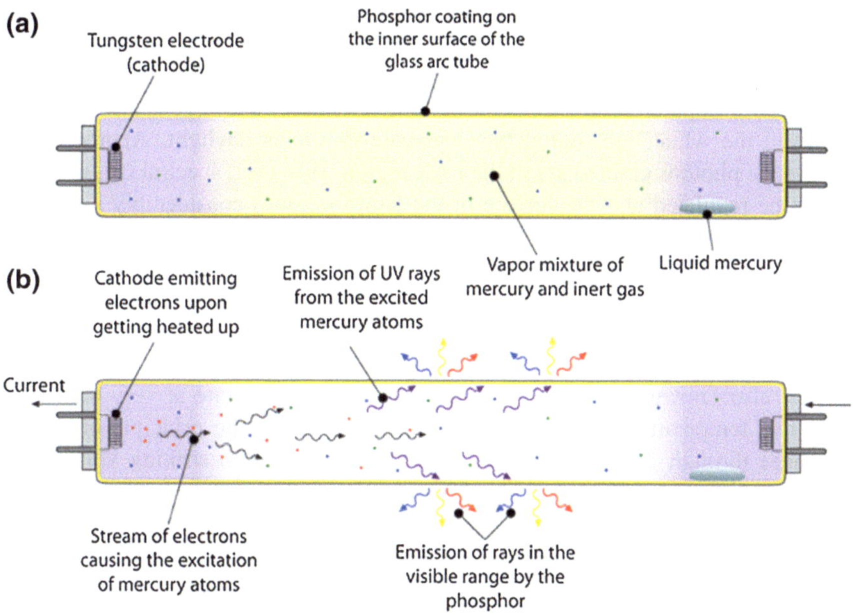


**Fig. 1.3** Components (**a**) and functioning (**b**) of a fluorescent lamp

vapors in a low-pressure environment (Fig. 1.3a). The inert gas present in the arc tube promotes the ionization of the gaseous metal (mercury) atoms. The two ends of the tube have electrodes composed of tungsten filaments projecting into the vapor mixture. Upon the passage of electricity, the filament gets heated up and starts emitting electrons (Simpson 2003). Since FLs work on alternating current, the two electrodes alternately emit electrons every half cycle. The electrons get accelerated toward the opposite electrode through the mercury vapor mixture due to the applied voltage. The electrons collide with the valence electrons of the mercury atoms causing electron impact ionization which leads to the release of more free electrons into the vapor mixture, a condition also referred to as breakdown. At this stage, the vapor starts conducting electricity freely. The mobile electrons cause the excitation of the other electrons in the outer orbitals of the mercury atoms. The excited electrons fall back to the ground state and in the process emit radiations in the UV range (Fig. 1.3b). These high-energy UV photons are absorbed by the phosphor coating which fluoresces or starts emitting photons of lower energy, i.e., within the visible range. Since the emission spectrum of an FL entirely depends upon the phosphor coating, a wide variety of phosphors have been used for developing white and colored FLs.

Energy losses in an FL occur in the ballast which supplies a pulse of high voltage to initiate the discharge. However, a significantly higher amount of energy is lost during the conversion of UV rays into visible light where almost half the energy of each photon is lost as heat. Since they were first launched commercially, fluorescent lamps (FLs) have been modified significantly for improving the luminous efficacy and reducing the cost of production. However, the overall energy conversion efficiency of the FLs is still below 30% (Shur and Žukauskas 2005). FLs have been a popular source of plant lighting in small- and large-scale operations owing to the white light output that appositely mimics daylight. Approximately 90% of the photons emitted are in the PAR region. However, spectral output of FLs cannot be regulated and the surface of the lamp becomes considerably hot during operation.

#### 1.2.2.2 High-Intensity Discharge Lamps (HIDLs)

HIDLs, also known as high-pressure discharge lamps, operate at very high pressures and temperatures. Like FLs, HIDLs also work on the principle of electric discharge through a gas and require ballasts for creating a striking voltage and maintaining the arc. However, the high operating pressure and temperature of HIDLs plays an important role in improving the spectral output and increasing the luminous efficacy. This is due to the fact that vaporized metals conduct electricity better under high pressure leading to higher number of electron excitations and more thermionic emissions (Kitsinelis 2011). HIDLs may be broadly classified into three types depending upon the "fill-gas" or vapor used—mercury, sodium, and metal halide (Fig. 1.4). It is worthy to note that all HIDLs essentially contain mercury in the fill-gas along with the other vapors.

As in FLs, the HPMLs contain a mixture of mercury and argon vapors, but at almost 200,000 times the pressure in an FL. The vapors are maintained in a quartz arc tube to withstand the high pressure and operating temperature (Fig. 1.5). The arc tube is housed inside an outer envelope made of borosilicate glass filled with nitrogen. The ionization of mercury atoms is triggered by the emission of electrons from the tungsten electrodes. However, due to the high pressure, the frequency of

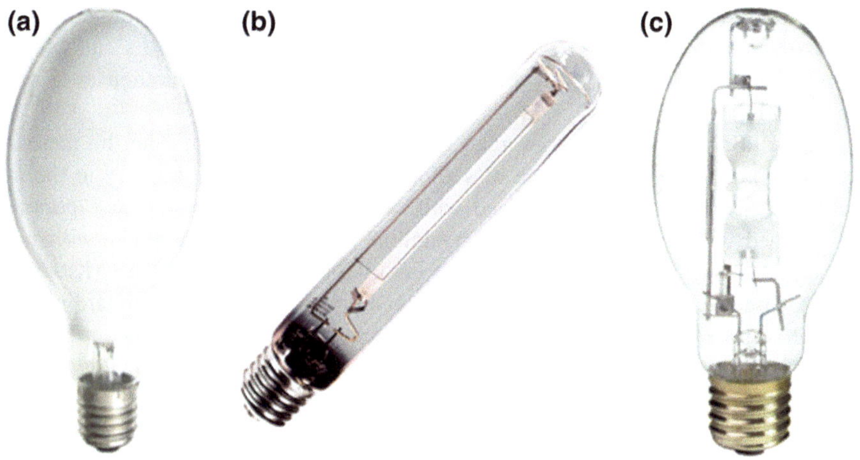


**Fig. 1.4** High-pressure mercury lamp (**a**), high-pressure sodium lamp (**b**), and metal-halide lamp (**c**)

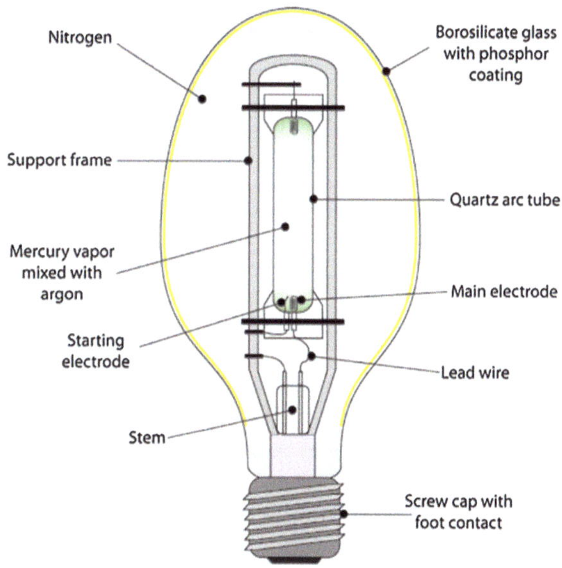


**Fig. 1.5** Design features of a high-pressure mercury vapor lamp

electron impacts on the mercury atoms becomes very high. This leads to the generation of a huge amount of heat. As a result, the mercury electrons get ionized to higher excitation states, leading to the emission of radiations at certain wavelengths in the visible range along with the UV radiations. A phosphor coating provided on the outer envelope converts the UV radiations into different visible wavelengths, resulting in white light (Kitsinelis 2011).

Despite the high-pressure discharge conditions, HPMLs have a luminous efficacy of around 60 lm/W. High lumen output has made the HPMLs suitable for various applications such as overhead lighting in factories and warehouses as well as street lighting.

The HPSLs have greater coverage over the visible spectrum than the mercury vapor lamps due to the presence of sodium vapors along with mercury in the arc tube. Further, the tube is pressurized with xenon instead of argon. The vapors are maintained within a ceramic or polycrystalline alumina tube which can withstand the corrosive nature of sodium vapors at high temperature and pressure (Kitsinelis 2011). The excitation of mercury and sodium atoms occurs by the bombardment of electrons from the tungsten electrodes. The electron impact ionization coupled with thermal ionization results in electrons jumping to various higher energy states, while falling back to the ground state, the electrons emit electromagnetic radiations covering a wide range in the visible spectrum. The components of HPSLs are depicted in Fig. 1.6.

Higher luminous efficacy (80–125 lm/W) and broad emission spectrum of HPSLs have made them a popular source of electrical lighting in public spaces and industrial buildings. A high emission peak in the 560–610 nm range renders a distinct yellow coloration to the light produced which limits its applications.

**Fig. 1.6** Structure of a high-pressure sodium vapor lamp

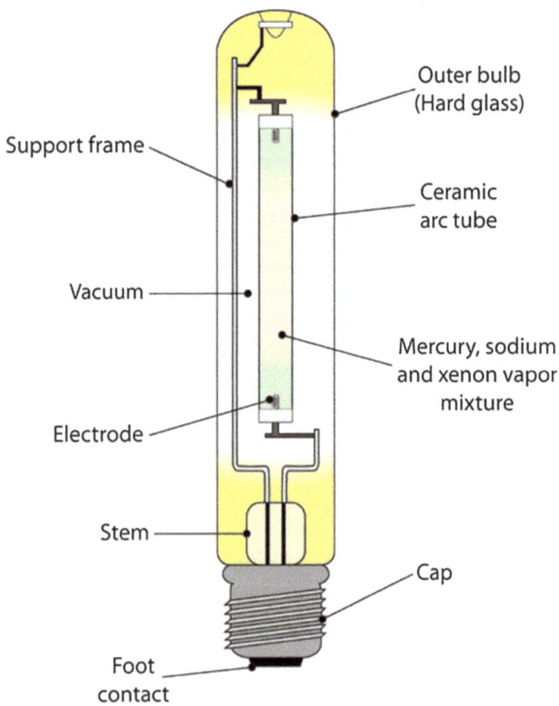

Further, the unbalanced spectral quality in relation to the absorption peaks of chlorophyll a, b and β-carotene makes them unsuitable for promoting photosynthesis and photomorphogenesis. Compared to other conventional light sources, HPSLs with high electrical efficiencies of 30–40% are the most energy-efficient light sources used in plant growth.

The metal-halide lamp is a modified version of high-pressure mercury vapor discharge lamp. The inclusion of metal halides along with the mercury vapor and inert gas permits the optimization of the spectral quality of the emitted radiation to a certain extent. Metals such as sodium, scandium, indium, thallium, and dysprosium are used in MHLs because of their characteristic emission spectra in the visible range. Generally, iodides, and sometimes bromides, of these metals are chosen because they are easier to vaporize and ionize than the pure metals as such. Like the other HIDLs, the pressurized gas is maintained within the arc tube and the same mechanism of operation is followed for electron excitation and light emission (Fig. 1.7). However, the outer casing is made of UV-filtering quartz glass to block the UV radiations of mercury. Since the light emitted by the lamp is a mixture of the radiations by the individual metals present in the vapor mixture, changing the combination of the metal halides allows the production of MHLs with various emission spectra (Simpson 2003). MHLs have an evenly distributed spectral output and produce white light with a high luminous efficacy of 100–120 lm/W. MHLs can be used in plant growth applications due to its high PAR, relative high percentage of blue radiation, and energy efficiency of approximately 25%.

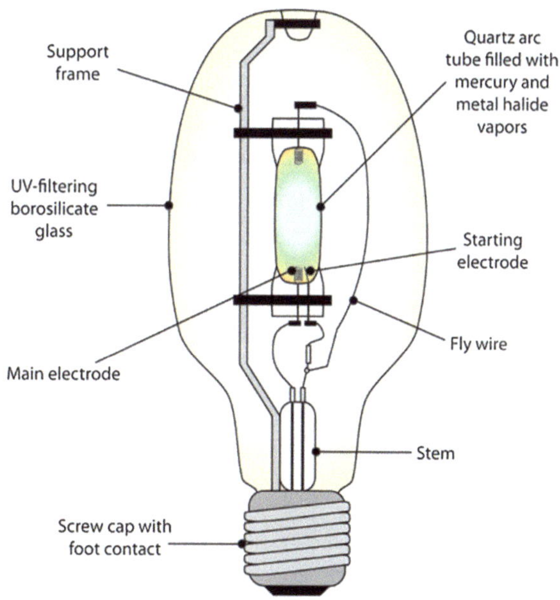


**Fig. 1.7** Structure of a metal-halide lamp

## 1.3 Light-Emitting Diodes (LEDs)

LEDs are known as solid-state light sources because they emit light from a semiconductor diode chip. Although the emission of light from ILs also occurs from a solid (filament), the cause of electromagnetic radiations is quite different from the LEDs. The ILs emit radiations due to the heating up of the filament, whereas LEDs emit light due to the transition of electrons from higher to lower energy orbital's. GDLs emit radiations due to release of excess energy from electrons too, but the source of energy is thermionic excitation due to the electric arc. In LEDs, the electrons are not impelled into higher excitation states but simply driven by the electrical potential difference from a higher energy orbital to a lower one. In this section, the major landmarks in the development of LEDs have been briefly outlined and the basic working principle of LEDs pertinent to plant scientists has been discussed.

### *1.3.1 Development of LED Technology*

A LED is a solid-state semiconductor device that emits light upon the flow of electricity (Fig. 1.8), following the principle of electroluminescence. Electroluminescence is the emission of light when electrons driven by an electrical or magnetic field enter a lower energy orbital and release the excess energy in the form of electromagnetic radiations. The phenomenon was first observed by H.J. Round in 1907 while working with silicon carbide (SiC). In 1927, Oleg Losev proposed a theory behind the phenomenon and outlined various practical applications of the technology (Zheludev 2007). Later, in 1955, R. Braunstein reported the emission of infrared radiations from various semiconductor alloys. James Biard and Gary Pittman (1961) of Texas Instruments accidentally discovered the emission of infrared radiations from gallium arsenide (GaAs) semiconductor upon the passage of electricity, while working on solar cells. They patented the design as "semiconductor radiant diode" in 1962, and that was the world's first light-emitting diode (LED). In the same

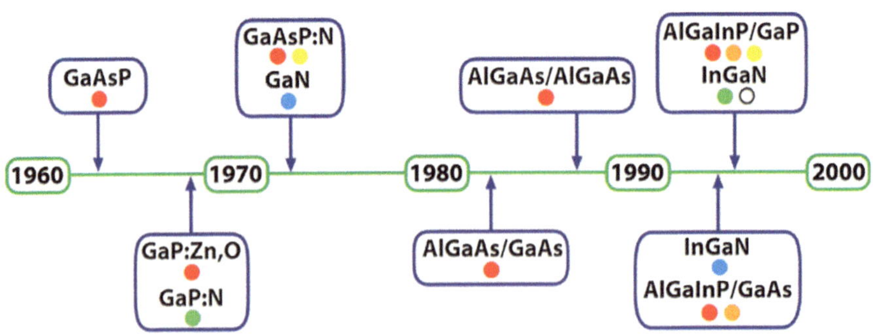


**Fig. 1.8** Historical development of semiconductor materials used for LED fabrication

year, Nick Holonyak Jr. designed the world's first LED producing visible light (red) using a gallium arsenide phosphide (GaAsP) diode. Ten years later, Holonyak's student M.G. Craford designed the GaAsP-based yellow LED and high-brightness red and red-orange LEDs. However, the LEDs being produced were too costly and were bright to an extent to be used only as indicators. In 1970, improvements in semiconductor fabrication and packaging techniques by Jean Hoerni and Thomas Brandt led to the drastic reduction in the cost of manufacturing LEDs. Initially, the development of light-emitting semiconductor technology was associated with red and infrared radiations. The lack of a viable blue LED hindered the utilization of this technology to plant growth applications. H.P. Maruska designed the first blue LEDs based on gallium nitride (GaN) in 1972. However, Maruska's LEDs had limited applications due to its low level of brightness. In 1994, Shuji Nakamura presented the design for a high-brightness blue LED employing an indium gallium nitride (InGan) diode. The newly developed LED with a peak emission wavelength of 450 nm was found to be suitable for use in studies on plant growth and development. The wavelength matches with the maximum absorption peak of plant photoreceptors of carotenoids. For this revolutionary invention of efficient blue LEDs which has enabled energy-efficient bright white light sources, the Nobel Prize in Physics 2014 was awarded jointly to Isamu Akasaki, Hiroshi Amano, and Shuji Nakamura. Over the years, gradual advancements in diode fabrication techniques have resulted in further reduction in the cost and significant increase in the luminous (lm/W) as well as photon (μmol/J) efficiencies.

Various semiconductor materials have been used since Holonyak's GaAsP-based model for fabricating red, green, blue, and white LEDs. Choice of the semiconductor alloy was guided by the need to increase the range of emission wavelength and luminous efficacy of the new LED as compared to its predecessors. The historical development of semiconductor material systems associated with improved performance of LEDs in terms of luminous efficacy is shown in Fig. 1.8. Further enhancement in luminous output and power efficiency could be attained by increasing the efficiency of radiative recombination (electron–hole pairing leading to photon emission) within the LEDs. This was achieved via bandgap engineering by the use of heterostructures and quantum wells. Advancements in epitaxial crystal growth techniques enabled the formation of customized heterostructures and quantum wells in LED chips (Schubert 2003). The technology led to the development of power-efficient high-brightness LEDs that have sufficient luminous output with desired wavelength to sustain optimal plant growth. Such LEDs are made from binary direct bandgap alloys from groups III–V elements of the periodic table, namely aluminum gallium arsenide (AlGaAs), aluminum gallium indium phosphide (AlInGaP), and aluminum indium gallium nitride (AlInGaN). Availability of high-brightness LEDs with spectral output matching with the action spectra of photosynthesis and photomorphogenesis created the platform for the LED-based plant illumination system (Tamulaitis et al. 2005).

In the near future, LED luminaries can become the smart solutions for sustaining plant growth in controlled environment agriculture and regulating morphogenic responses in plant tissue culture.

### *1.3.2 Structure and Working Principle of LED*

The LED comprises a semiconductor chip housed within an epoxy or plastic lens, with connecting wires for directing the electrical current. The dual in-line package (DIP) LED (Fig. 1.9a) has been the most commonly used LED design. The newly developed high-power LEDs (Fig. 1.9b) produce higher luminosity due to higher current flow than the DIP-LEDs. The components of the DIP and high-power LEDs have been depicted in Fig. 1.10. The chip is a small (approximately 1 $mm^2$ in size) semiconductor wafer that has been impregnated with specific impurities or dopants. There are two types of dopants: *n*-type, i.e., elements having a high number of valence electrons, and *p*-type, i.e., elements having a high number of empty slots or "holes" in the valence shell. The *p*-type- and *n*-type-doped semiconductor crystals are fused together to form a "*p–n* heterojunction." As the electric current moves across the diode from the *p*-side to the *n*-side, electrons from the *n*-side cross over

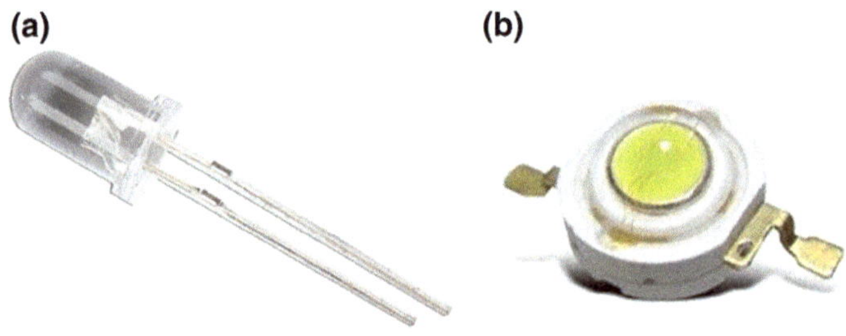


**Fig. 1.9** Dual in-line package (**a**) and high-power (**b**) light-emitting diodes

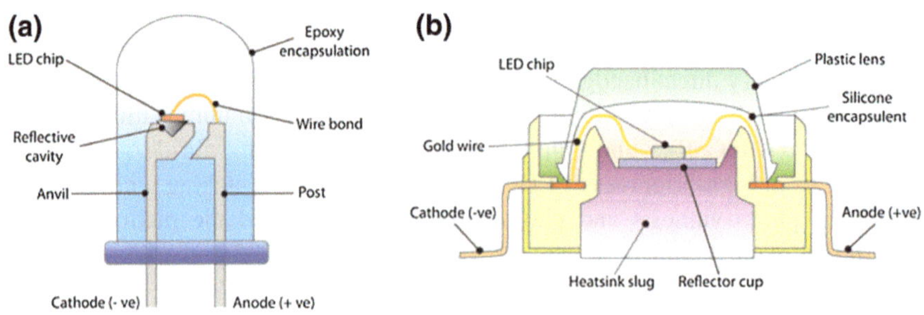


**Fig. 1.10** Components of conventional DIP (**a**) and modern high-power (**b**) light-emitting diodes

to the *p*-side. These electrons now fall into the vacant spaces in the orbitals of the *p*-type dopant resulting in "electron-hole pairing."

As the energy of the newly acquired orbital is lower than the energy possessed by the electron, the excess energy is liberated as electromagnetic radiation having a specific wavelength or color. This wavelength corresponds to the difference in valence shell energies of the *p* and *n* dopants (Fig. 1.11). The phenomenon can be mathematically expressed as $\Delta E = (hc)/\lambda$ ($\Delta E$ = change in energy of an electron, $h$ = Planck's constant, $c$ = velocity of light, $\lambda$ = wavelength of light). By virtue of its constituent dopants, an LED is capable of emitting light at a fixed wavelength only.

The application of red and blue monochromatic LEDs alone or in combinations has been reported for plant morphogenesis both in vivo and in vitro over the decades (Bula et al. 1991; Kim et al. 2005; Massa et al. 2008; Dutta Gupta and Jatothu 2013; Agarwal and Dutta Gupta 2016). However, such LED lighting suffers from the waveband mismatch with the photosynthetic action spectrum and the high fabrication cost of the complicated circuit. Application of white LEDs eliminates the likelihood of such an event since they have a broad spectral output. Moreover, constructing a circuit with only white LEDs is relatively simpler than making a red-blue mixed LED panel because the voltage requirements of red and blue LEDs differ significantly. White LEDs can be fabricated by using a combination of red, green, and blue LED units in the same fixture (Fig. 1.12a). Such LEDs are called trichromatic or tetrachromatic depending upon the combination of monochromatic LEDs used (Lei et al. 2007). White LEDs made by red, green, and blue LED clusters have a tunable spectral output controlled by the drive current through individual red, green, and blue LED units (He and Zheng 2010). Phosphor-coated

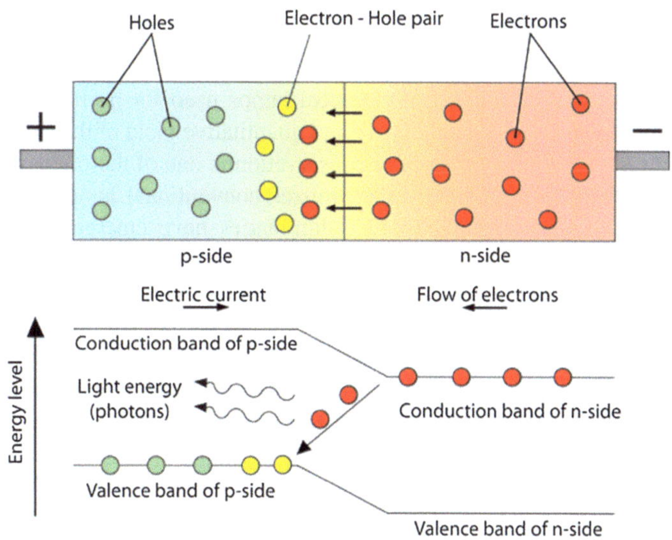


**Fig. 1.11** Working principle of an LED

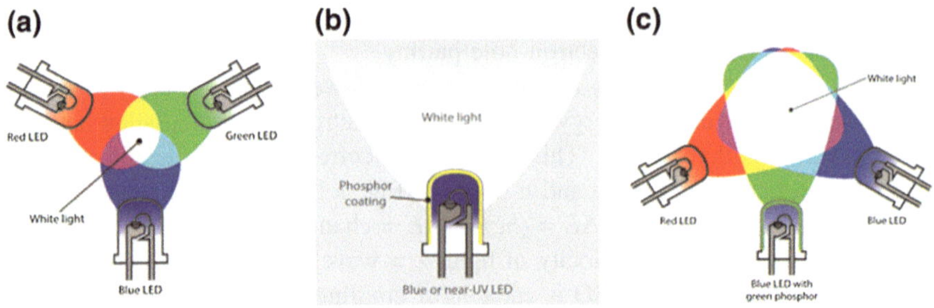


**Fig. 1.12** Types of white LEDs produced by RGB color mixing (**a**), phosphor coating in UV and blue LEDs (**b**) and a hybrid of phosphor and color mixing (**c**)

blue and UV-LEDs are the preferred source of white light owing to their common availability at low cost (Fig. 1.12b). However, the initial phosphor-coated LED models suffered from significant energy losses at the phosphor due to low energy conversion efficiency (Bourget 2008). Approaches are being made to develop high-efficacy white LEDs using a hybrid model which includes colored phosphors along with monochromatic LEDs (Fig. 1.12c). A decrease in total internal reflection within the chip and device encapsulation with multicolor-emitting phosphors could enhance the luminous efficiency (Pattison et al. 2016). Recently, Chen et al. (2016) proposed the potential of $Eu^{+}$-doped fluorophosphate in fabricating white LEDs for application in plant growth.

## 1.4 Comparative Assessment of the Different Artificial Lighting Systems

Although the various conventional electrical lamps used for horticultural lighting have the capacity to boost the qualitative and quantitative yield of the plants, they all suffer from certain limitations. Energy conservation is one of the major concerns in controlled environment agriculture that utilizes conventional lamps, especially in northern latitudes. New-generation LED luminaries have emerged as potentially viable and promising plant lighting system to be used in controlled environment agriculture. Emergence of solid-state lighting has not only offers the energy-efficient interior agriculture but has also opened up new frontiers for studying plant response to a specific wavelength and/or radiation quantity. A detailed comparison of the attributes of LEDs and conventional lamps used for plant lighting is essential for comprehensively assessing the benefits of using LEDs in indoor cultivation setups and plant research laboratories. Lamp features such as spectral quality, luminous efficacy, power requirement, life span, heat emission, robustness, and ease of disposal are discussed in the following section for assessing the performance of each lighting system.

### 1.4.1 Spectral Quality

Availability of a proper light environment is pivotal for plant growth. Incident spectrum and photon flux density (PFD) are two major factors that govern plant development in response to the lighting conditions. Plants essentially utilize the infrared, red, and blue portions of the incident spectrum for conducting photosynthesis and regulating numerous developmental and adaptive processes. The typical absorption spectra of the most common photosynthetic and photomorphogenic photoreceptors are shown in Fig. 1.13. Chlorophylls absorb photons and utilize the energy for photosynthesis (Anderson et al. 1995). The main absorption peaks of chlorophyll are located in the red (625–675 nm) and blue regions (425–475 nm). Carotenoids, the auxiliary photoreceptors of chlorophyll, absorb light mainly in the blue region. Photomorphogenic responses including germination, phototropism, leaf expansion, flowering, stomatal development, chloroplast migration, and shade avoidance are regulated by three types of photoreceptors, viz. phytochromes, cryptochromes, and phototropins (Smith 1995; Sancar 2003; Briggs and Christie 2002). Interconvertible forms of Pr and Pfr in the red at 660 nm and in the far-red at 730 nm, respectively, constitute the phytochrome photoreceptor system. Phytochrome-mediated photomorphogenic responses are critically regulated by the sensing of R/FR ratio (Shinomura et al. 2000). The pigments absorbing blue light include both cryptochromes (cry1, cry2) and phototropins (phot1, phot2). The cryptochrome system controls several aspects of morphological responses, such as germination, leaf expansion, stem elongation, and stomatal opening. It also regulates the circadian rhythm in flowering plants (Cashmore et al. 1999).

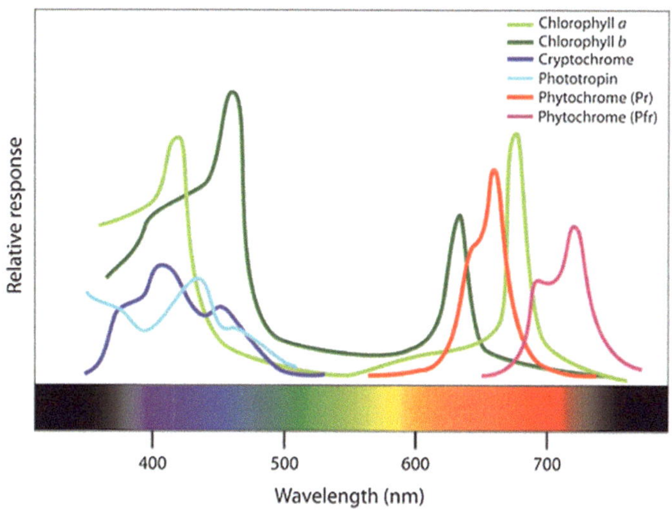


**Fig. 1.13** Utilization of various wavebands of light for photosynthesis by chlorophylls, and for photomorphogenesis by phytochrome, cryptochrome and phototropin

Phototropins are involved in the regulation of pigment content and the positioning of photosynthetic organelles in order to optimize the harvesting of light and to prevent photoinhibition (Spalding and Folta 2005).

Insolation (incoming solar radiation that reaches the earth's surface) contains all the regions of the visible spectrum along with radiations in the infrared and UV regions. Intensity of solar radiations is relatively higher in the blue-yellow (460–580 nm) range. Like sunlight, all conventional electric lamps, viz. ILs, FLs, and HIDLs, are broad-spectrum light sources. The spectral quality of light (wavelengths) produced by different artificial lighting sources along with sunlight is depicted in Fig. 1.14. The ILs have a continuous emission spectrum having high proportions of photons in the infrared and red ranges, the PFD gradually reducing toward blue. Due to the presence of phosphor coating, white FLs also have a continuous visible spectrum with peaks near 400–450 nm (violet-blue), 540–560 nm (green-yellow), and 620–630 nm (orange-red) that results in a balanced white color rendition. HPMLs employing phosphor coatings also feature a similar emission spectrum but with sharper peaks than FLs. Spectral emission of HPS lamps exhibits peaks in the 560–610 nm (yellow-orange) region which imbues these lamps with a predominantly yellow light output. MHLs emit a continuous visible light spectrum with several peaks distributed evenly across the entire spectrum. FLs, HPMLs, and MHLs are capable of delivering bright white light and are hence also referred to as "daylight lamps."

LEDs are essentially monochromatic light sources and have a specific emission wavelength which is determined by the constituent elements of the LED chip. Since LEDs for all wavelengths in the visible range are available, a wide variety of light spectra can be obtained from LED-based luminaries by simply embedding specific LEDs for the desired wavelengths. All conventional artificial light sources have significant emissions in regions of the visible spectrum that plants simply do not require. Since electrical lamps produce light at the expense of electrical energy, delivering wavelengths of light that are not utilized by the plants becomes impractical and a costly affair. With LEDs, it is possible to produce artificial light with selected peak wavelength emission that closely matches the absorption peak of a known important photoreceptor. Furthermore, the designs of ILs and GDLs do not allow the regulation of operating light intensity. Intensity of emission from LED lamps can be easily regulated by altering the electrical current. Thus, it is possible to construct LED panels with specific peak emission that are utilized by plants, having intensity control for adjusting the PFD most suited for the plants being raised. In this way, customized LED luminaries would allow a versatile control of radiation intensity and spectrum.

### *1.4.2 Luminous Efficacy and Power Requirement*

Efficient conversion of electrical energy to light energy is an important factor for selecting the light source for indoor plant cultivation. The reason being that if the

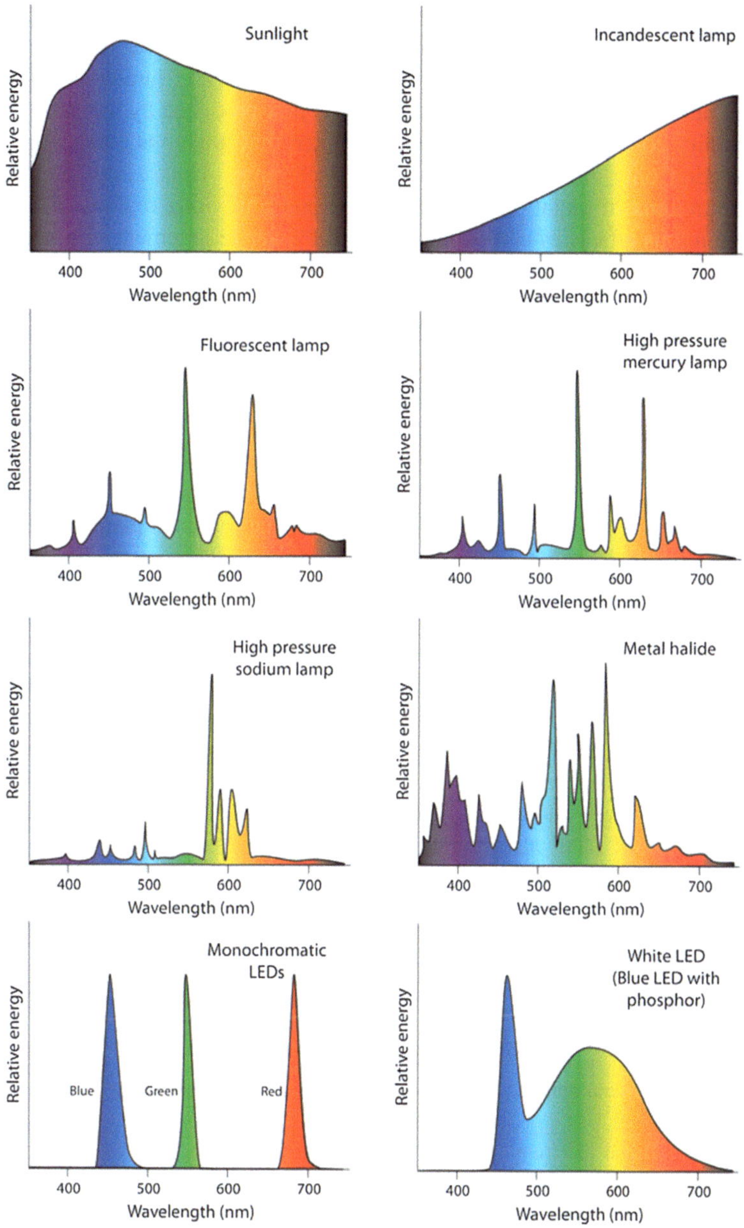


**Fig. 1.14** Spectral outputs of the various light sources

electricity consumption for providing appropriate lighting conditions to the crops becomes very high then the procedure may become economically unfeasible for overall management of the crop production system. The luminous efficacy of artificial light sources is a measure of luminous flux produced by the lamp per watt of electricity consumed (lm/W). It must be noted that luminous efficacy takes into consideration only the spectral output in visible range. Thus, lamps emitting significant amounts of radiations in the infrared and ultraviolet regions tend to have lower luminous efficacies compared to others. Power requirement of a lamp refers to the wattage supply required for operating the lamp. The lower the power requirement, the cheaper and easier it is to run any electrical lamp.

Among all artificial light sources, HPS and MH lamps have the highest luminous efficacies (Table 1.1). However, if we consider the lumens utilized by plants, the value gets reduced significantly since only the blue and red regions must be considered for plant use. Thus, the useful luminous output of even the most power-efficient electrical lamps may be considered to be quite low for plant growth. Although the luminous efficacies of conventional light sources have improved significantly since their initial development, the values attained plateau in the range of 80–125 lm/W (Fig. 1.15). LEDs with luminous efficacy of 80–150 lm/W are already available in the market. Combinations of monochromatic LEDs can also be used to produce specific spectra that may be completely utilized by the plants, thus making the useful luminous output equivalent to the total luminous output. Further, due to rapid advancements in LED lighting technology, it is expected that LEDs with an efficacy of >200 lm/W will be developed within the next few years (Fig. 1.15; US Department of Energy 2016). The power requirement of a typical LED is 10–100 times less than most conventional lamps, thus making LED lamps highly cost-effective (Table 1.1). Since LEDs consume less electricity, application of this technology shall also reduce the pressure on fossil fuel reserves used for generating electricity.

**Table 1.1** Features of various electrical lamps used for plant lighting

| Lamp type | Spectral output | Luminous efficacy (lm/W) | Power requirement (W) | Life span (h) |
|---|---|---|---|---|
| Incandescent | Broad spectrum | 20 | 15–1000 | 1000 |
| Fluorescent | Broad spectrum | 100–120 | 5–125 | 1000–30,000 |
| HPM | Broad spectrum | 60 | 100–250 | 10,000–20,000 |
| HPS | Broad spectrum | 80–125 | 35–1000 | 10,000–30,000 |
| Metal halide | Broad spectrum | 100–120 | 35–400 | 10,000–20,000 |
| LED | Specific wavelengths | 80–150 | 0.1–5 | >50,000 |

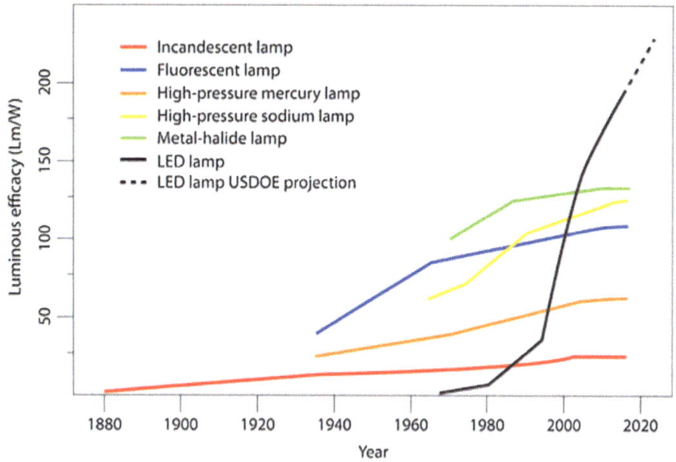


**Fig. 1.15** Timeline of improvement in the luminous efficacies of various artificial lights

## *1.4.3 Heating of the Lamp*

Dissipation of heat from the lamp is undesirable for indoor farming as well as for in vitro propagation from various aspects. Artificial light sources generating a lot of heat tend to raise the ambient temperature, a situation which may affect the quality of crops and the process of morphogenesis during in vitro culture. Additionally, this increases the load on the cooling system used for maintaining the temperature, leading to an increment in the electricity consumption. Furthermore, such light sources need to be placed at a safe distance from the crops/cultures as direct exposure to the heat may prove to be fatal. In vertical farming models where the crops are grown in tiers, using light sources having lower surface temperatures allows the placing of crops closer to the light source, thus giving more space for constructing more tiers and obtaining a higher yield per volume of the farming space. This notion is also applicable for in vitro culture. Dissipation of heat to the surroundings during any form of energy conversion has been considered as a loss of energy from the system. Since light sources having cool operating temperatures lose lesser energy to the surroundings in the form of heat, they are able to convert electrical energy to light energy more efficiently.

All conventional lamps involve heating up of the conducting medium as an essential step for operation. Inelastic collisions of electrons occurring in ILs and GDLs liberate a lot of heat energy, a condition absent in LEDs. Like all other devices conducting electricity, LEDs also generate heat due to their intrinsic resistance at the p-n junction. However, the heat generated is negligible as compared to that in the conventional lamps. Furthermore, incorporation of heat sinks in modern high-power LED designs (Fig. 1.10b) allows the LED to keep on operating at cool temperatures even while conducting significantly higher electrical currents.

LEDs have a cool surface temperature and are safer for growing plants as they practically do not emanate any heat as compared to the ILs and GDLs (Mitchell et al. 2012).

### *1.4.4 Life Span, Dimming, Directionality, Robustness, and Safety*

Life span of the luminaire affects the overall operating cost because frequent replacement of a large number of lamps on a commercial scale involves a huge capital input on a regular basis. Conventional lamps gradually wear out from within owing to the extremely high operating temperatures. Due to the low working temperature, LED components do not wear out easily and that extended its life span by several thousand hours (Table 1.1).

As the IL and GDL illumination units grow old, precipitations on the inner surface tend to make the lamp dim. Hence, despite the lamp functioning optimally, the luminosity produced by it gets reduced. LEDs are solid-state light sources that do not contain any vapors or gases nor involve vaporization of elements, thus eliminating the chances of dimming due to precipitations.

All conventional artificial light sources, by virtue of their design, emit light in all directions. The use of reflective coatings in fixtures reduces the loss of light within the fixture. However, the luminous flux or the total useful light obtained in the desired direction becomes significantly lower than the total light produced by the lamp. An artificial light source with directionality of light emission can be used to provide greater luminous flux to the plants with significantly lower fixture losses. An LED contains a reflective cavity housed within the epoxy cover that concentrates all the photons in a single direction. Furthermore, half-isotropic spatial pattern of LEDs makes them directional emitters. LEDs with a small viewing angle and the use of secondary optics such as collimator lenses can improve the luminous efficacy by directing the light toward the plant canopy.

Small size and robustness of lamps also increase their desirability. Small illumination units occupy a small volume and provide more space for growing crops, especially in vertical farms. Further, lamps made of durable materials are easy to handle and thus more user friendly. Artificial light sources devoid of hazardous materials such as mercury are preferable from the point of view of disposal. ILs and GDLs are made up of different types of glass filled with various gases. Users have to exercise caution while handling such lamps. GDLs contain mercury which is highly toxic when released in the environment, making the disposal of spent GDLs a matter of concern. HIDLs get highly pressurized during operation, thus making them quite unsafe in case of any manufacturing defects. Luminaires for such lamps are often large and make them uneconomical in terms of space. On the contrary,

LEDs are small, solid-state lamps housed within epoxy or plastic lens. LEDs are, thus, not only more robust and easy to handle, but also occupy a very small portion of the space being utilized for growing plants. Advantages of LEDs over conventional electrical light sources from the perspective of plant growth and development are listed as follows:

- Choice of the peak emission for customized plant growth and development,
- Versatile control of the flux emission and the light spectrum,
- High luminous efficacy,
- Small size and directional light emission,
- Long life expectancy,
- Negligible heat emission,
- Does not get dim with age,
- Economical in terms of space and power (wattage) requirement,
- Plastic body, hence more robust and easy to handle, and
- Easy to dispose without any environmental hazards.

## 1.5 Conclusions

Application of artificial lighting as supplemental and sole light sources for growing plants has been in practice for almost a century. Advancements in lighting technology have allowed the implementation of electrical lamps at large scales for controlled environment agriculture and in vitro transplant production. Conventionally, filament- and gas-based electrical lamps, viz. ILs, FLs and the different HIDLs, have been employed in greenhouses and controlled environment plant production units. High-power requirement and relatively short life span of these lamps made such crop production systems highly uneconomical. Furthermore, the lack of intelligent control and risks in handling and disposal reduced the usefulness of ILs and GDLs for large-scale interior agriculture. The development of power-efficient high-brightness LEDs has been a major breakthrough in lighting technology that has dramatically changed the scenario of plant lighting for both commercial and research endeavors. LEDs are semiconductor light sources that have the ability to deliver photons more precisely than all other contemporary electrical lamps. LEDs have been recognized as a new artificial lighting source to promote photosynthesis, to regulate photomorphogenesis, and to enhance nutritional quality of leafy vegetables due to its several aforesaid advantages. Advancements in the LED technology over time including packaging, current drop, phosphor coatings, intelligent control of light distribution, intensity and spectral quality along with the reduction in prices will make LED-based illumination system a smart choice for novel open as well as closed plant production systems.

## References


Agarwal A, Dutta Gupta S (2016) Impact of light emitting-diodes (LEDs) and its potential on plant growth and development in controlled-environment plant production system. Curr Biotechnol 5:28–43

Anderson JM, Chow WS, Park YI (1995) The grand design of photosynthesis: acclimation of the photosynthetic apparatus to environmental cues. Photosynth Res 46:129–139

Bourget CM (2008) An introduction to light-emitting diodes. HortScience 43(7):1944–1946

Boyle G (2004) Renewable energy: power for a sustainable future, 2nd edn. Oxford University Press, UK

Briggs WR, Christie JM (2002) Phototropins 1 and 2: versatile plant blue light receptors. Trends Plant Sci 7(5):204–210

Bula RJ, Morrow RC, Tibbitts TW, Barta DJ, Ignatius RW, Martin TS (1991) Light-emitting diodes as a radiation source for plants. HortScience 26(2):203–205

Bula RJ, Tibbitts TW, Morrow RC, Dinauer WR (1992) Commercial involvement in the development of space-based plant growing technology. Adv Space Res 12(5):5–10

Cashmore AR, Jarillo JA, Wu YJ, Liu D (1999) Cryptochromes: blue light receptors for plants and animals. Science 284(5415):760–765

Chen J, Zhang N, Guo C, Pan F, Zhou X, Suo H, Zhao X, Goldys EM (2016) Site-dependent luminescence and thermal stability of $Eu^{2+}$ doped fluorophosphate toward white LEDs for plant growth. ACS Appl Mater Interfaces 8:20856–20864

Dutta Gupta S, Jatothu B (2013) Fundamentals and applications of light emitting-diodes (LEDs) in vitro plant growth and morphogenesis. Plant Biotechnol Rep 7:211–220

Harvey RB (1922) Growth of plants in artificial light. Bot Gaz 74:447–451

He G, Zheng L (2010) Color temperature tunable white-light light-emitting diode clusters with high color rendering index. Appl Opt 49(24):4670–4676

Kim HH, Wheeler RM, Sager JC, Yorio NC, Goins GD (2005) Light-emitting diodes as an illumination source for plants: a review of research at Kennedy Space Center. Habitat (Elmsford) 10:71–78

Kitsinelis S (2011) Light sources: technologies and applications. CRC Press, Florida

Lei Z, Xia G, Ting L, Xiaoling G, Ming LQ, Guangdi S (2007) Color rendering and luminous efficacy of trichromatic and tetrachromatic LED-based white LEDs. Microelectron J 38:1–6

Massa G, Kim H, Wheeler RM, Mitchell CA (2008) Plant productivity in response to LED lighting. HortScience 43:1951–1956

Mitchell CA, Both AJ, Bourget CM, Burr JF, Kubota C, Lopez RG, Morrow RC, Runkle ES (2012) LEDs: the future of greenhouse lighting! Chron Hortic 52:6–10

Mpelkas CC (1980) Light sources for horticultural lighting. Inst Electr Electron Eng Trans Ind Appl IA-16(4):557–565

Nakamura S, Fasol G (1997) The blue laser diode: GaN based light emitters and lasers. Springer, Berlin

Nakamura S, Senoh M, Nagahama SI, Iwasa N, Matsushita T, Mukai T (2000) Blue InGaN-based laser diodes with an emission wavelength of 450 nm. Appl Phys Lett 76(1):22–24

Pattison PM, Tsao JY, Krames MR (2016) Light-emitting diode technology status and directions: opportunities for horticultural lighting. Acta Hortic 1134:413–426

Pfeiffer NE (1926) Microchemical and morphological studies of effect of light on plants. Bot Gaz 81:173–195

Pinho P, Halonen L (2014) Agricultural and horticultural lighting. In: Karlicek R, Sun CC, Zissis G, Ma R (eds) Handbook of advanced lighting technology. Springer, Switzerland, pp 1–14

Round HJ (1907) Discovery of electroluminescence—blue light emission from silicon carbide (SiC). Electron World 19:309

Sancar A (2003) Structure and function of DNA photolyase and cryptochrome blue-light photoreceptors. Chem Rev 103(6):2203–2238

Schubert EF (2003) Light-emitting diodes. Cambridge University Press, UK

Shinomura T, Uchida K, Furuya M (2000) Elementary processes of photoperception by phytochrome A for high-irradiance response of hypocotyl elongation in Arabidopsis. Plant Physiol 122(1):147–156

Shur MS, Žukauskas A (2005) Solid-state lighting: toward superior illumination. Proc Inst Electr Electron Eng 93(10):1691–1703

Siemens CW (1880) On the influence of electric light upon vegetation, and on certain physical principles involved. Proc R Soc Lond 30:210–219

Simpson RS (2003) Lighting control—technology and applications. Focal Press, Oxford

Smith H (1995) Physiological and ecological function within the phytochrome family. Annu Rev Plant Biol 46:289–315

Spalding EP, Folta KM (2005) Illuminating topics in plant photobiology. Plant, Cell Environ 28:39–53

Tamulaitis G, Duchovskis P, Bliznikas Z, Breive K, Ulinskaite R, Brazaityte A, Novickovas A, Zukauskas A (2005) High-power light-emitting diode based facility for plant cultivation. J Phys D Appl Phys 38:3182–3187

US Department of Energy (2016) Solid-state lighting: R & D plan. http://energy.gov/sites/prod/files/2016/06/f32/ssl_rdplan_%20jun2016_2.pdf. Accessed 11 Sept 2016

Zheludev N (2007) The life and times of the LED—a 100-year history. Nat Photonics 1:189–192

Zissis G, Kitsinelis S (2009) State of art on the science and technology of electrical light sources: from the past to the future. J Phys D Appl Phys 42:173001

# Chapter 2
# LED Supplementary Lighting

**Yasuomi Ibaraki**

## 2.1 Introduction

Recently, the use of light-emitting diode (LED) lighting systems has rapidly spread to various fields due to the improved luminous efficacy and reduced production cost and market price of LED lamps. LEDs have also begun to be used in plant production due to various advantages such as flexibility in controlling lighting conditions, including the wavelength, irradiated portion, and timing. LEDs are used not only for plant production under a controlled environment such as a plant factory but also as light sources for supplementary lighting. During plant production, supplementary lighting, defined as irradiation that is additional to sunlight, is used to improve the light environment. Supplementary lighting is used to improve plant growth, i.e., to compensate for a shortage of sunlight for photosynthesis; to control plant morphogenesis, including flowering; to protect the plants from diseases; and to improve plant quality. LED technology has several merits for use in supplementary lighting systems. In this chapter, the advantages of using LEDs in supplementary lighting systems are discussed along with their applications for several purposes in plant production. The methods adopted for evaluating the efficiency of supplementary lighting have also been illustrated.

---

Y. Ibaraki (✉)
Faculty of Agriculture, Yamaguchi University 1677-1 Yoshida,
Yamaguchi 753-8515, Japan
e-mail: Ibaraki@yamaguchi-u.ac.jp



S. Dutta Gupta (ed.), *Light Emitting Diodes for Agriculture*,
DOI 10.1007/978-981-10-5807-3_2

## 2.2 Advantages of LED for Supplementary Lighting Systems

The advantages of using LEDs as a light source in plant production include their potential for miniaturization and reduction of weight, low radiant heat emittance, and flexibility in wavelength, intensity, light distribution. The primary advantage of using LEDs in supplementary lighting systems is their small size. Small lighting devices can minimize the interception of sunlight, i.e., shading, thereby maximizing the sunlight received by plants and in turn the production efficiency. In addition, the small size of LEDs contributes to enhanced portability, which is one of the desirable features of supplementary lighting devices because the requirement of supplementary lighting depends on the stage of plant growth and development, and the lighting devices may be moved to avoid a reduction in the lighting efficiency with plant growth (Ibaraki 2016). In addition, LED supplementary lighting systems can be set close to plants due to the low emittance of heat; therefore, they are suitable for intracanopy lighting, in which lighting devices are set inside the plant canopy. For example, LEDs have previously been set inside the canopy of tomato plants being grown in a greenhouse (Tokuno et al. 2012; Deram et al. 2014; Tewolde et al. 2016).

Research into the spectral effects of light on plant growth and development has progressed due to the use of LEDs, which can provide light of a narrow bandwidth at a relatively high intensity. LEDs can be used in supplementary lighting systems with light of a specific wavelength. For example, blue-violet LED supplementary lighting systems have been used for protecting plants from diseases (Tokuno et al. 2012) and red LEDs have been applied for controlling flowering (Liao et al. 2014). The intensity of light in LEDs can be easily controlled by regulating the electrical current or duty ratio. This would be helpful for a dynamic control of supplementary lighting according to the variation in solar radiation. Moreover, LEDs with high directivity may be useful for controlling the lighting direction or site, which may be effective in improving the lighting efficiency. Thus, LEDs provide flexibility in controlling the light environment and are suitable for supplementary lighting.

Several types of LED lighting devices for supplementary lighting have been developed. One type can be installed in existing general lighting equipment as an alternative to fluorescent lights and electric lamps. Another type is specially designed for plant production. These include line (bar), flat (plate), and small unit types. The flat-type LED lighting system is easy to handle and provides uniform irradiation but more shading. The line-type LED lighting system is equipped with LED lamps arrayed in a linear fashion and is easy to handle and can be used not only for downward lighting but also for sideward lighting. In addition, tape-type systems consisting of a flexible material also exist. The small unit LED lighting system consists of multiple small units equipped with one or several LED lamps (Fig. 2.1). This type of lighting system can be used as a line type or flat type depending on the arrangement of units. Moreover, light intensity is controlled by regulating the distance between the units.

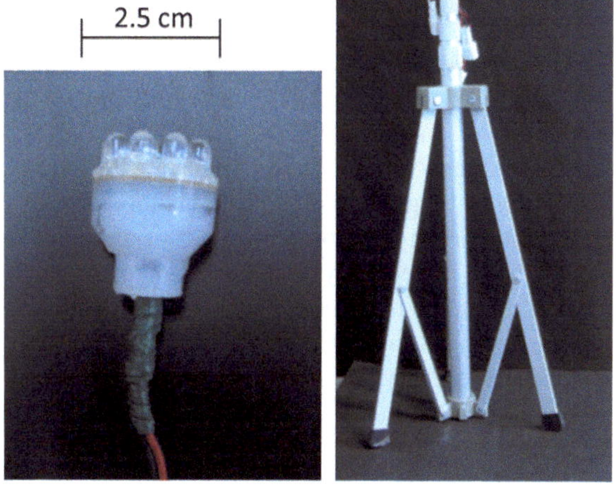


**Fig. 2.1** Example of a small unit LED supplementary lighting system (*left*) consisting of a small LED unit (*right*)

## 2.3 Supplementary Lighting for Photosynthesis

One of the main purposes of supplementary lighting is to compensate for a shortage of light for photosynthesis under conditions of low sunlight. LED supplementary lighting has reportedly been used for this purpose in tomatoes (Deram et al. 2014; Tewolde et al. 2016), lettuces (Wojciechowska et al. 2015), cucumbers (Trouwborst et al. 2010), strawberries (Hidaka et al. 2013), and peppers (Li et al. 2016). Supplementary lighting is used for irradiation during the daytime to improve the light intensity or for nighttime irradiation to prolong the photoperiod.

It should be noted that the spectral properties of LEDs are very important, even for supplementary lighting for photosynthesis. White light with a spectrum similar to that of sunlight is suitable for this purpose. However, there are several types of white LEDs with different spectral properties. Their effects on photosynthesis depend on the spectral properties. Although the ratio of red and blue lights is one of the indices used for evaluating the spectral properties of light for plant growth, absolute blue light intensity was also reported to be important, affecting the photosynthetic rate (Cope and Bugbee 2013).

The irradiated portion is also important because the photosynthetic properties depend on the leaf age and/or position, and light intensity distribution exists inside the canopy. The effect of supplementary lighting depends on the position of the leaves to be irradiated. For example, the degree of improvement of photosynthetic rates by increasing the photosynthetic photon flux density (PPFD) may differ between the upper leaves that have already received light with a high PPFD and lower leaves that receive light with a low PPFD.

## 2.4 Supplementary Lighting for Controlling Morphogenesis

In some species of plants, flowering can be controlled by regulating the photoperiod with supplementary lighting, allowing flower growers to control flowering according to market demand. In general, night breaking with supplementary lighting inhibits floral differentiation in short-day plants and promotes flower bud formation in long-day plants. This photoperiodism is thought to be caused by the photoreceptor phytochrome, which has two forms that mainly absorb red and far-red light, respectively, and so red and/or far-red light irradiation is important for controlling flowering. Incandescence lamps with a relatively high proportion of red and far-red light have previously been used for this purpose, but their use has been limited in recent years due to their low luminous efficacy and high electricity consumption. Consequently, the use of LED lamps has been now tested for controlling flowering.

The use of LEDs with a narrow range of wavelengths has revealed that the response of plants to the quality of light during the night break varies between species and that light other than red light may also affect flowering. For example, the optimal spectral properties of supplementary lighting for controlling flowering in chrysanthemum differ between varieties (Liao et al. 2014; Ochiai et al. 2015). In addition, the light quality supplied during the daily photoperiod might affect the light quality required for effective night break (Higuchi et al. 2012). Plants may be roughly divided into four groups based on their flowering response to different spectra of irradiation during night: inhibition mainly by red light, promotion mainly by red light, promotion mainly by far-red light, and no effect (Hisamatsu 2012).

Plant growth retardants are commonly used to regulate morphogenesis, but it is preferable to limit their use due to their potential negative effects on human health and the environment (Islam et al. 2015). Therefore, environmental control is a promising alternate method for controlling morphogenesis, which includes controlling the temperature difference between the daytime and nighttime (the DIF) and manipulating supplementary lighting regimes. Blue, red, and far-red lights are generally effective for controlling morphogenesis, associating with different photoreceptors (phototropins and cryptochromes for blue, and phytochromes for red and far red), and LED lamps can be used to control these spectral properties of light being irradiated on plants.

## 2.5 Supplementary Lighting for Other Purposes

### 2.5.1 Protection from Plant Disease

It has recently been reported that irradiation with specific wavelengths of light has the potential for suppressing plant disease. For example, irradiation with ultraviolet B (UV-B) light could suppress disease in strawberry (Kanto et al. 2009) and in rose (Kobayashi et al. 2013) and supplementary lighting devices using fluorescent lamps that provide UV-B irradiation are now commercially available ("Tahunarei," Panasonic). Green light irradiation during the nighttime has also been reported to have suppressive effects on plant disease (Kudo et al. 2011), and supplementary blue-violet LED lighting, which has an emission peak at around 405 nm, has been used to suppress disease in house-grown tomato (Tokuno et al. 2012). It is believed that blue-violet LED lighting induces resistance to plant disease (Ito et al. 2013), as well as having direct suppressive effects (Imada et al. 2014). It has also been reported that red light induces resistance to plant disease (Wang et al. 2010; Suthaparan and Torre 2010), although there are no reports of supplementary red LED lighting being used for this purpose. The protection of plants from disease through the use of lights is an emerging technology and so some issues remain to be resolved, including the dependence of different plant species and optimization of the lighting conditions.

### 2.5.2 Improving the Concentration of Functional Components

Light irradiation can induce the production of some secondary metabolites and so supplementary lighting has the potential for improving their contents as functional ingredients. Many studies have reported on the spectral effects of light on secondary metabolites, including phenolic acid and flavonoids, which are used as defense mechanisms under stressful conditions (Shetty et al. 2011). Furthermore, control of light quality has often been reported as enhancing antioxidant capacity (e.g., Ebisawa et al. 2008; Li and Kubota 2009; Shiga et al. 2009; Samuoliene et al. 2012; Carvalho et al. 2016) as a result of increasing the concentration of metabolites that serve as antioxidants, such as ascorbic acid and flavonoids. In an investigation on the effects of the proportion of blue/red light on rose, chrysanthemum, and campanula, Ouzounis et al. (2014) found that a high blue light ratio increased the concentrations of phenolic acid and flavonoids, although the effects differed between species.

LED supplementary lighting is effective for controlling the spectral properties of light being irradiated on plants. Furthermore, LEDs with various peak wavelengths are now available, which can be used to modify the light spectrum during the daytime and to irradiate plants with specific wavelengths of light during the night time.

## 2.6 Evaluation of the Efficiency of Supplementary Lighting

### *2.6.1 Methods for Evaluation of the Efficiency of Supplementary Lighting*

Although the use of artificial lighting in plant production has increased, little attention has been paid to the efficiency of lighting (Ibaraki and Shigemoto 2013). As artificial lighting, including supplementary lighting, consumes energy, thereby increasing the cost of production, it is critical that plant growers improve the efficiency of their lighting systems (Ibaraki 2016). An adequate evaluation of lighting efficiency is essential for determining methods of improving it.

As previously described, supplementary lighting is implemented for various purposes. The direct evaluation of supplementary lighting involves estimating the benefit/return corresponding to the objectives of an endeavor. This could be obtained by calculating the supplementary lighting per unit cost or per unit energy consumption required for the lighting. For example, a method to evaluate the efficiency of supplementary lighting for photosynthesis is to estimate the amount of biomass produced per unit of energy used to irradiate the plants (Ibaraki 2016). The possible benefits of supplementary lighting depend on the purposes of the lighting, e.g., increase in the concentration of the target component or control of period of flowering.

These parameters can be estimated by comparing results of a cultivation utilizing supplementary lighting with that not utilizing supplementary lighting. However, this approach may not be realistic as it requires cultivation without supplementary lighting for every comparison. Moreover, comparison or quantification of the results might be difficult in some cases, such as supplementary lighting for controlling morphogenesis or protection from plant disease. Alternatively, a possible important index is determining to what degree the distribution of light intensity is changed by the supplementary lighting. Thus, lighting efficiency can also be evaluated by interpreting the extent of light intensity that can be improved on leaf surfaces.

A method of evaluating the efficiency of supplementary lighting based on light intensity distribution on a canopy surface, expressed as a PPFD histogram, was previously developed (Ibaraki and Shigemoto 2013). This is based on the method developed by Ibaraki et al. (2012a, b), wherein the reflection images of plant canopy surfaces at specific ranges of wavelength acquired with a digital camera are used for the estimation of PPFD on leaf surfaces. The pixel value of the image is converted into a PPFD by a regression model determined from the PPFD measured at one point on the canopy with simultaneous imaging, following which a PPFD histogram is constructed. To characterize the PPFD distribution, an average PPFD, a median PPFD, and the coefficient of variance of the PPFD over the illuminated canopy surface can be calculated from the PPFD histogram. In addition, the fraction of leaf area with a PPFD value greater than a certain threshold can be calculated. This method has been applied for analyzing PPFD distribution on the canopy surface in tomatoes grown in a greenhouse (Ibaraki et al. 2012a) and lettuces

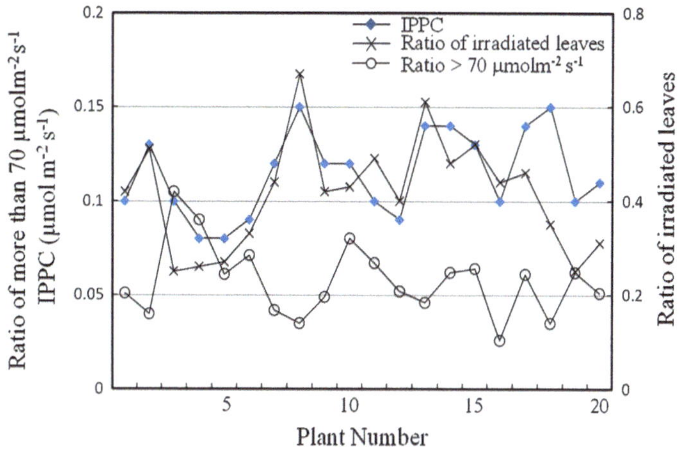


**Fig. 2.2** Example of the variation of efficiency of supplementary lighting with line-type LED lighting systems for tomato plants. Parameters were estimated for evaluating the efficiency in each plant ($n = 20$)

cultivated under artificial lighting (Miyoshi et al. 2016). Integrated PPFD over all illuminated leaves per unit power consumption (IPPC) was then proposed as a criterion for evaluating the efficiency of supplementary lighting (Ibaraki and Shigemoto 2013). IPPC was calculated by the following equation:

$$\text{IPPC}\ \left(\mu\text{mol}\,\text{s}^{-1}\,\text{W}^{-1}\ \text{or}\ \mu\text{mol}\,\text{J}^{-1}\right) = \frac{\text{Averaged PPFD}\ \left(\mu\text{mol}\,\text{m}^{-2}\,\text{s}^{-1}\right) \times \text{Projected leaf area}\ \left(\text{m}^2\right)}{\text{Power consumption for lighting}\ (\text{W})}.$$

Ibaraki and Shigemoto (2013) reported that the histogram pattern of PPFD on a tomato plant canopy surface under supplementary lighting depended on the canopy structures, types of light sources, and distance between lamps and the canopy surfaces, along with the difference in the efficiency, IPPC. Figure 2.2 shows an example of variation of the efficiency, IPPC, when each plant was irradiated by the same type of supplementary devices set at the same distance from plants in tomato cultivation in a greenhouse (Fig. 2.2). It should be noted that lighting efficiency depended on the canopy structure (leaf distribution pattern); therefore, lighting efficiency changed with time corresponding to plant growth.

### 2.6.2 *Practices to Increase the Efficiency of Supplementary Lighting*

Figure 2.3 shows a strategy for improving the efficiency of supplementary lighting. An effective method is to use lamps with high luminous efficacy. However, lighting

Fig. 2.3 Practices for improving the efficiency of supplementary lighting

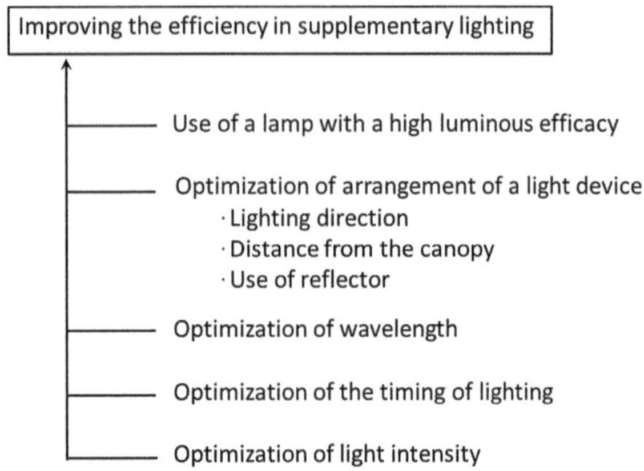


efficiency also depends on the arrangement of the lamps and/or the plant canopy structure being irradiated (Ibaraki 2016). Minimizing unnecessary irradiation, such as irradiation that excludes plants, should be considered. This might depend on several factors, including lighting direction, light distribution of lamps, and the distance from plants. The timing of lighting is also important for supplementary lighting (Ibaraki 2016): Lighting at night is effective for promoting the growth of lettuce (Fukuda et al. 2004), and end-of-day lighting, which is irradiation just before the onset of darkness, is effective in controlling plant morphological events (e.g., Yang et al. 2012). The efficiency of supplementary lighting can be improved in regard to irradiation position (Ibaraki 2016). Moreover, the parts of plants to be irradiated should be considered based on the physiological properties of the plants.

## 2.7 Conclusion

LED supplementary lighting is a promising technology for improving crop productivity and quality by controlling plant growth and development. The most significant merit of LED technology as a light source for supplementary lighting is its flexibility for controlling the light environment. By using LEDs, the effects of light quality, i.e., light spectrum, on plant growth and development can be revealed in detail, thereby allowing effective supplementary lighting guidelines to be developed. To take full advantage of LED supplementary lighting, it is essential to adequately evaluate the efficiency of LED lighting. Therefore, for evaluating the efficiency of supplementary lighting, it is important to determine how much of the light distributed on the plant canopy surface is changed by the supplementary lighting. Moreover, this contributes to improve stability and reproducibility in both research and application of supplementary lighting.

# References

Carvalho SD, Schwieterman ML, Abrahan CE, Colquhoun TA, Folta KM (2016) Light quality dependent changes in morphology, antioxidant capacity, and volatile production in sweet basil (*Ocimum basilicum*). Front Plant Sci 7:1328

Cope KR, Bugbee B (2013) Spectral effects of three types of white light-emitting diodes on plant growth and development: absolute versus relative amounts of blue light. HortScience 48:504–509

Deram P, Lefsrud MG, Orsat V (2014) Supplemental lighting orientation and red-to-blue ratio of light-emitting diodes for greenhouse tomato production. HortScience 49:448–452

Ebisawa M, Shoji K, Kato M, Shimomura K, Goto F, Yoshihara T (2008) Supplementary ultraviolet radiation B together with blue light at night increased quercetin content and flavonol synthase gene expression in leaf lettuce (*Lactuca sativa* L.). Environ Control Biol 46:1–11

Fukuda N, Nishimura S, Fumiki Y (2004) Effect supplemental lighting during the period from middle of night to morning on photosynthesis and leaf thickness of lettuce (*Lactuca sativa* L.) and tsukena (*Brassica campestris* L.). Acta Hortic 633:237–244

Hidaka K, Dan K, Imamura H, Miyoshi Y, Takayama T, Sameshima K, Kitano M, Okimura M (2013) Effect of supplemental lighting from different light sources on growth and yield of strawberry. Environ Control Biol 51:41–47

Higuchi Y, Sumitomo K, Oda A (2012) Daylight quality affects the night break response in the short-day plant chrysanthemum, suggesting differential phytochrome-mediated regulation of flowering. J Plant Physiol 169:1789–1796

Hisamatsu T (2012) Chapter 13. In: Goto E (ed) Agri-photonics II, the latest technology of plant factories with LED lighting. CMC, Tokyo, pp 114–124 (In Japanese)

Ibaraki Y (2016) Lighting efficiency in plant production under artificial lighting and plant growth modeling for evaluating the lighting efficiency. In: Fujiwara K, Runkle ES (eds) Kozai T. LED lighting for urban agriculture, Springer, pp 151–161

Ibaraki Y, Shigemoto C (2013) Estimation of supplemental lighting efficiency based on PPFD distribution on the canopy surface. J Agric Meteorol 69:47–54

Ibaraki Y, Kishida T, Shigemoto C (2012a) Image-based estimation of PPFD distribution on the canopy surface in a greenhouse. Acta Hortic 956:577–582

Ibaraki Y, Yano Y, OkuharaH Tazuru M (2012b) Estimation of light intensity distribution on a canopy surface from reflection images. Environ Control Biol 50:117–126

Imada K, Tanaka S, Ibaraki Y, Yoshimura K, Ito S (2014) Antifungal effect of 405-nm light on *Botrytis cinerea*. Lett Appl Microbiol 59:670–676

Islam MA, Gislerød HR, Torre S, Olsen JE (2015) Control of shoot elongation and hormone physiology in poinsettia by light quality provided by light emitting diodes-a minireview. Acta Hortic 1104:131–136

Ito S, Yoshimura K, Ibaraki Y (2013) Chapter 14. In: Goto E (ed) Agri-Photonics II. The latest technology of plant factories with LED lighting, CMC, Tokyo, pp 125–132 (In Japanese)

Kanto T, Matsuura K, Yamada M, Usami T, Amemiya Y (2009) UV-B radiation for control of strawberry powdery mildew. Acta Hortric 842:68

Kobayashi M, Kanto T, Fujikawa T (2013) Supplemental UV radiation controls rose powdery mildew disease under the greenhouse conditions. Environ Control Biol 51:157–163

Kudo R, Ishida Y, Yamamoto K (2011) Effects of green light irradiation on induction of disease resistance in plants. Acta Hortic 907:251–254

Li Q, Kubota C (2009) Effects of supplemental light quality on growth and phytochemicals of baby leaf lettuce. Environ Exp Bot 67:59–64

Li X, Lu W, Hu G, Wang XC, Zhang Y, Sun GX, Fang Z (2016) Effects of light-emitting diode supplementary lighting on the winter growth of greenhouse plants in the yangtze river delta of China. Bot Stud 57. doi:10.1186/s40529-015-0117-3

Liao Y, Suzuki K, Yu W, Zhuang D, Takai Y, Ogasawara R, Shimazu T, Fukui H (2014) Night break effect of LED light with different wavelengths on floral bud differentiation of

*Chrysanthemum morifolium* Ramat 'Jimba' and 'Iwa no hakusen'. Environ Control Biol 52:45–50
Miyoshi T, Ibaraki Y, Sago Y (2016) Development of an android-tablet-based system for analyzing light intensity distribution on a plant canopy surface. Comput Electron Agric 122:211–217
Ochiai M, Liao Y, Shimazu T, Takai Y, Suzuki K, Yano S, Fukui H (2015) Varietal differences in flowering and plant growth under night-break treatment with LEDs in 12 *Chrysanthemum* cultivars. Environ Control Biol 53:17–22
Ouzounis T, Frette X, Rosenqvist E, Ottosen CO (2014) Spectral effects of supplementary lighting on the secondary metabolites in roses, chrysanthemums, and campanulas. J Plant Physiol 171:1491–1499
Samuolienė G, Sirtautas R, Brazaitytė A, Viršilė A, Duchovskis P (2012) Supplementary red-LED lighting and the changes in phytochemical content of two baby leaf lettuce varieties during three seasons. J Food Agric Environ 10:701–706
Shetty R, Fretté X, Jensen B, Shetty NP, Jensen JD, Jørgensen HJ, Newman MA, Christensen LP (2011) Silicon-induced changes in antifungal phenolic acids, flavonoids, and key phenyl-propanoid pathway genes during the interaction between miniature roses and the biotrophic pathogen *Podosphaera pannosa*. Plant Physiol 157:2194–2205
Shiga T, Shojil K, Shimada H, Hashida S, Goto F, Yoshihara T (2009) Effect of light quality on rosmarinic acid content and antioxidant activity of sweet basil, *Ocimum basilicum* L. Plant Biotechnol 26:255–259
Suthaparan A, Torre A (2010) Specific light-emitting diodes can suppress sporulation of *Podosphaera pannosa* on greenhouse roses. Plant Dis 94:1105–1110
Tewolde FT, Lu N, Shiina K, Maruo T, Takagaki M, Kozai T, Yamori W (2016) Night time supplemental LED inter-lighting improves growth and yield of single-truss tomatoes by enhancing photosynthesis in both winter and summer. Front Plant Sci 7:448. doi:10.3389/fpls. 2016.00448.e-Collection
Tokuno A, Ibaraki Y, Ito S, Araki H, Yoshimura K, Osaki K (2012) Disease suppression in greenhouse tomato by supplementary lighting with 405 nm LED. Environ Control Biol 50:19–29
Trouwborst G, Oosterkamp J, Hogewoning SW, Harbinson J, Ieperen WV (2010) The responses of light interception, photosynthesis and fruit yield of cucumber to LED-lighting within the canopy. Physiol Plant 138:289–300
Wang H, Jiang YP, Jing H (2010) Light quality affects incidence of powdery mildew, expression of defense-related genes and associated metabolism in cucumber plants. Eur J Plant Pathol 127:125–135
Wojciechowska R, Długosz-Grochowska O, Kołton A, Żupnik M (2015) Effects of LED supplemental lighting on yield and some quality parameters of lamb's lettuce grown in two winter cycles. SciHortic 187:80–86
Yang ZC, Kubota C, Chia PL, Kacirac M (2012) Effect of end-of-day far-red light from a movable LED fixture on squash rootstock hypocotyl elongation. Sci Hortic 136:81–86

# Chapter 3
# Influence of Light-Emitting Diodes (LEDs) on Light Sensing and Signaling Networks in Plants

**T. Pocock**

## 3.1 Introduction

In 2016, an estimated 54% of the world's population lived in urban centers, and by 2030, this will increase to 60% (United Nations 2016). In addition to the demographic shift, 50% more food will need to be produced by 2030 resulting in an estimated 45% increase in energy and 30% increase in water consumption (U.S. Global Change Research Program (USGCRP) 2015). Simultaneously, agricultural areas are becoming increasingly climate-challenged suffering from drought, heat waves, flooding, hurricanes, and pollution, and this is physically threatening our food security (IPCC 2014; USGCRP 2015). As a response to the above concerns, many entrepreneurs are investing in the intensive cultivation methods of controlled environment agriculture (CEA) to ensure the year-round availability of food. Initially, CEA referred to greenhouses, but a more recent CEA system is the plant factory (PF), where light is supplied solely by electric lamps (sole-source lighting; SSL; Kozai et al. 2015). Both types of facilities are equipped with heating, ventilation, and air conditioning (HVAC) systems, $CO_2$ enrichment systems, and light (Kozai et al. 2015). Out of the 35,000 plant species under cultivation, approximately 7,000 are used for food; while in CEA, the major food crops are tomatoes, leafy greens, cucumbers, pepper, and eggplant, each of which contains many different cultivars (Khoshbakht and Hammer 2008). Understanding and unifying plant evolution, plant physiological processes and acclimation responses will accelerate the development of lighting programs for CEA crops growing under light-emitting diode (LED) light sources.

T. Pocock (✉)
Center for Lighting Enabled Systems and Applications (LESA),
Rensselaer Polytechnic Institute, 110 8th Street, CII 7015,
Troy, NY 12180, USA
e-mail: pococt@rpi.edu



S. Dutta Gupta (ed.), *Light Emitting Diodes for Agriculture*,
DOI 10.1007/978-981-10-5807-3_3

Plants have survived in the natural world through their ability to sense and integrate at least fifteen different environmental variables (light quantity, quality and duration, temperature, humidity, $CO_2$, soil moisture, and nutritional status, among others) simultaneously with high sensitivity, and they manage these large data sets through sophisticated and complex sensing and signaling networks (Trewavas 2002;

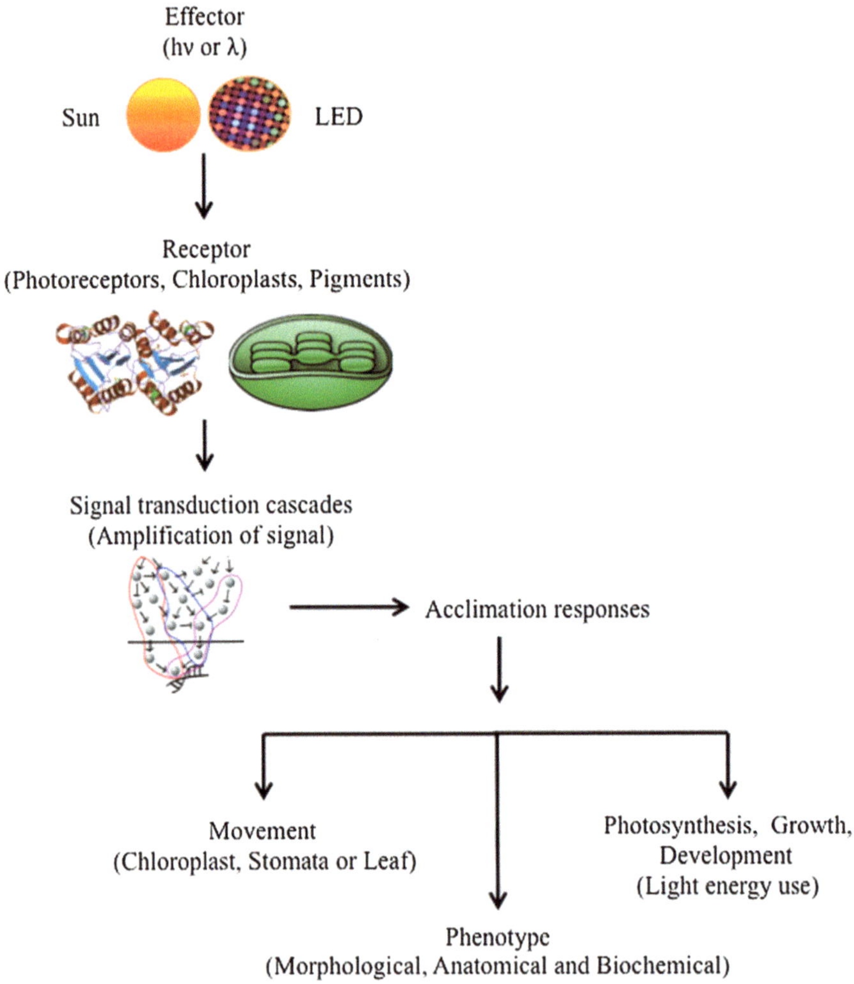


**Fig. 3.1** Acclimation paradigm. Both the photon flux density (PFD) (hν) and the spectral distribution (λ) are primary effectors in plant acclimation responses. The receptors consist of families of photoreceptors as well as the photosynthetic apparatus and pigments within the chloroplast. Once activated by narrow, broad, or intense bands of light, they transmit signal molecules that turn on acclimation or housekeeping genes primarily in the nucleus. The acclimation responses include optimizing photosynthetic efficiency, altering development and growth, avoidance (movement) and photoprotection, and repair (phenotype). The acclimation paradigm has traditionally been examined with respect to stress, but this fundamental knowledge is transformative and can be used to influence CEA grown crops using the quantity (PFD), quality (λ), duration and timing (Pulse Width Modulation (PWM) or photoperiod)

Karpinski and Szechynska-Hebda 2010). They utilize light not only as an energy source for photosynthesis, but also as an information source that enables them to predict and respond to imminent changes in their environment. The evolution of plant sensing and signaling networks has made the plant kingdom one of the richest on Earth and has enabled them to dominate every terrestrial environment (Trewavas 2002; Scheffers et al. 2012). The early investigations on light signaling networks were performed under high and low photosynthetic photon flux densities (PPFD, 400–700 nm) or darkness and the steps in signal transduction during light-mediated development (photomorphogenesis) have been elucidated. Signal transduction cascades are three-step biochemical events (receive–transduce–respond) that serve to amplify and coordinate signals that induce the acclimation paradigm (Fig. 3.1).

The first step is sensing and processing a signal imposed by an effector such as light. Plants sense changes in the photon flux density (PFD between 380 and 750 nm), daily light integral (DLI), quality (wavelength and spectral ratios), duration (clouds, sunflecks, frequency, and pulse width modulation; PWM), and timing (photoperiod) by way of organelles (chloroplasts) and families of individual photoreceptors. In contrast to animals, whose photoreceptors are limited to specialized organs such as the eyes, plants have light sensors embedded in every tissue (Galvão and Frankhauser 2015). In addition, photoreceptors are expressed in different tissues at different developmental stages (Goosey et al. 1997; Sakamoto and Briggs 2002). To the human eye, light responses in plants often appear subtle due to their sessile nature, but internally light has a dynamic and strong influence on the regulation of plant processes. The second step of signal transduction amplifies the signal through biochemical cascades involving the phosphorylation or conformational change of second messengers that ultimately result in the activation or repression of gene expression and/or the modification of proteins and pigments (Clark et al. 2001; Giraudat and Schroeder 2001; Taiz and Zeiger 2010). Lastly, a cellular, long distance, or whole plant response is realized. Plants have the ability to acclimate, and this is often confused with adaptation, the inherited genetic capacity to survive environmental pressures over many generations. Acclimation responses involve either short-term (seconds to minutes) or long-term (hours, weeks, seasons) reversible reprogramming of molecular and biochemical events that result in altered plant physical and chemical attributes (Taiz and Zeiger 2010; Chory 2010; Alter et al. 2012; Demmig-Adams et al. 2012; Schottler and Toth 2014; Ruban 2015). Acclimation products protect plants under stress indoors or in nature where it serves to maintain photosynthesis and growth; however, these acclimation products can play a concomitant protective function in human health and can add value to the consumer and grower, respectively. Positive acclimation responses include nutrient density (functional foods), stress resistance (shelf life), increased pathogen and herbivore defense metabolites (aroma and taste), and pigmentation (appearance).

Plants are resilient, and another unique plant quality is the level of plasticity that many species possess. Phenotypic plasticity is defined as the range of phenotypes (morphology, anatomy, development, and nutrient density) expressed by a single genotype as a response of changes in the environment (Gratani 2014). Phenotypic responses to light include the control of leaf mass per unit area, stomata size and density, height, flowering time, seed size, water use efficiency, leaf size, shape and thickness, root-to-shoot ratio, specific root length, plant chemical defense, pigmentation, energy capture within the chloroplast, thylakoid membrane dynamics, and photosynthesis (Nicotra et al. 2010; Ruban 2015). In short, the interaction of a crop plant with light and the ability of a crop to manifest different and desirable phenotypes are genetically determined and will depend on the selection of the crop species and cultivar (Fig. 3.2).

Conversely, growth conditions, particularly light, have been shown to be able to push genetic limitations further (Rosevear et al. 2001; Marin et al. 2015). Another interesting aspect of light acclimation of plants is their ability to store and use PFD and spectral compositions for several days and even months in anticipation (Karpinski and Szechynska-Hebda 2010; Thelier and Luttge 2012). Hardening, conditioning, or priming are common terms used to describe the acclimation process. Farmers often place field transplants in cold frames where they are exposed to sunlight and low temperatures before planting them in the field due to the increased stress tolerance as a result of acclimation. The effects of crop acclimation are not new, but the idea of the maintenance and induction of acclimation responses through stored light memory is just the beginning to be understood at the physiological and molecular levels (Crisp et al. 2016). Fundamental plant acclimation research is primarily focused on advancing the understanding of controlled stress responses at the cellular and molecular levels in the model plant organism, *Arabidopsis thaliana*. Crop physiology needs to be matched with LED electronic characteristics to achieve acclimation and vice versa. The development of light strategies using LEDs in CEA is in its infancy, and there have been rapid advances in the electronics and spectrum control. Learning how to use LEDs to influence crops can progress further through

**Fig. 3.2** Acclimation of the red pigmentation in red lettuce Rouxai is spectrum dependent. Pictorial representation of 28-day-old red lettuce Rouxai plants grown under phosphor-converted LEDs (*left*) and cool white fluorescent (*right*) light under similar cultivation and environmental conditions. Growth conditions were 200 μmol $m^{-2}$ $s^{-1}$ PPFD, 16 h photoperiod, and day/night temperatures 25/20 °C in isolated hydroponic unit areas containing 53 plants, EC 1.6 (Pocock data unpublished)

the understanding how, what, and when plants sense and process narrow bands of the spectrum. This review describes the characteristics of natural and electric light and how plants sense and respond to fluctuations on time scales from femtoseconds (light capture) to months (development). Understanding and integrating fundamental plant physiology research into applied CEA through advanced solid-state technology has a large potential to further enhance crop production. How to activate or reverse light acclimation through sensing and signaling networks in plants using the characteristics of LEDs is discussed.

## 3.2 Characteristics of LED Systems that Can Activate Plant Networks

Electric lamps have been used in CEA for almost 150 years, and horticulture lighting technologies closely follow those used for human applications (Wheeler 2008). Until now, the best light environment for specific crops or different growth stages (propagation, vegetative, flowering, and fruiting) in greenhouses was limited to incandescent lamps (1920s), fluorescent lamps (1930s), and high-intensity discharge lamps such as high-pressure sodium (HPS) and metal halide (MH) (1950s) (Wheeler 2008). Plant factories require a different light strategy as they consist of closely spaced shelves of crops, usually leafy greens, where fluorescent lamps or, more recently, LED systems are being used. The first technological development since the 1950s is solid-state lighting such as the light-emitting diode (LED). The first visible-light LED produced in 1962 was red, and since the 1990s, they have become brighter and many more direct-emission wavelengths have become available through materials and deposition improvements (Krames 2016). Phosphor conversion of light has been employed for over 50 years, for instance, fluorescing natural earth elements or synthetic materials are mixed and applied to fluorescent and metal halide lamps to block much of the UV light while emitting 'white' light within the human visible range (380–720 nm) (Van Broekhoven 2001). In 2014, the Nobel Prize was awarded for the invention of an efficient blue LED, and this led to the creation of the white LED through the phosphor down conversion of high-energy blue wavelengths (George et al. 2013). The wide availability of adjustable narrow band and phosphor-converted wavelengths offers almost limitless light algorithms; however, due to crop and wavelength specificity, generalizations are difficult to make. The physical benefits of using LEDs in horticultural settings include their long lifetime, rapid cycling (on/off), the variety of spectral distributions (color mixing), the lack of infrared radiation (heat), the potential for new designs, and in some cases energy savings. Three characteristics of LED systems can be used to activate beneficial acclimation responses are quantity (PFD), quality (wavelengths), duration (milliseconds to days), and timing of light.

Some plants function better under low light such as leafy greens ($\leq 250$ µmol $m^{-2}$ $s^{-1}$ or $\sim 12$ DLI), where they are able to balance photosynthesis with respiration (Kiang et al. 2007). Others do well at higher light (>600 µmol $m^{-2}$ $s^{-1}$ or $\sim 26$ DLI) with the upper limit determined by resource limits (water, nutrients, and $CO_2$). The PFDs at crop level under LED systems can be adjusted through fixture engineering and design such as increased number of LEDs, adjusting current, good thermal management, and the use of optics to concentrate photons or simply, physically moving luminaires closer to the crop. The intensity levels of light (PFD) required by plants are far greater than that for human vision and light output is a challenge for horticultural fixtures. The spectral distributions and the peak maxima ($\lambda_{max}$) of a horticultural light system had significant effects on plant growth, development, and metabolism. Johkan et al. (2012) not only confirmed that green light participates in photosynthesis and plant growth, but the results also showed that 10 nm differences in $\lambda_{max}$ of the green LEDs can have significant negative effects on photosynthesis and photomorphogenesis of lettuce. This has also been observed in the blue region of the spectrum, red lettuce grown under LEDs with similar PPFDs, and blue (10–17%) to red (90–87%) ratios had significantly less fresh weight and lower anthocyanin concentration when the $\lambda_{max}$ of the blue LEDs was 434 nm compared to 470 nm (Pocock 2015a). Thus, crop species and their cultivars are differentially sensitive to narrow bands of light emitted by LED systems. In addition to the effects of the $\lambda_{max}$ for green LEDs, this could potentially be the same for the red region of the spectrum, as a broad band between 580 and 630 nm has been correlated to photosystem II photodamage (Takahashi et al. 2010). The only other region that had high photodamage efficiency was UV, below 420 nm (Takahashi et al. 2010). Another characteristic of LED systems that can be used to maintain, inhibit, or activate sensing and signaling networks is dimming through pulse width modulation (PWM). Much like photosynthesis, dimming LEDs using PWM consists of light and dark phases. The duty cycle is the fraction of one period the LED is on and is frequently expressed as a percentage. Low PFDs are achieved with lower duty cycles, and the highest PFD is achieved at duty cycles of 100% (on all of the time). Another light modulation aspect of LED systems is the frequency in Herz (Hz) or cycles per second; for HPS and fluorescent light, it is between 50 and 60 Hz. Although PWM and frequency are not usually adjustable or described in horticultural LED data sheets, it could be used to modulate and acclimate CEA crops. The effects of pulsed and continuous light on plant and algal photochemistry and growth have been examined historically using rotating shutters in front of incandescent and HID light sources, but the results were not consistent and were often contradictory, although recently LEDs have been used as a pulsed light source. It took a few billion years for photosynthesis to evolve under a fluctuating Sun, and it will take time to explore, understand, and improve it using programmable narrow bands of electric light.

## 3.3 Mimicking the Sun

Advantages to growing high value and perishable crops (greens, tomatoes, cucumbers, and peppers) in CEA compared to field production are: year-round production, higher yields, higher visual quality, and sometimes, better nutritional quality, and thus higher market value (Gruda 2005). There is increased awareness and demand for nutritional or functional foods, and the superior aromas and flavor of field grown crops are suggested to be in large part due to the fluctuating solar light quantity (PFD), quality (wavelength), and distribution (time and space) (Bian et al. 2014). Greenhouse crops can obtain approximately 73% of their required light energy from the Sun, and the light transmitted through most greenhouse glazings scatter the light while filtering out the UV-B and most of the UV-A. Lighting requirements and the spectral control of crops in greenhouses and under sole-source lighting in PFs needs to be examined separately, as greenhouse crops are exposed to the natural fluctuations of sunlight. Better understanding on both the dynamics of the Sun and the dynamics of plant sensing and signaling networks will be beneficial for both scenarios.

Solar fluctuations are occurring constantly on long-term time scales such as the 11-year cycle, annual, or monthly (seasonal), and on the short-term including daily, minutes, and even sub-second time scales (Tomson 2010; Kopp and Lean 2011). LED technology has now enabled horticultural lighting to mimic both the spatial, spectral, temporal, and fluctuating properties of the Sun. In nature or in a greenhouse, both the PPFD and the spectrum are in constant flux and plants respond or acclimate immediately. A fourteen-year study of the DLI in Ithaca, New York, indicated yearly variances of 15 (winter) to 60 mol $m^{-2}$ $day^{-1}$ (summer) on a given day during the summer months (Albright et al. 2000). On a shorter timescale, fluctuations in solar radiation can be significant under fast moving clouds and can oscillate up to 600 W/$m^2$ (2742 μmol/$m^2$/s) within seconds (Tomson 2010). The spectral quality of light also changes significantly over time such as under different cloud types moving across the sky, the growth and shading of neighboring plants or rapid sunflecks. Solar spectral distributions were monitored under different sky conditions for 15 months in Miami, Florida, and significant diurnal variations were observed (Lee and Downum 1991). Large increases in the blue-to-red ratios occurred at sunrise and sunset, there was an early dawn drop in the blue-to-red ratio followed by a large increase at sunrise and a decrease an hour later and, R:FR ratios were significantly high immediately after sunrise (Lee and Downum 1991; Lee and Hernandez-Andres 2005). Atmospheric water content, aerosols, and cloud cover are dynamic and will also affect the amount and type of light reaching the plant canopy. Under thinly cloudy skies, blue light was enhanced by 5–15%, red light was diminished by 6–11%, daily light integral was 85% lower, and the light reaching the plant was diffuse (Reinhardt et al. 2010). Clearly, the natural light environment is dynamic and the benefits of dynamic lighting on crops can now be theorized and tested. The ability to rapidly control all LED characteristics is opening up new possibilities for targeting specific sensing and signaling pathways in plants (Pocock 2015b). The acclimation processes that plants use in nature are often beneficial and can now be mimicked using LEDs.

## 3.4 Integration of Plant Biology into Engineered Lighting Systems

There is a large potential to increase crop yield and crop quality in CEA using solid-state lighting, although the precise knowledge of which wavelengths to use for how long, at what time, and at what PPFD for specific crop plants is still largely unknown. Crop quality is a subjective and complex term. It has different meanings depending on the desired outcome; for producers, it can refer primarily to high yield, stress tolerance, shelf life, appearance, and smell. Consumers would more likely describe quality as freshness, appearance, color, smell, taste, texture, nutritional value, and health benefits. Photosynthesis underlies all plant functions including the energy required for growth, acclimation processes, and the respective improvement of crop qualities. Therefore, an important function of any horticultural light system is the ability to drive photosynthesis. McCree (1972) published the action spectra for 22 different crop plants in order to determine the spectral boundaries of photosynthetic active radiation (PAR), and it was concluded and subsequently confirmed to be between 400 and 700 nm (Hogewoning et al. 2010). Isolated chlorophyll absorbs in the blue and red regions of the spectrum; however, in the plant cell, chlorophyll is bound to a protein and embedded in the thylakoid membrane, and this broadens its absorption properties across the solar spectrum (Blankenship 2002). Light capture is fundamental to photosynthesis, and apart from chlorophyll, accessory pigments are important components of the light harvesting complexes where they absorb in the blue, green, and red regions. Accessory pigments include the carotenoids (i.e., β-carotene, zeaxanthin, and lutein), and in addition to light harvesting, they protect the photosynthetic apparatus from over

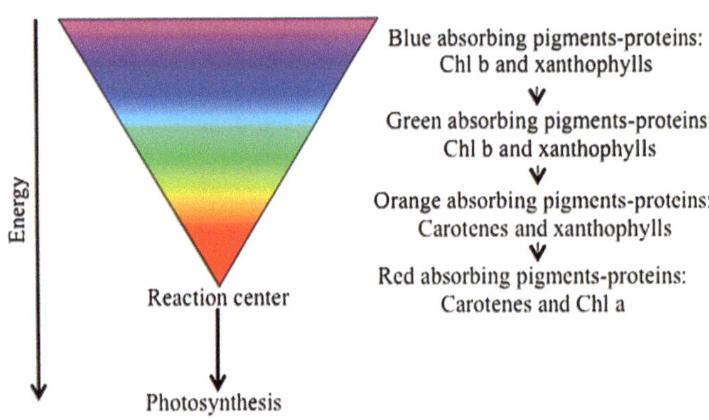


**Fig. 3.3** Funnel concept of light absorption in photosynthetic antennae systems. The distal parts of the antennae (furthest from the reaction centers) absorb maximally at shorter wavelengths (higher energy) than do the pigment–protein antennae complexes proximal to the reaction centers of the photosystems. Although not all energy transfer events are downhill, this schema depicts the organization of light harvesting and the need for a broad spectrum of light for maximal efficiency. Redrawn from Blankenship (2002), Ruban (2015)

excitation through quenching excited chlorophyll and the thermal dissipation of excess light energy (Esteban et al. 2015; Ruban 2015). Pictorially, light harvesting can be seen as funnelling light down an energy gradient to the reaction centers (Fig. 3.3). Blue and red light are absorbed preferentially at the adaxial side (top) of leaves and are more efficient at driving photosynthesis in this region compared to green light (Sun et al. 1998; Nishio 2000; Terashima et al. 2009).

As a consequence, green light is transmitted deeper into the leaf and is more efficient than either blue or red light at driving $CO_2$ fixation at the abaxial sides (bottom) (Sun et al. 1998; Terashima et al. 2009). Green light is absorbed and utilized by plants in photosynthesis and should be present in CEA light environments (Vogelmann and Han 2000). Unlike conventional lighting, a unique feature of LEDs is that their narrow bands can be simultaneously and rapidly used to activate endogenous sensing and signaling networks used in acclimation processes. Typically, sensing and signaling in plants refer to the five families of photoreceptors containing at least 12 different photoreceptors; however, there is another primary sensing and signaling network that works daily to maintain photosynthetic efficiency and is known as photosynthetic control (Pfannschmidt 2003; Möglich et al. 2010; Pfalz et al. 2012; Snowden and Inzé 2016). There is a lot of redundancy and cross talk between the plant sensing signaling networks, and this increases the complexity of programming LED light programs to yield desired crop qualities. Different plant species, different spectral combinations, different photoperiods and irradiances, and different modes of growth have been used in the published literature which makes it difficult to draw general or definitive conclusions.

Another LED characteristic that can be engineered into horticultural LED light systems is PWM (Sager and Giger 1980; Tennessen et al. 1995; Jao and Fang 2004; Olvera-Gonzalez et al. 2013). Only a few reports of the effect of pulsing conventional or LED light on photochemistry and plant health have been published, and more evidence needs to be collected before this opportunity can be used. There was no general consensus with respect to the benefits of pulsing light in the literature cited above; however, one common thread was observed. The length of the dark period of PWM had the primary influence on the photosynthetic rates of plants grown or exposed to pulsed LED light. From an energy perspective, Jao and Fang (2004) examined the interactive effects of duty cycles and frequency in the growth of potato plantlets and concluded that using blue and red LEDs at 180 Hz and a 50% duty cycle over a 16 h photoperiod reduced energy consumption without significantly affecting yield even when taking heat removal from the growth area into account. *A. thaliana* plants moved from growth under constant light in the laboratory to fluctuating solar conditions in the field developed higher maximum quantum efficiencies of PSII, had lower NPQ and zeaxanthin values and higher photosynthetic capacities (Wituszynska et al. 2013). In contrast, when light from undisclosed light sources fluctuated from 200 to 2000 μmol $m^{-2}$ $s^{-1}$, a decrease in quantum yields of photosynthesis but higher NPQ and zeaxanthin concentrations was observed (Kromdijk et al. 2016). These latter results confirm findings from a sunfleck simulation study, where short pulses of light (20 s) in contrast to long pulses (40 min) increased NPQ (Alter et al. 2012). When using more than one wavelength (color), the timing of the pulses of the different LEDs can be in or out of phase which can alter plant metabolism. The effect of timing of pulsed and direct current (100% duty cycle) of blue and red LEDs was examined on the model organism *A. thaliana* (Shimada and Taniguchi 2011). The pulsed red and blue LEDs were driven at 2.5 kHz with a 45% duty cycle in phase ($\phi = 0°$) and out of phase ($\phi = 180°$), and both the leaf area and chlorophyll concentration were negatively affected when grown out of phase (Shimada and Taniguchi 2011). Different spectral regions were illuminating the plant canopy in a nonuniform pattern and therefore were 'affecting' the sensing and system networks sequentially; the biology was also out of phase.

## 3.5 Sensing and Signaling Networks: What Are They and How Do They Work?

### *3.5.1 Photosynthetic Sensing and Signaling Networks: Photosynthetic Control*

Photosynthetic sensing and signaling networks, also referred to as photosynthetic control (PSC), are predominant in plants and are used in rapid day-to-day operations (Pfannschmidt et al. 1999; Zachgo et al. 2013). Light is a requirement for plants, but paradoxically, it is damaging, and without acclimation mechanisms, plants would not have survived four million years of evolution in terrestrial environments. Beneficial light acclimation responses initiated by PSC include increased photosynthetic efficiency, stomatal opening, nutrient density, germination, plant height, specific leaf weight, and plant defense (Pocock et al. 2001; Pfannschmidt 2003; Zhu et al. 2004; Buchanan and Balmer 2005; Potters et al. 2010; Wang et al. 2009; Hüner et al. 2012). The chloroplast, and more precisely the energy balance of the different components of the electron transport chain, is considered a global plant sensor (Huner et al. 1998; Murchie and Lawson 2009; Biswal et al. 2011). The complex set of photosynthetic reactions can be divided into three phases based on their time constants: (1) primary photochemistry, (2) electron shuttling, and (3) carbon metabolism (Ruban 2015). Primary photochemistry involves light capture, energy transfer, and charge separation within the photosystems that subsequently induces the electron transfer reactions (Blankenship 2002). The two photosystems, photosystem II (PSII) and photosystem I (PSI), work electrochemically in series within the thylakoid membrane, they absorb different spectral regions. For efficient photosynthesis, they must receive similar amounts of light energy as imbalances will cause limitations in the electron transport rate (Durnford and Falkowski 1997; Huner et al. 1998; Hüner et al. 2012; Ruban 2015). Primary photochemistry is typically very efficient and rapid, light capture by the light harvesting complexes occurs within femtoseconds ($10^{-15}$) while charge separation occurs within pico- and nanoseconds ($10^{-12}$–$10^{-9}$) (Tennessen et al. 1995;

Biswal et al. 2011). The second phase involves the extraction of electrons from water, and the subsequent transfer from PSII to PSI through a series of electron carriers. This occurs within micro- to milliseconds ($10^{-6}$–$10^{-3}$) unless the light energy absorbed exceeds the utilization of the chemical energy in plant metabolism (Huner et al. 1998; Biswal et al. 2011; Ruban 2015). The third phase, carbon metabolism, utilizes ATP and NADPH that are products of electrochemical gradients and electron transport, and it occurs on the order of seconds to minutes (Blankenship 2002; Ruban 2009; Biswal et al. 2011; Ruban 2015). The rate constants for the three phases of photosynthesis span orders of magnitude and imbalances are constantly occurring either through changes in the environment (light, temperature, $CO_2$) or, in some cases, circadian rhythms (Hennessey and Field 1991; Dodd et al. 2014; Hüner et al. 2012). The balance of energy flow between the three photophysical and photochemical phases of photosynthesis is called photostasis and has been expressed mathematically as follows:

$$\sigma_{\text{PSII}} E_k = n\tau^{-1},$$

where $\sigma_{\text{PSII}}$ is the effective absorption cross section of PSII (probability of a photon being absorbed and used in photochemistry), $E_k$ the irradiance, $n$ the number of metabolic electron sinks, and $\tau^{-1}$ is the rate that photosynthetic electrons are utilized by metabolic sinks (carbon, nitrogen, sulfur) (Huner et al. 1998; Falkowski and Chen 2003; Öquist and Huner 2003; Ensminger et al. 2006; Hüner et al. 2012). The left side of the equation is affected by the quantity (high light), quality (wavelengths), and timing (PWM or longer fluctuations) of light, whereas the right side consists of enzymatic reactions and is temperature dependent. Therefore, the timing and duration of light can affect changes in crop characteristics using PSC, but temperature must also be taken into account.

LEDs can activate PSC in three ways (quantity, quality, and timing), and this elevates the signals within the chloroplast (Murchie and Lawson 2009). These signals derived from chloroplasts regulate plastid gene expression, but they are also considered retrograde in that they convey information to the nucleus which in turn turns on nuclear gene expression necessary for the acclimation paradigm (Chi et al. 2013). A benefit of increasing crop quality using PSC is that the onset can be measured in situ before visual acclimation responses occur using chlorophyll *a* fluorescence (CF) (Maxwell and Johnson 2000). This noninvasive technique measures the redox status of the chloroplast (plant health and stress) and is used regularly in research laboratories to rapidly measure photochemical efficiencies ($F_V/F_M$, (Y(II)) and nonphotochemical quenching (NPQ), the latter an indicator of nutrient density (Müller et al. 2001; Kopsell and Sams 2013; Cohu et al. 2014; Ouzounis et al. 2014). In short, using PSC and CF to optimize the crop can speed up the development of LED light programs by estimating the chloroplastic energy balance under a specified light source.

LED systems can be used to disrupt photostasis at any time during crop growth. Plastoquinone (PQ) shuttles electrons out of PSII into the electron transport chain, and it is well accepted that PQ is a primary sensor and signal in PSC

(Huner et al. 1998; Pfannschmidt 2003; Pfalz et al. 2012; Foyer et al. 2012). The redox state of PQ is measured as the photochemical quenching of CF and is expressed as 1-$q_P$ or 1-$q_L$, and growth under conditions where the PQ pool is reduced (high 1-$q_P$) has practical applications as it can increase stress tolerance, modulate plant height, and increase nutritive qualities among others (Huner et al. 1998; Pocock et al. 2001; Cohu et al. 2014). Spectral activation of PSC on crop plants using LED rather than conventional systems is less well understood. The PQ pool can sense the spectrum and is 'activated' (reduced) when the spectral distribution is below 680 nm and 'de-activated' (oxidized) under a spectral distribution that contains far-red light (>700 nm) (Piipo et al. 2006). How this can be used to influence crop growth and development is unknown. The acidification of the lumen (ΔpH) during photosynthesis also senses changes in light and regulates electron transport through the activation of nonphotochemical quenching (NPQ) in PSII (Foyer et al. 2012; Kono and Terashima 2014). The photoprotective xanthophyll cycle is a significant component of NPQ and involves the increase in zeaxanthin concentrations within the plant, an important dietary nutrient for eye health (Niyogi 1999; Semba and Dagnelie 2003; Kopsell and Sams 2013; Cohu et al. 2014). ΔpH and NPQ are reversed in the dark, and this sensing system is considered to be primarily a response to fluctuating and high light rather than light quality (Tikkanen et al. 2010). However, it has been recently reported that orchids (*Phalaenopsis* hybrid 'Vivien' and 'Purple Star') growing under natural shaded light with supplemental 40% blue and 60% red LEDs resulted in higher NPQ with subsequent higher levels of zeaxanthin and another beneficial nutrient, lutein (Ouzounis et al. 2014). Adjusting the intensity, spectrum, and timing of LEDs offers a way to increase crop nutritive value through the regulation of the redox state of PQ, ΔpH, and the resulting NPQ (Hüner et al. 2012; Cohu et al. 2014). Two other lesser-known redox signals are thioredoxin and reactive oxygen species (ROS). The ferredoxin/thioredoxin (F/T) system is a light sensor that is linked to PSI. In plants, it has been found to regulate 43 proteins and 15 physiological processes (Buchanan and Balmer 2005). The F/T system is activated by high light or preferential excitation of PSI (>700 nm) relative to PSII (<680 nm) and is primarily involved in the regulation of photosynthesis. For a long time, reactive oxygen species (ROS) were solely associated with light-induced oxidative stress (Mittler et al. 2011; Baxter et al. 2014). It is now accepted that ROS is a PSC redox signaling molecule that regulates protective responses against pathogens in plants (Lehmann et al. 2015).

Another beneficial nutrient that is under PSC is the red/purple pigment anthocyanin. Similarly to zeaxanthin and lutein, their dietary uptake is correlated with human health such as in the treatment of vision disorders, protection against neurological disorders, reduced incidence of cardiovascular disease, increased cognitive ability, and, lastly, they act to enhance antioxidant defences (Lila 2004). Anthocyanins are water-soluble pigments belonging to the flavonoid group of polyphenols that are found in all plant tissues, including flowers, berries, and leaves (Davies 2004). They are deposited in the leaf epidermis where they act as a sunscreen to protect the photosynthetic apparatus from over excitation and damage

(Huner et al. 1998; Davies 2004). An example of using PSC to increase anthocyanin concentrations in red lettuce Rouxai plants was observed after a 24 h shift from a spectrum that resulted in high biomass but green leaves to a spectrum where the plants had lower biomass but significantly higher anthocyanin red leaves (Pocock 2015a). Lettuce plants were grown under a red (660 nm) and blue (460 nm) or phosphor-converted white LEDs (high biomass) and shifted to cool white fluorescent light (high anthocyanin), and this resulted in a 15-fold increase in anthocyanin concentration within 24 h (Pocock 2017). The reddening of the lettuce crop was also captured with time-lapse photography in parallel with in situ CF measurements (Fig. 3.4).

The effective PSII quantum yields (Y(II)) increased and NPQ decreased significantly and was detected within 1 min after the spectral shift, 4 h before visible reddening was becoming apparent. Functionally, photochemical energy conversion became more efficient while thermal dissipation of absorbed light energy decreased after the shift. In experiments where the plants were shifted back to their original growth light, acclimation was completely reversed (data unpublished). Spectral analyses revealed that the wavelengths responsible for this acclimation response has $\lambda_{max}$ at 402, 530 nm, or 485 nm, and this is currently being teased out in the laboratory. Acclimation responses under PSC have been studied extensively from a stress perspective. However, as described above, light can be used in multiple ways to elicit desired acclimation responses on crop plants grown in CEA.

**Fig. 3.4** Time-lapse photography of red lettuce Rouxai was shifted at day 24 from a phosphor-converted horticultural LED environment where the plants had high biomass but poor biochemistry (pigmentation) to a cool white fluorescent environment that produced small plants with high biochemistry (pigmentation). Numbers in lower right of frames represent number of hours post shift

### 3.5.2 Photoreceptor Sensing and Signaling Networks: Photoreceptor Control

Photoreceptor (PR) sensing and signaling networks, referred to here as photoreceptor control (PRC), involve a multitude of photoreceptors that sense light across the spectrum from UVB (280–315 nm) to far-red (700–750 nm). Long-term development and the generation of new tissue are controlled by PRC; however, it operates in parallel and coordination with the rapid operational PSC (Kami et al. 2010; Pfalz et al. 2012). The genes encoding two of the photoreceptor classes, phytochrome and cryptochrome, are diurnally regulated indicating circadian control as well (Fankhauser and Staiger 2002). This suggests that tuning narrow bands of light in coordination with the circadian rhythm of photoreceptor expression could be another way to optimize crop growth and development, although more research is required.

All PRs contain organic, nonprotein components known as chromophores that serve as the primary site of photon absorption (Möglich et al. 2010). The best studied PR class is the five-member phytochrome (PHY) family that sense the ratio of red and far-red light through its chromophore, phytochromobilin (Casal 2000; Franklin and Quail 2010; Kami et al. 2010; Chen and Chory 2011). The biologically inactive $P_r$ form is synthesized in the dark and is converted to the active $P_{fr}$ form within 1–2 min in the presence of red light (660–670 nm) (Chen and Chory 2011). The conversion to $P_{fr}$ is reversible by far-red light (725–735 nm), and therefore, R and FR light act as rapid molecular light switches (Smith and Whitelam 1997). The use of mutants has led to the elucidation of many of the specific functions attributed to each of the PHYA-E variants including a large number of plant developmental processes such as seed germination (Casal and Sanchez 1998), de-etiolation (Franklin and Quail 2010), shade avoidance and plant height (Franklin 2008), branching (Leduc et al. 2014), stomatal development (Casson and Hetherington 2010), photoperiodic responses and flowering (Thomas 2006), circadian clock (Devlin and Kay 2000), plant immunity (herbivores/pathogens) (Ballare 2014), and freezing tolerance (Franklin and Whitelam 2007). The individual phytochrome species are controlled by red and far-red light as well as the fluence rate, they have unique and overlapping functions throughout the life cycle of plants (Franklin and Quail 2010; Li et al. 2011; Casal 2013). The most abundant phytochrome in the dark is phyA that functions as a molecular switch that regulates germination, de-etiolation, height, leaf architecture, stomatal index, circadian clock entrainment, and photoperiod perception (Franklin and Quail 2010). Green light is also sensed by the plant as 'shade,' and it induces shade morphology similar to PHY suggesting that the typical red/far-red response could in fact be broader with a greater redundancy (Zhang et al. 2011). Finely tuned and timed light programs could be developed using LEDs to impose beneficial crop qualities throughout the crop lifecycle through PHY control.

There are three distinct classes of specific UV-A/blue light sensors: (1) cryptochromes (cry1, cry2, and cry3), (2) phototropins (phot1 and phot2), and

(3) Zeitlupes (ZTL, FKF1, and LKP2) (Lin and Shalitin 2003; Christie 2007; Demarsy and Fankhauser 2009; Somers and Fujiwara 2009). The first class cryptochromes (CRY) are ubiquitous photoreceptors across all kingdoms (Lin and Todo 2005; Li and Yang 2007). There are three CRYs in the model plant, *A. thaliana*, cry1 is involved in photomorphology (e.g., plant height and phytochemicals), cry2 is involved in regulating photoperiodic flowering, and cry3 is localized to the mitochondria and chloroplast where it plays a role in repairing UV-induced DNA damage (Koorneef et al. 1998; Lin 2000; Liu et al. 2011; Selby and Sancar 2006; Pokorny et al. 2008). CRYs contain two chromophores, flavin adenine dinucleotide (FAD) that functions as the primary light sensor and a pterin derivative that harvests and transfers additional light energy to FAD from the near UV region (370–390 nm) (Hoang et al. 2008). CRYs show maximal activity within microseconds when exposed to photons between 400 and 500 nm with $\lambda_{max}$ at 450 nm and shoulders at 430 and 470 nm while the less active pterin absorbs maximally at 380 nm (Ahmed et al. 2002; Möglich et al. 2010; Chaves et al. 2011; Christie et al. 2015). The absorption and activation properties are similar for all plant blue light receptors (Conrad et al. 2014). In addition to the above examples, CRYs are involved in the control of seed dormancy and germination (Barrero et al. 2014), de-etiolation (Ahmed and Cashmore 1993), circadian clock (Somers et al. 1998; Devlin and Kay 2000), anthocyanin biosynthesis (Ahmed et al. 2002), branching (Leduc et al. 2014), and stomatal opening (Sellaro et al. 2010). Similarly to PHYs, there is a photocycle for CRYs where blue light activates and darkness or green light (500–600 nm) reverses or balances the activity. CRYs not only sense blue light, but they sense and respond to changes in the blue-to-green ratio with higher biological impact under high PFDs (Bouly et al. 2007; Sellaro et al. 2010). However, it has also been reported that green light responses were enhanced under low light conditions as is found in northern greenhouses during the darker months (Wang and Folta 2013). The second class of specific UV-A/blue light sensors is the phototropin (PHOT) and the Zeitlupe (ZTL) families that share an FMN chromophore and light oxygen voltage domains (LOV) (Christie et al. 2015). PHOTS and ZTLs absorb and are activated by the same spectral region as the CRYs; however in contrast to CRYs, once activated they are not reversed by green light. In fact, their absorbance spectrum gets shifted down to between 350 and 450 nm when in the active form (Christie et al. 2015). In *A. thaliana*, there are two PHOTs, phot 1 and phot2, which regulate a wide range of relatively rapid responses to optimize photosynthetic efficiency and growth under low light conditions (Takemiya et al. 2005). These include phototropism, chloroplast accumulation or avoidance responses, stomatal opening, and leaf flattening (Suetsugu and Wada 2013). ZTL serve a different function where they are primarily involved in the control of slower light responses such as entrainment of the circadian clock and the onset of flowering (Nelson et al. 2000; Demarsy and Fankhauser 2009). The dark inactivation of the three ZTL members, ZTL, FKF1, and LPK2, is slow, approximately 62.5 h compared to tens of seconds for PHOTs and therefore do not revert fully to their inactive state during the night (Zikihara et al. 2006; Demarsy and Fankhauser 2009; Suetsugu and Wada 2013). The last and most recently discovered class of plant

photoreceptor is the UV RESISTANCE LOCUS 8 (UVR8) which absorb in the UVB range (280–315 nm) eliciting photoprotective and repair mechanisms in response to UVB damage such as the movement and anchoring of cell nuclei to the bottom of plant cells as well as other as yet unidentified acclimation processes (Heijde and Ulm 2012; Iwabuchi et al. 2016).

## 3.6 Conclusion

It is not hard to imagine futuristic CEA with advanced LED lighting automation whose spectral combinations are programmed to drive crop physiological processes either through sensing technology or manually on command. Light is the biggest effector in plant growth and development, and when LED systems are fully integrated into CEA control systems, both electrical and biological efficacies will be accelerated. With respect to engineering, LED arrays can be static or used dynamically, and almost every spectral ratio within the horticultural relevant range can be delivered how and when you want it. LEDs participation in the acclimation paradigm will influence crop plants; however, the level of individual phenotypic plasticity (flexibility) is important to consider. The continuous spectral distribution of the Sun cannot be achieved without LED down conversion using phosphors, but the fluctuations and the PFD of the spectrum can be mimicked. Can this improve crop productivity with respect to yield, robustness, or nutrient density? Programming the light environment is one way to influence crops, but using the plant to modify the environment is another strategy. For instance, basic research has shown that the number and aperture of the stomata is highly regulated by blue and green light at ambient (350–400 ppm) $CO_2$ concentrations. Would it be beneficial to shut down crop transpiration when the relative humidity gets too elevated in high-density CEA, such as PFs or when photosynthesis is down-regulated through circadian rhythms or stress? Plant acclimation responses are well described as are PSC and PRC, and once a particular crop and the desired crop attributes are identified, an LED program can be specified and validated. Horticultural light programs can be manufactured solely from an engineering point of view, and the information provided in this chapter will help light manufacturers, growers, and consumers wade through the lighting maze. The relationship of plants with light is complex, and there is a lot yet to be discovered.

**Acknowledgements** This work was supported primarily by the Engineering Research Centers Program (ERC) of the National Science Foundation under NSF Cooperative Agreement No. EEC-0812056 and in part by New York State under NYSTAR contract C090145 and NYSERDA contract J50562.

# References

Ahmed M, Cashmore AR (1993) HY4 gene of *A. thaliana* encodes a protein with characteristics of a blue-light photoreceptor. Nature 366:162–166

Ahmed M, Grancher N, Heil M, Black RC, Giovani B, Galland P, Lardemer D (2002) Action spectrum for cyptochrome-dependent hypocotyl growth inhibition in *Arabidopsis*. Plant Physiol 129:774–785

Albright LD, Both A-J, Chiu AJ (2000) Controlling greenhouse light to a consistent daily integral. T Am Soc Agri Eng 43:421–431

Alter P, Dreissen A, Luo F-L, Matsubara S (2012) Acclimatory responses of *Arabidopsis* to fluctuating light environment: comparison of different sunfleck regimes and accessions. Photosynth Res 113:221–237

Ballare C (2014) Light regulation of plant defense. Annu Rev Plant Biol 65:1–790

Barrero JM, Downie AB, Xu Q, Gubler F (2014) A role for barley CRYTOCHROME1 in light regulation of grain dormancy and germination. Plant Cell 26:1094–1104

Baxter A, Mittler R, Suzuki N (2014) ROS as key players in plant stress signaling. J Exp Bot 65:1229–1240

Bian ZH, Yang QC, Liu WK (2014) Effects of light quality on the accumulation of phytochemicals in vegetables produced in controlled environments: a review. J Sci Food Agri 95:869–877

Biswal B, Joshi PN, Raval MK, Biswal UC (2011) Photosynthesis, a global sensor of environmental stress in green plants: stress signaling and adaptation. Curr Sci 101:47–56

Blankenship RE (2002) Molecular mechanisms of photosynthesis. Blackwell Science Ltd, Oxford

Bouly JP, Schleicher E, Dionisio-Sese M, Vandenbussche F, Van der Straeten D, Bakrim N, Meier S, Batschauer A, Galland P, Bitti R, Ahmad M (2007) Cryptochrome blue light photoreceptros are activated through interconversion of flavin redox states. J Biol Chem 282:9383–9391

Buchanan B, Balmer Y (2005) Redox regulation: a broadening horizon. Annu Rev Plant Biol 56:187–220

Casal J, Sanchez R (1998) Phytochromes and seed germination. Seed Sci Res 8:317–329

Casal J (2000) Phytochromes, cryptochromes, phototropin: photoreceptor interactions in plants. Photchem Photobiol 71:1–11

Casal J (2013) Photoreceptor signaling networks in plant responses to shade. Annu Rev Plant Biol 64:1–863

Casson S, Hetherington A (2010) Environmental regulation of stomatal development. Curr Opin Plant Biol 13:90–95

Chaves I, Byrdin M, Hoang N, van der Horst TJ, Batschauer A, Ahmad M (2011) The cryptochromes: blue light photoreceptors in plants and animals. Annu Rev Plant Biol 62:335–364

Chen M, Chory J (2011) Phytochrome signaling mechanisms and the control of plant development. Trends Cell Biol 21:664–671

Chi W, Sun X, Zhang L (2013) Intracellular signaling from plastid to nucleus. Annu Rev Plant Biol 64:559–582

Chory J (2010) Light signal transduction: an infinite spectrum of possibilities. Plant J 61:982–991

Christie JM (2007) Phototropin blue-light receptors. Annu Rev Plant Biol 58:21–45

Christie JM, Blackwood L, Petersen J, Sullivan S (2015) Plant flavoprotein photoreceptors. Plant Cell Physiol 56:401–413

Clark GB, Thompson G Jr, Roux SJ (2001) Signal transduction mechanisms in plants: an overview. Curr Sci 80:170–177

Cohu CM, Lombardi E, Adams WW III, Demmig-Adams B (2014) Increased nutritional quality of plants for long-duration spaceflight missions through choice of plant variety and manipulation of growth conditions. Acta Astronaut 94:799–806

Conrad KS, Manahan CC, Crane BR (2014) Photochemistry of flavoprotein light sensors. Nat Chem Biol 10:801–809

Crisp P, Ganguly D, Eichten SR, Borevitz O, Pogson BJ (2016) Reconsidering plant memory: intersections between stress recovery, RNA turnover, and epigenetics. Sci Adv 2:1–14

Davies KM (2004) An introduction to plant pigments in biology and commerce. In: Davies K (ed) Plant pigments and their manipulation. Blackwell Publishing, Oxford, p 1

Demarsy E, Fankhauser C (2009) Higher plants use LOV to perceive blue light. Curr Opin Plant Biol 12:69–74

Demmig-Adams B, Cohu CM, Muller O, Adams WW III (2012) Modulation of photosynthetic energy conversion efficiency in nature: from seconds to seasons. Photosynth Res 113:75–88

Devlin P, Kay S (2000) Cytochromes are required for phytochrome signaling to the circadian clock but not for rhythmicity. Plant Cell 12:2499–2509

Dodd AN, Kusakina J, Hall A, Gould PD, Hanaoka M (2014) The circadian regulation of photosynthesis. Photosynth Res 119:181–190

Durnford D, Falkowski P (1997) Chloroplast redox regulation of nuclear gene transcription during photoacclimation. Photosynth Res 53:229–241

Ensminger I, Busch F, Huner NPA (2006) Photostasis and cold acclimation: sensing low temperature through photosynthesis. Physiol Plantarum 126:28–44

Esteban R, Moran JF, Beccerril JM, García-Plazaola I (2015) Versatility of carotenoids: an integrated view on diversity, evolution, functional roles and environmental interactions. Environ Exp Bot 119:63–75

Falkowski PG, Chen YB (2003) Photoacclimation of light harvesting system in eukaryotic algae. In: Green BR, Parson WW (eds) Advances in photosynthesis and respiration: light harvesting antennas in photosynthesis, vol 13. Kluwer, The Netherlands, p 423

Fankhauser C, Staiger D (2002) Photoreceptors in *Arabidopsis thaliana*: light perception, signal transduction and entrainment of the endogenous clock. Planta 216:1–16

Foyer CH, Neukermans J, Queval J, Noctor G, Harbinson J (2012) Photosynthetic control of electron transport and the regulation of gene expression. J Exp Bot 63:1637–1661

Franklin K, Whitelam G (2007) Light-quality regulation of freezing tolerance in *Arabidopsis thaliana*. Nat Genet 39:1410–1413

Franklin KA (2008) Shade avoidance. New Phytol 179:930–944

Franklin K, Quail P (2010) Phytochrome functions in *Arabidopsis* development. J Exp Bot 61:11–24

Galvão VC, Frankhauser C (2015) Sensing the light environment in plants: photoreceptors and early signaling steps. Curr Opin Neurobiol 34:46–53

George NC, Denault KA, Seshadri R (2013) Phosphors for solid-state white lighting. Annu Rev Mater Res 43:481–501

Giraudat J, Schroeder J (2001) Cell signaling and gene regulation: plant signal transduction pathways: greying of the black boxes. Curr Opin Plant Biol 4:379–381

Goosey L, Palecanda L, Sharrock RA (1997) Differential patterns of expression of the *Arabidopsis* PHYB, PUYD, and PHYE phytochrome genes. Plant Physiol 115:959–969

Gratani L (2014) Plant phenotypic plasticity in response to environmental factors. Adv Bot 2014:1–17

Gruda N (2005) Impact of environmental factors on product quality of greenhouse vegetables for fresh consumption. Crit Rev Plant Sci 24:227–247

Heijde M, Ulm R (2012) UV-B photoreceptor-mediated signaling in plants. Trends Plant Sci 17:230–237

Hennessey TL, Field CB (1991) Oscillations in carbon assimilation and stomatal conductance under constant conditions. Plant Physiol 96:831–836

Hoang N, Bouly JP, Ahmad M (2008) Evidence of a light-sensing role for folate in *Arabidopsis* cryptochrome blue-light receptors. Mol Plant 1:68–74

Hogewoning SW, Trowborst G, Maljaars H, Poorter H, van Ieperen W, Harbinson J (2010) Blue light dose-response of leaf photosynthesis, morphology, and chemical composition of *Cucumis sativus* grown under different combinations of red and blue light. J Expt Bot 61:3107–3117

Huner NPA, Öquist G, Sarhan F (1998) Energy balance and acclimation to light and cold. Trends Plant Sci 3:224–230

Hüner NPA, Bode R, Dahal K, Hollis L, Rosso D, Krol M, Ivanov AG (2012) Chloroplast imbalance governs phenotypic plasticity: the "grand design of photosynthesis" revisited. Front Plant Sci. doi:10.3389/fpls.2012.00255

IPCC (2014) Climate change 2014: synthesis report. In: Pachauri RK, Meyer LA (eds) Contribution of working groups I, II and III to the fifth assessment report of the intergovernmental panel on climate change. IPCC, Geneva, p 151

Iwabuchi K, Hidema J, Tamura K, Takagi S, Hara-Nishimura I (2016) Plant nuclei move to escape ultraviolet-induced DNA damage and cell death. Plant Physiol 170:678–685

Jao RC, Fang W (2004) Effects of frequency and duty ratio on the growth of potato plantlets in vitro using light emitting diodes. HortScience 39:375–379

Johkan M, Shoji K, Goto F, Hahida S, Yoshihara T (2012) Effect of green light wavelength and intensity on photomorphogenesis and photosynthesis in *Lactuca sativa*. Environ Exp Bot 75:128–133

Kami C, Lorrain S, Hornitschek P, Fankhauser C (2010) Light-regulated plant growth and development. Curr Top DevBiol 91:29–66

Karpinski S, Szechynska-Hebda M (2010) Secret life of plants from memory to intelligence. Plant Signal Behav 5:1391–1394

Khoshbakht K, Hammer K (2008) How many plant species are cultivated? Genetic Res Crop Evol 55:925–928

Kiang N, Sieffert J, Govindjee B (2007) Spectral signatures of photosynthesis I: review of earth organisms. Astrobiol 7:222–251

Kono M, Terashima I (2014) Roles of the cyclic electron flow around PSI (CEF-PSI) and $O_2$-dependent alternative pathways in regulation of the photosynthetic electron flow in short-term fluctuating light in *Adrabidopsis thaliana*. Plant Cell Physiol 55:990–1004

Koorneef M, Alonso-Blanco C, Peeters AJM, Soppe W (1998) Genetic control of flowering time in *Arabidopsis thaliana*. Annu Rev Plant Biol 49:345–370

Kopp G, Lean J (2011) A new, lower value of total solar irradiance: evidence and climate significance. Geophys Res Lett 38:1–7

Kopsell DA, Sams CE (2013) Increases in shoot tissue pigments, glucosinolates, and mineral elements in sprouting broccoli after exposure to short-duration blue light from light emitting diodes. J Am Soc Hortic Sci 138:31–37

Kozai T, Niu G, Takagaki M (eds) (2015) Plant factory: an indoor vertical farming system for efficient quality food production. Academic Press

Krames M (2016) 6-1: invited paper: status and future prospects for visible-spectrum light-emitting diodes. SID Symp Dig Tech Pap 47:39–41

Kromdijk J, Glowacka K, Leonelli L, Gabilly ST, Iwai M, Niyogi KK, Long SP (2016) Improving photosynthesis and crop productivity by accelerating recovery from photoprotection. Science 354:857–861

Leduc N, Roman H, Barbier F, Péron T, Huché-Thélier L, Lothier J, Demotes-Mainard S, Sakr S (2014) Light signaling in bud outgrowth and branching in plants. Plants 3:223–250

Lee D, Downum K (1991) The spectral distribution of biologically active solar radiation at Miami, Florida, USA. Biometeorology 35:48–54

Lee R, Hernandez-Andres J (2005) Colors of the daytime overcast sky. Appl Optic 44:5712–5722

Lehmann S, Haridon FL, Tjamos SE, Metraus JP (2015) Reactive oxygen species and plant resistance to fungal pathogens. Phytochem 112:54–62

Li J, Li G, Wang H, Deng XW (2011) Phytochrome signaling mechanisms. *Arabidopsis Book* 9: e0148. doi:10.1199/tab.0148

Li QH, Yang HQ (2007) Cryptochrome signaling in plants. Photochem Photobiol 83:94–101

Lila MA (2004) Anthocyanins and human health: an in vitro investigative approach. J Biomed Biotechnol 5:305–313

Lin C (2000) Photoreceptors and regulation of flowering time. Plant Physiol 123:39–50

Lin C, Shalitin D (2003) Cryptochrome structure and signal transduction. Annu Rev Plant Biol 54:469–496

Lin C, Todo T (2005) The cryptochromes. Genome Biol 6:220.1–220.9

Liu H, Liu B, Zhao C, Pepper M, Lin C (2011) The action mechanisms of plant cryptochromes. Trend Plant Sci 16:684–691
Marin A, Ferreres F, Barberà GG, Gil MI (2015) Weather variability influences color and phenolic content of pigmented baby leaf lettuces throughout the season. J Agri Food Chem 63:1673–1681
Maxwell K, Johnson GN (2000) Chlorophyll fluorescence-a practical guide. J Exp Bot 51:659–668
McCree K (1972) The action spectrum, absorptance and quantum yield of photosynthesis in crop plants. Agri Meteorol 9:191–216
Mittler R, Vanderauwera S, Suzuki N, Miller G, Tognetti VB, Vanderpoele K, Gollery M, Shulaev V, Van Breusegem F (2011) ROS signaling: the new wave? Trend Plant Sci 16:300–309
Möglich A, Yang X, Ayers RA, Moffat K (2010) Structure and function of plant photoreceptors. Annu Rev Plant Biol 61:1–720
Murchie EH, Lawson T (2009) Agriculture and the new challenges for photosynthesis research. New Phytol 181:532–552
Müller P, Li XP, Niyogi KK (2001) Non-photochemical quenching. A response to excess light energy. Plant Physiol 125:1558–1566
Nelson DC, Lasswell J, Rogg LE, Cohen MA, Bartel B (2000) FKF1, a clock-controlled gene that regulates the transition to flowering in *Arabidopsis*. Cell 101:331–340
Nicotra AB, Atkin OK, Bonser SO, Davidson AM, Finnegan EJ, Mathesius U, Poot P, Purugganan MD, Richards CL, Valadares F, van Kleunen M (2010) Plant phenotypic plasticity in a changing climate. Trend Plant Sci 15:684–692
Nishio JN (2000) Why are higher plants green? Evolution of the higher plant photosynthetic pigment complement. Plant Cell Environ 23:539–548
Niyogi KK (1999) Photoprotection revisited: genetic and molecular approaches. Annu Rev Plant Biol 50:333–359
Olvera-Gonzalez E, Alaniz-Lumbreras D, Torres-Argüelles González-Ramirez E, Villa-Hernández J, Araiza-Esquivel M, Ivanov-Tsonchev R, Olvera-Olvera C, Castaño VM (2013) A LED-based smart illumination system for studying plant growth. Light Res Technol 46:128–139
Öquist G, Huner NPA (2003) Photosynthesis of overwintering evergreen plants. Annu Rev Plant Biol 54:329–355
Ouzounis T, Fretté Z, Ottosen C-O, Rosenqvist E (2014) Spectral effects of supplementary lighting on the secondary metabolites in roses, chrysanthemums, and campanulas. J Plant Physiol 175:192
Pfalz J, Liebers M, Hirth M, Grübler B, Schröter H, Dietzel B (2012) Environmental control of plant nuclear gene expression by chloroplast redox signals. Front Plant Sci 3:257–266
Pfannschmidt T, Nilsson A, Allen JF (1999) Photosynthetic control of chloroplast gene expression. Nature 397:625–628
Pfannschmidt T (2003) Chloroplast redox signals: how photosynthesis controls its own genes. Trends Plant Sci 8:33–41
Piipo M, Allahverdiyeva Y, Paakkarinen V, Suoranto UM, Battchikova N, Aro EM (2006) Chloroplast-mediated regulation of nucleargenes in *Arabidopsis thaliana* in the absence of light stress. Physiol Genomics 25:142–152
Pocock T, Hurry V, Savitch LV, Hüner NPA (2001) Susceptibility to low-temperature photoinhibition and the acquisition of freezing tolerance in winter and spring wheat: the role of growth temperature and irradiance. Int J Plant Biol 113:499–506
Pocock T (2015a) Advanced lighting technology in controlled environment agriculture. Light Res Technol. doi:10.1177/1477153515622681
Pocock T (2015b) Light-emitting diodes and the modulation of specialty crops: light sensing and signaling networks in plants. HortScience 50:1281–1284
Pocock T (2017) Energy, LEDs and enhanced plant yield in controlled environment agriculture. NYSERDA report number 31793 (revised Sept 2014). New York State Energy and Research Development Authority

Pokorny R, Klar T, Hennecke U, Carell T, Batschauer A, Essen LO (2008) Recognition and repair of UV lesions in loop structures of duplex DNA by DASH-type cryptochrome. Proc Natl Acad Sci USA 105:21023–21027
Potters G, Horemans N, Jansen MAK (2010) The cellular redox state in plant stress biology. Plant Physiol Biochem 48:292–300
Reinhardt K, Smith WK, Carter GA (2010) Clouds and cloud immersion alter photosynthetic light quality in a temperate mountain cloud forest. Botany 88:462–470
Rosevear MJ, Young AJ, Johnson GN (2001) Growth conditions are more important than species origin in determining leaf pigment content of British plant species. Func Ecol 15:474–480
Ruban AV (2009) Plants in light. Commun Integrat Biol 2:50–55
Ruban A (2015) Evolution under the sun: optimizing light harvesting in photosynthesis. J Exp Bot 66:7–23
Sager J, Giger W (1980) Re-evaluation of published data on the relative photosynthetic efficiency of intermittent and continuous light. Agric Meteorol 22:289–302
Sakamoto K, Briggs WR (2002) Cellular and subcellular localization of Phototropin 1. Plant Cell 14:1723–1735
Selby CP, Sancar A (2006) Acryptochrome/photolyase class of enzymes with single-stranded DNA-specific photolyase activity. Proc Natl Acad Sci USA 103:17696–17700
Sellaro R, Crepy M, Trupkin SA, Karayekov E, Buchovsky AS, Rossi C, Casal JJ (2010) Cryptochrome as a sensor of the blue/green ratio of natural radiations in *Arabidopsis*. Plant Physiol 154:401–409
Semba RD, Dagnelie G (2003) Are lutein and zeaxanthin conditionally essential nutrients for eye health? Med Hypothes 61:465–472
Scheffers B, Joppa LN, Pimm SL, Laurance WF (2012) What we know and don't know about Earth's missing biodiversity. Trend Ecol Evol 27:501–510
Schottler M, Toth S (2014) Photosynthetic complex stoichiometry dynamics in higher plants: environmental acclimation and photosynthetic flux control. Frontier Plant Sci 5:1–15
Shimada A, Taniguchi Y (2011) Red and blue pulse timing control for pulse width modulation light dimming of light emitting diodes for plant cultivation. J Photochem Photobiol 104:399–404
Snowden KC, Inzé D (2016) Editorial overview: cell signaling and gene regulation: the many layers of plant signaling. Curr Opin Plant Biol 33:iv–vi
Smith H, Whitelam GC (1997) The shade avoidance syndrome multiple responses mediated by multiple phytochromes. Plant Cell Environ 20:840–844
Somers DE, Devlin PF, Kay SA (1998) Phytochromes and cryptochromes in the entrainment of the *Arabidopsis* circadian clock. Science 282:1488–1490
Somers DE, Fujiwara S (2009) Thinking outside the F-box: novel ligands for novel receptors. Trend Plant Sci 14:206–213
Suetsugu N, Wada M (2013) Evolution of three LOV blue light receptor families in green plants and photosynthetic stramenopiles: phototropin, ZTL/FKF1/LKP2 and autochrome. Plant Cell Physiol 54:8–23
Sun J, Nishio JN, Vogelman TC (1998) Green light drives $CO_2$ fixation deep within the leaves. Plant Cell Physiol 39:1020–1026
Taiz L, Zeiger E (2010) Plant physiology, 5th edn. Sinauer Associates Inc., Sunderland, p 782
Takahashi S, Milward SE, Yamori W, Evans JR, Hillier W, Badger MR (2010) The solar action spectrum of photosystem II damage. Plant Physiol 153:988–993
Takemiya A, Inoue S, Doi M, Kinoshita T, Shimazaki K (2005) Phototropins promote plant growth in response to blue light in low light environments. Plant Cell 17:1120–1127
Tennessen DJ, Bula RJ, Sharkey TD (1995) Efficiency of photosynthesis in continuous and pulsed light emitting diode irradiation. Photosynth Res 44:261–269
Terashima I, Fujita T, Inoue T, Chow WS, Oguchi R (2009) Green light drives leaf photosynthesis more efficiently than red light in strong white light: revisiting the enigmatic question of why leaves are green. Plant Cell Physiol 50:684–697
Thelier M, Luttge U (2012) Plant memory: a tentative model. Plant Biol 15:1–12
Thomas B (2006) Light signals and flowering. J Exp Bot 57:3387–3393

Tikkanen M, Grieco M, Kangasjärvi S, Aro EM (2010) Thylakoid protein phosphorylation in higher plant chloroplasts optimizes electron transfer under fluctuating light. Plant Physiol 152:723–735

Tomson T (2010) Fast dynamic process of solar radiation. Sol Energy 84:318–323

Trewavas A (2002) Plant intelligence: mindless mastery. Nature 515:841

United Nations, Department of Economic and Social Affairs, Population Division (2016) The World's Cities in 2016—Data Booklet (ST/ESA/SER.A/392)

U.S. Global Change Research Program (2015) Our changing planet: the U.S. Global Change Research Program for Fiscal Year 2016, Washington, DC, USA

Van Broekhoven J (2001) Lamp Phosphors. In: Kane R, Sell H (eds) Revolution in lamps: a chronicle of 50 years of progress. Fairmont Press Inc., GA, p 288

Vogelmann TC, Han T (2000) Measurements of gradients of absorbed light in spinach leaves from chlorophyll fluorescence profiles. Plant Cell Environ 23:1303–1311

Wang H, Gu M, Cui J, Shi K, Zhou Y, Yu J (2009) Effects of light quality on $CO_2$ assimilation, chlorophyll-fluorescence quenching, expression of Calvin cycle genes and carbohydrate accumulation in *Cucumis sativus*. J Photochem Photobiol 96:30–37

Wang Y, Folta K (2013) Contributions of green light to plant growth and development. Am J Bot 100:70–78

Wituszynska W, Galazka K, Rusaczonek A, Vanderauwera S, Van Breusegem F, Karpinski S (2013) Multivariable environmental conditions promote photosynthetic adaptation potential in *Arabidopsis thaliana*. J Plant Physiol 170:548–559

Wheeler R (2008) A historical background of plant lighting: an introduction to the workshop. HortScience 43:1942–1943

Zachgo S, Hanke GT, Scheibe R (2013) Plant cell microcompartments: a redox-signaling perspective. Biol Chem 394:203–216

Zhang T, Maruhnich SA, Folta KM (2011) Green light induces shade avoidance symptoms. Plant Physiol 157:1528–1536

Zhu XG, Ort DR, Whitmarsh J, Long SP (2004) The slow reversibility of photosystem II thermal energy dissipation on transfer form high to low light may cause large losses in carbon gain by crop canopies: a theoretical analysis. J Exp Bot 55:1167–1175

Zikihara K, Iwata T, Matsuoka D, Kandori H, Todo T, Tokutomi S (2006) Photoreaction cycle of the light, oxygen, and voltage domain in FKF1 determined by low-temperature absorption spectroscopy. Biochem 45:10828–10837

# Chapter 4
# Optimizing LED Lighting in Controlled Environment Agriculture

**Marc W. van Iersel**

## 4.1 Introduction

Controlled environment agriculture encompasses a range of production systems, including high tunnels, greenhouses, and indoor production facilities (often referred to as vertical farms or plant factories). The main goal of controlled environment agriculture is to provide growers with some level of control over the environmental conditions that crops are exposed to, thereby extending the growing season and increasing production as well as crop quality. The focus of this chapter is specifically on greenhouses and indoor production, because other forms of controlled environment agriculture (e.g., high tunnels) rarely employ supplemental lighting.

The lighting situation and requirements in greenhouses and indoor production facilities are different. In greenhouses, sunlight is typically the primary source of light. Supplemental electric light can be provided to increase crop yield and quality. This can be especially beneficial at higher latitudes, where seasonal fluctuations in daily light integral (DLI, photosynthetic photon flux density (PPFD) integrated over 24-h periods) are large and where DLI is low in winter time (Albright et al. 2000). Indoor production systems receive, at most, a small fraction of the required light from sunlight and are highly reliant on electric lighting.

The cost of electricity to provide electric light in both greenhouses and indoor production facilities is high. Supplemental lighting in greenhouses is often the second-highest operating expense, after labor. For example, a single 1000-W high-pressure sodium (HPS) light consumes about 1075-W, including both the bulb and ballast (Wallace and Both 2016). If 600 1000-W HPS lights are used per hectare for 180 days per year and 16 h per day, and at an electricity cost of US $0.10 kWh,

M.W. van Iersel (✉)
Department of Horticulture, University of Georgia, Athens, GA 30602, USA
e-mail: mvanier@uga.edu



S. Dutta Gupta (ed.), *Light Emitting Diodes for Agriculture*,
DOI 10.1007/978-981-10-5807-3_4

the annual electricity cost to operate the lights is ~$180,000 ha. The average farm gate value for greenhouse vegetables in the USA is ~$700,000 ha year (USDA-NASS 2014), so reducing operating expenses with more efficient lighting strategies will have a major impact on the profitability of greenhouses.

Despite the high cost of electric light, light levels in greenhouses are often poorly controlled. The amount of light greenhouse crops receive from the sun is highly variable and can change within seconds, because of shading by the greenhouse structure or changing weather conditions. Likewise, day-to-day changes in weather conditions can result in large variability in DLI, while seasonal changes in the position of the earth relative to the sun result in large differences in DLI over the course of a year (Albright et al. 2000). These changes depend on latitude and are greater further away from the equator.

Electric light is generally the only form of light available for crop growth in indoor production facilities, so the cost of providing that light is even more important than in greenhouses. Zeidler et al. (2013) published a comprehensive assessment of the technical and financial aspects of building and operating large skyscraper-like vertical farms. They estimated that the vertical farm they designed can increase production per unit area 1,115× compared to field production. However, the cost of the LED lighting system accounts for ~30% of the initial capital cost of the vertical farm and electricity accounts for ~60% of the annual operating costs (Zeidler et al. 2013). Much of this electricity is needed to provide lighting for the crops and for air-conditioning needed to remove the heat generated by the lights. General estimates are that 40–50% of total operating costs for vertical farms are associated with lighting (Zeidler et al. 2013; Watanabe 2011). Because of the high capital and operating costs, production of staple crops in vertical farms is not likely to be financially feasible in the foreseeable future (Banerjee and Adenauer 2014). More efficient lighting techniques are essential to improve the sustainability and profitability of vertical farms.

Developing optimal lighting strategies for controlled environment agriculture is complicated because of the number of factors that should be included in such a strategy. These factors include the capital cost of the lighting system, the efficiency of the lights, the cost of electricity (which may vary both short-and long-term), the ability of the crop to intercept and use the provided light to produce salable yield, the effect of this light on crop quality, and the value of the crop. The goal of this chapter is to outline how physiological information regarding plant responses to light can be used to develop more efficient and cost-effective approaches for lighting in controlled environment agriculture.

## 4.2 Benefits of LEDs as Grow Lights

LEDs are gradually gaining popularity for use in controlled environment agriculture. They have long been promoted as being more energy-efficient than other lights, but this has not been backed up by data. Nelson and Bugbee (2014) and

Wallace and Both (2016) recently compared the energy efficiency of various LED and HPS grow lights, expressed as mole of photosynthetic photons produced per joule of electricity used. They reported that the best LED and HPS lights had similar efficiency. Recent improvements in LED technology have increased the efficiency of some of the newest LED lights, exceeding that of the most efficient HPS lights (see Chap. 5). However, LED lights are still not universally more efficient than HPS lights.

### 4.2.1 Lack of Radiant Heat

Another advantage of LED lights over HPS or metal halide lights is that they produce little radiant heat, which allows them to be placed close to, or even within, the canopy without damaging the crop. Increasing the proximity of lights to the canopy ensures that the light can be efficiently delivered to leaves. For indoor production system, it decreases space requirements, allowing for multi-layer production systems with the different shelves relatively close together (Fig. 4.1). LED fixtures can also be designed, by using the appropriate lenses and reflectors, to ensure that most of the light emitted is indeed delivered to the canopy. Overall, this allows for a more efficient use of energy, light, and space.

**Fig. 4.1** A multi-tiered, LED-lit production system for *leafy greens* at AeroFarms (Newark, NJ, USA). Photograph by Marc van Iersel

### 4.2.2 Control of Light Spectrum

Generally, LEDs produce light within a fairly narrow spectral range, often expressed as the full width at half maximum (FWHM), i.e., the wavelength range within which the PPFD is at least half of the maximum PPFD (Fig. 4.2). LEDs are available with peaks at many different wavelengths, providing much flexibility in which spectra can be produced. White light can be produced either by combining LEDs with different wavelengths (i.e., blue, green, and red or blue and yellow) or, most common for LED grow lights, by coating blue LEDs with a phosphor (For more details see Chap. 1). The phosphor will absorb some fraction of the photons emitted by the blue LEDs and re-emit light with longer wavelengths through luminescence, generating white light (Chen et al. 2010). The emitted spectrum can be adjusted based on the type and amount of phosphor coating. Adding a phosphor coating reduces the efficiency of the LED, and white LEDs are thus less efficient than blue or red LEDs (Nelson and Bugbee 2014). Because chlorophyll has absorption peaks in the blue and red regions of the spectrum, many LED grow lights use only red and blue LEDs (red/blue LEDs). However, the idea that plants cannot efficiently use light with wavelengths other than at chlorophyll absorption peaks is incorrect: Higher plants have a variety of carotenoids that are part of the light-harvesting complex surrounding photosystem I and II. Those carotenoids efficiently absorb much of the light not absorbed by chlorophyll *a* and *b*. As a result, plants can use most of the light with wavelengths of 400–700 nm quite efficiently for photosynthesis (Ouzounis et al. 2015). It is currently not clear what the optimal spectrum for maximum photosynthetic efficiency is (see Sect. 4.3.3 for more in-depth discussion). Beyond driving photosynthesis, light spectrum can have distinct effects on plant morphology and the production of secondary compounds. Manipulating the light spectrum to improve plant quality is an important tool that has become available to controlled environment agriculture with the advent of LED lighting.

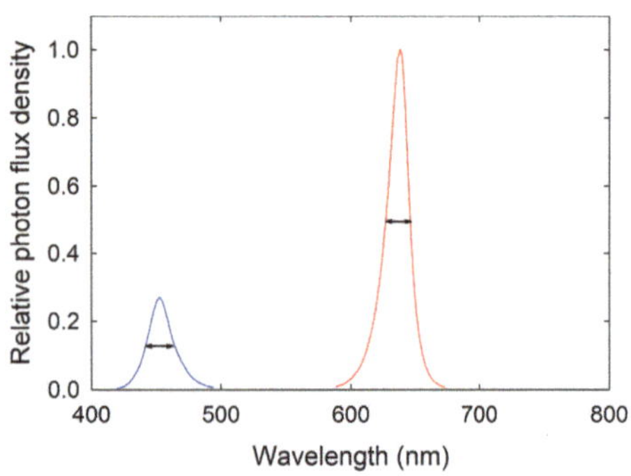


**Fig. 4.2** Spectral distribution of a grow light with *blue* and *red* LEDs. *Horizontal arrows* indicate the full width at half maximum (FWHM), a measure of how sharp the spectral peaks are

### 4.2.3 Controllability of LEDs

An underutilized property of LED lights is the ability to quickly and precisely control the intensity of their light output. This is commonly accomplished using one of two methods: current control or pulse width modulation. Pulse width modulation allows for control of the frequency at which LED lights are turned on and off (typically thousands of times per second) as well as the duty cycle (fraction of time the LEDs are energized during each short on/off cycle). Decreasing the duty cycle of LED lights reduces the PPFD. Adjusting the light output from LED grow lights is technically easy and cheap (van Iersel and Gianino 2017). Control ability allows for the rapid adjustment of the PPFD provided by the LED lights based on the ability of the crop to use that light efficiently. However, there has been little research into optimizing methods for controlling the PPFD provided by LED lights. As a matter of fact, there has been little research into optimal lighting strategies in general, despite the high cost associated with supplemental lighting.

## 4.3 Optimizing Lighting Control

Heuvelink and Challa (1989) developed guidelines for supplemental lighting based on a simple economic principle: the cost of providing more supplemental light needs to be lower than or equal to the value of the additional yield that results from adding that supplemental light. They predicted the increase in salable yield that could be achieved with supplemental light using a mechanistic crop photosynthesis model to predict carbohydrate production and the crop conversion efficiency (gram of salable product per gram of carbohydrate). Taking into account the price of electricity needed to operate the lights and the sales price of the harvested product, they were able to calculate the break-even point for supplemental lighting (Heuvelink and Challa 1989). It is not clear whether these guidelines were ever implemented by the greenhouse industry.

Clausen et al. (2015) used a similar approach to develop a supplemental lighting control system based on a leaf photosynthesis model. Their approach takes into account weather forecasts and real-time electricity pricing, preferentially providing supplemental light when electricity prices are low. The system can be programmed to achieve a specific ‘daily photosynthesis integral,’ calculated from the leaf photosynthesis model. Using the developed DynaLight desktop software to implement this strategy resulted in 25% energy savings in the production of *Campanula* without notable reductions in crop quality (Kjaer et al. 2011; Clausen et al. 2015).

Albright et al. (2000) used a different approach to control light in lettuce production. Based on multiple greenhouse trials, they determined that lettuce biomass accumulation was tightly correlated with the cumulative amount of PPFD the crop received during the production cycle. They developed a system called ‘light and shade system implementation’ (LASSI) that can control the DLI inside a greenhouse by applying shade and supplemental light as needed (Albright et al. 2000; Mathieu et al. 2004). LASSI makes hourly decisions regarding the need for shade or supplemental light, and supplemental light is preferentially provided when electricity prices are low. Controlling DLI inside the greenhouse results in steady and predictable year-round lettuce production.

It is important to note that all of these lighting control approaches are based on the use of HPS lights. Algorithms developed for HPS lights may not be optimal for LED lights. The light output from HPS lights cannot be precisely controlled, and the lights cannot be turned on and off rapidly. LED lights, on the other hand, can be precisely controlled to make rapid adjustments and can thus be programmed to instantaneously respond to environmental or crop physiological parameters (van Iersel and Gianino 2017). This is a significant improvement over the slow and coarse control of HPS lights. The cost-effectiveness of LED lighting can be increased by taking full advantage of the controllability of LED lights. Specifically, supplemental light should be provided only when crops can use the supplemental light efficiently.

### *4.3.1 How Much Light Is Optimal?*

To develop optimal lighting strategies, it is important to have a quantitative understanding of how crops use light. Crop light use efficiency can be divided into two components: light absorption and utilization of absorbed light. Light absorption is largely a function of canopy size: Small plants will intercept a relatively small fraction of the provided light. Such inefficiencies can be reduced by growing the plants closer together, as is common in seedling production. For example, ornamental seedlings are commonly grown at densities of up to 4000 plants/$m^2$. Likewise, leafy greens in plant factories are normally grown at very high densities (Fig. 4.3). Although the main goal of these high plant densities is to use the available space as efficiently as possible, it also increases light interception. This can be beneficial when supplemental lighting is used, since a greater plant density will increase light interception and thus the economic benefits of supplemental lighting.

It may also be possible to increase canopy light interception by manipulating light quality (i.e., the spectral composition of the light). Phytochrome is a pigment-protein complex that plays an important role in controlling cell and leaf elongation, and its activity depends on the phytochrome photo-equilibrium, which can be manipulated by altering the ratio of red to far-red light (Sager et al. 1988). A high proportion of far-red light generally triggers shade responses in plants. Many plants respond by producing larger thinner leaves (an increase in the specific leaf area), although this response is species-dependent. The molecular mechanisms of phytochrome regulation of plant growth and development were recently reviewed by Demotes-Mainard et al. (2016). For many crop species, using light with a high proportion of far-red light can increase canopy size and thus light interception. This can be especially valuable during the seedling stage, when light interception is

**Fig. 4.3** High-density production of microgreens under LED lighting at AeroFarms (Newark, NJ, USA). The high plant density optimizes space use and ensures that the crop intercepts most of the provided light. Photograph by Marc van Iersel

typically low. Kubota et al. (2012) found that providing far-red light at the end of the photoperiod increases light interception and growth of baby leaf lettuce grown solely under LED light. Interestingly, they did not see such an increase in growth in greenhouse-grown lettuce, possibly because sunlight is already rich in far-red light, especially at the start and end of the photoperiod.

LED lights have some advantages over HPS lights with regard to light interception. The use of appropriate reflectors and/or lenses allows for focusing the provided light toward the crop canopy (Nelson and Bugbee 2014). In tall vine crops, such as bell peppers, cucumbers, and tomatoes, it is possible to use LEDs for intra-canopy lighting. Placing LED lights inside the canopy results in efficient light interception, while providing light to the part of canopy that typically receives little sunlight (Gómez and Mitchell 2016).

The other component of crop light use efficiency is the utilization of absorbed light by the leaves. Leaves of most plants typically absorb about 84% of the light that reaches the leaf (Björkman and Demmig 1987). The absorbed light can then be used for electron transport in the light reactions of photosynthesis (photochemistry). Photochemistry results in the production of reduced ferredoxin and, subsequently, NADPH. At the same time, photochemistry produces a proton gradient across the thylakoid membrane of the chloroplasts that is used to produce ATP. The NADPH and ATP produced by photochemistry can subsequently be used in the Calvin cycle to produce carbohydrates (Lawlor 2000; Ruban 2015). However, not all absorbed light is used for photochemistry; some of it is dissipated as heat, and a small fraction, typically around 1–2%, is re-emitted as chlorophyll fluorescence (Maxwell and Johnson 2000).

Photosystem II is generally considered to be the slowest part of the linear electron transport chain in chloroplasts of higher plants. When excitation energy from a photon reaches the reaction center of photosystem II, that energy is used to move an electron from the reaction center chlorophyll to pheophytin and subsequently to the plastoquinone pool. The reaction center chlorophyll subsequently receives an electron from a water molecule that is split into hydrogen ions and oxygen at the oxygen evolving complex. As the electron moves from the reaction center chlorophyll to the plastoquinone pool, the reaction center of photosystem II is briefly closed, i.e., it cannot accept additional excitation energy. Under high light conditions, more excitation energy will reach the reaction centers of photosystem II, resulting in more electron transport, but also in a higher proportion of closed reaction centers. As more reaction centers close, plants have to up-regulate processes to dissipate the excess light energy because excess absorbed energy can cause photoinhibition (damage to the photosynthetic machinery). Dissipation of excess light energy occurs through various processes that are collectively referred to as the non-photochemical quenching of chlorophyll fluorescence. Non-photochemical quenching is at least partly controlled by the pH of the chloroplast lumen. High rates of electron transport result in the accumulation of $H^+$ in the lumen, and the resulting low pH up-regulates processes that help dissipate excess light energy. Several in-depth reviews of these processes have been published in the last few years (Demmig-Adams et al. 2012; Horton 2012; Ruban 2015).

The quantum yield of photosystem II ($\Phi_{PSII}$) is the fraction of light absorbed by the photosystem II light-harvesting complex that is used for electron transport through photosystem II. At higher light levels, more PSII reaction centers are closed and $\Phi_{PSII}$ decreases. This has direct implications for controlling supplemental lighting, which would ideally be provided when it can be used most efficiently for electron transport, i.e., under low ambient light conditions when most reaction centers are open and $\Phi_{PSII}$ is high. Thus, supplemental lighting efficiency is high when ambient light levels are low (van Iersel and Gianino 2017). The principle of controlling supplemental lighting based on ambient light levels is consistent with photosynthesis model-driven supplemental lighting control methods (Heuvelink and Challa 1989; Clausen et al. 2015). These photosynthesis models show a non-linear increase in photosynthesis with increasing PPFD. Supplemental lighting is thus expected to have the greatest impact on photosynthesis under low ambient light conditions.

There is no generic answer to the question of how much supplemental light to provide. This depends on the expected economic yield increase that results from the supplemental light, the value of the harvested product, and the cost associated with providing the supplemental light. This makes it difficult, if not impossible, to provide generally applicable guidelines for supplemental lighting. Lighting control strategies need to be tailored to suit the needs of specific crops and production systems.

### 4.3.2 Chlorophyll Fluorescence as a Tool to Monitor Crop Performance

Knowing how efficiently plants use absorbed light can help determine whether supplemental light is likely to be used efficiently. Chlorophyll fluorescence measurements can be used to monitor crop photosynthetic light use efficiency, specifically $\Phi_{PSII}$. Pulse-amplitude-modulated chlorophyll fluorometry has become a common technique to monitor photochemical processes (Maxwell and Johnson 2000). By combining steady-state fluorescence with fluorescence measured under saturating light conditions, $\Phi_{PSII}$ can be calculated (Genty et al. 1989). Note that $\Phi_{PSII}$ (the ratio between electrons transported through PSII and absorbed photons) and the quantum yield of photosynthesis (molecules of $CO_2$ fixed or $O_2$ produced per photon) are distinctly different measures of how efficiently plants use light for photosynthetic processes. In principle, both chlorophyll fluorescence and $CO_2$ exchange measurements can be used to determine photosynthetic responses to environmental conditions, including light. However, chlorophyll fluorescence has distinct advantages over $CO_2$ exchange: It is cheaper, the equipment does not require regular calibration, and it responds instantaneously to changes in light conditions.

If $\Phi_{PSII}$ and the PPFD are known, the rate of electron transport through photosystem II can be estimated, and this is generally indicative of the overall rate of photosynthesis (Maxwell and Johnson 2000). Applying this principle to roses in a greenhouse, Schapendonk et al. (2006) found a strong correlation between the electron transport rate and leaf net photosynthesis. They concluded that chlorophyll fluorescence measurements can be used for real-time optimization control of greenhouse lighting to optimize growth and yield. However, they appear to have not yet implemented this approach. Janka et al. (2015) used measurements of $\Phi_{PSII}$ to monitor chrysanthemum (*Dendranthema grandiflora*) for high light and temperature stress. They monitored diurnal fluctuations in $\Phi_{PSII}$, as well as non-photochemical quenching. Using dark-adapted measurements of photosystem II, they also were able to detect damage to photosystem II induced by high light or temperature stress during the day.

Figures 4.4 and 4.5 outline how fluorescence data may be used to develop optimal lighting strategies. The PPFD inside a greenhouse can be highly dynamic, and $\Phi_{PSII}$ rapidly responds to changes in PPFD. Simply looking at the dynamic changes in $\Phi_{PSII}$ over the course of a day, it may seem as if these changes are too rapid and unpredictable to be of much practical use. However, when plotting $\Phi_{PSII}$ versus PPFD, it is clear that there is a strong relationship between the two. Quantum yield gradually decreases, while electron transport rate increases asymptotically as PPFD increases. These relationships vary from species to species (Fig. 4.5), thus quantitative, crop-specific information regarding the photosynthetic response to PPFD is needed to develop optimal lighting strategies. This task is complicated by the fact that species-specific light responses may depend on environmental conditions (Janka et al. 2015) and production practices (e.g., salinity, fertility). Little research has been done so far on real-time measurements of $\Phi_{PSII}$ or electron transport rate to optimize growing conditions for crops in controlled environments.

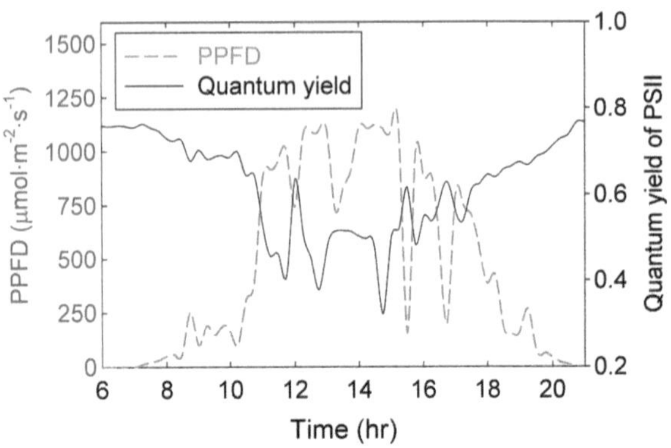


**Fig. 4.4** Changes in photosynthetic photon flux density (*PPFD*) and the quantum yield of photosynthesis II of geranium (*Pelargoniu* × *hortorum*) over the course of a 15-h period in a greenhouse

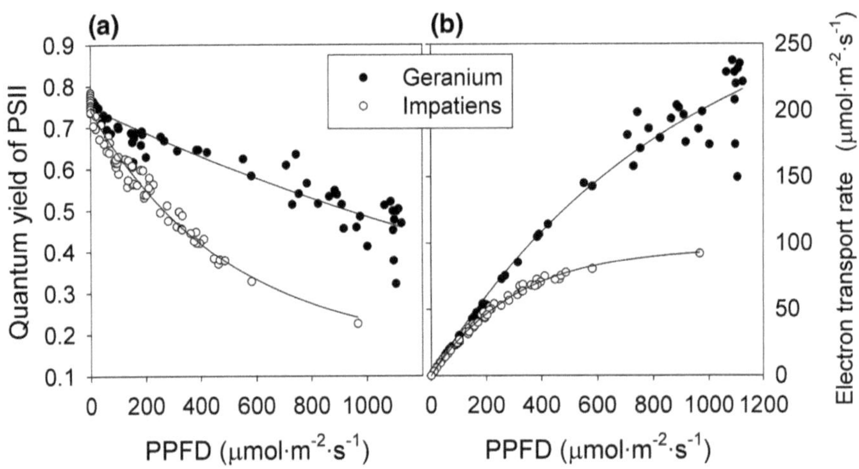


**Fig. 4.5** The relationship between the photosynthetic photon flux density (*PPFD*) and the quantum yield of photosystem II (*left*) and electron transport rate (*right*) for geranium (*Pelargonium* × *hortorum*) and impatiens (*Impatiens walleriana*)

However, the technique is promising, because it provides quantitative information about how efficiently plants use light. It is important to note an inherent tradeoff between $\Phi_{PSII}$ and electron transport: To achieve high electron transport rates, high PPFDs are needed, but that inherently will result in relatively low $\Phi_{PSII}$ (Fig. 4.5).

One limitation of chlorophyll fluorescence, as well as $CO_2$ exchange measurements, is that these are typically point measurements. Measuring a small part of one leaf of an entire canopy may not be representative of the entire crop. In principle, scanning the entire canopy would be preferable and this could provide information on spatial variability of photosynthesis. Currently, imaging chlorophyll fluorescence systems are expensive and limited to research applications, and most are restricted to individual leaves or small plants (Gorbe and Calatayud 2012; Takayama 2015). Remote sensing techniques to determine chlorophyll fluorescence at larger scales typically measure solar-induced chlorophyll fluorescence (Porcar-Castell et al. 2014). Passive solar-induced chlorophyll fluorescence measurements can monitor spatial differences and temporal changes in photochemical activity of canopies (Pinto et al. 2016), but do not allow for the quantification of $\Phi_{PSII}$ or the calculation of the electron transport rate. This is still a relatively new technique and there are serious challenges in the interpretation of these signals (see review by Porcar-Castell et al. 2014). Whether this technique will be suitable for large-scale application in commercial horticultural production remains to be seen.

### *4.3.3 Revisiting the Optimal Spectrum for Photosynthesis*

LEDs provide unique opportunities to tailor the light spectrum to the needs of the crop. There are multiple recent papers that review the effects of light spectrum on plant growth, development, and secondary metabolism (Ouzounis et al. 2015; Pocock 2015; Bugbee 2016). The goal here is not to add to those reviews, but rather to highlight some underappreciated or neglected aspects of the spectral composition of light on photosynthesis.

There has been a considerable amount of past work on the action spectrum of photosynthesis, starting 80 years ago (Hoover 1937). McCree (1972) did a series of comprehensive studies to quantify how efficiently leaves use light with different wavelengths for photosynthesis. He used different species, grown under different conditions, and his results are still considered the standard for the action spectrum of photosynthesis. He determined the spectral response of photosynthesis based on the energy of the photons reaching a leaf (action spectrum), as well as the number of absorbed photons (quantum yield of photosynthesis). His results indicate that red light (625–675 nm) is used most efficiently, with a lower peak in the blue part of the spectrum (450 nm, Fig. 4.6). These peaks roughly correspond to the absorption peaks of chlorophyll, the main photosynthetic pigment. However, numerous other photosynthetic pigments are involved in light harvesting in the thylakoid membranes (Ouzounis et al. 2015). These pigments can absorb much of the light that is poorly absorbed by chlorophyll *a* and *b* and transfer the excitation energy to chlorophylls, allowing plants to use much of the light with wavelengths from 400 to 700 nm for photosynthesis. McCree (1972) did indeed find that plants can use green and yellow lights quite efficiently (Fig. 4.6).

McCree's action spectrum has been replicated by other researchers (Inada 1976; Evans 1987; Hogewoning et al. 2012) with similar results. Despite our knowledge of the action spectrum of photosynthesis, the recent literature still contains references arguing that green and yellow lights are used inefficiently for photosynthesis (e.g., Singh et al. 2015).

One important limitation of the studies looking at spectral effects on photosynthetic efficiency is that different wavelengths of light were generally provided by

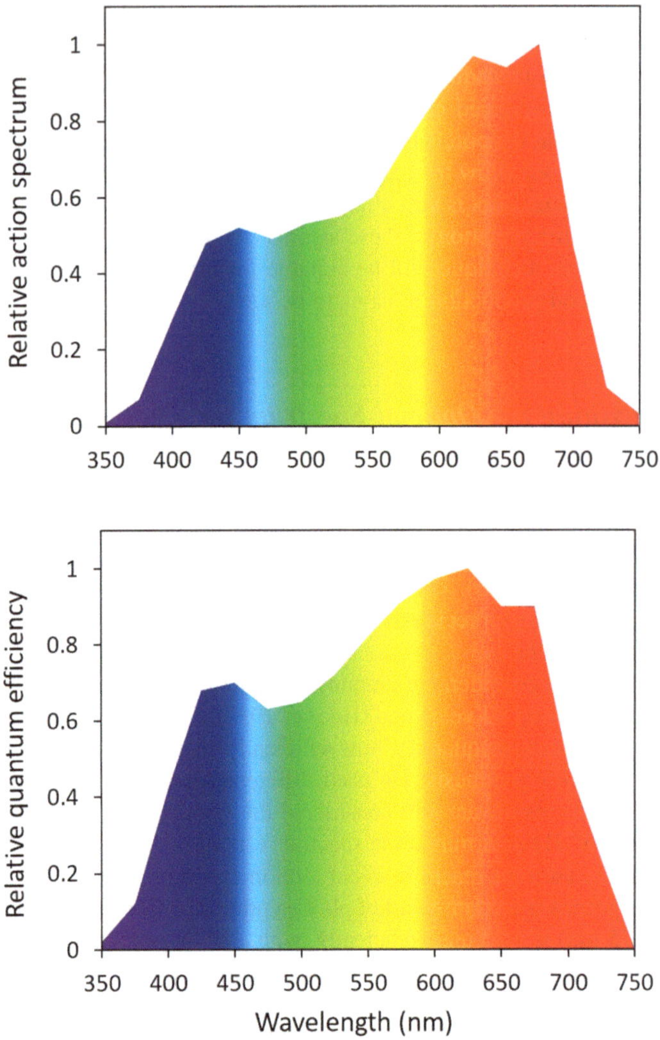


**Fig. 4.6** Action spectrum of photosynthesis, based on incident energy flux of the light reaching the leaf surface, (*top*) and relative quantum yield (*bottom*), based on number of absorbed photons. Based on data from McCree (1972, Table 6) of eight species of field-grown plants

either using a monochromator or by using filters. In both cases, the resulting light will not have a sharp peak at the wavelength of interest. Hogewoning et al. (2012), for example, reported that the FWHM for the filters they used ranged from 10 to 40 nm. McCree (1972) did not specify the exact spectral distribution of the different wavelengths of light he used. However, this information may be relevant to the interpretation of his results, especially at wavelengths where the quantum yield of photosynthesis changes rapidly. As initially described by Emerson and Lewis (1943), McCree (1972) found a rapid drop in quantum yield of photosynthesis as the wavelength of the light was increased from 675 to 700 nm. Assuming a symmetrical distribution of the provided light around the center peak, half of the light used for measurements at 700 nm would have had wavelengths below 700 nm. It seems likely that much of the observed photosynthetic activity of 700 nm light was the result of those shorter wavelength photons, rather than photons with wavelengths over 700 nm. That interpretation seems to be consistent with the findings by both Inada (1976) and Hogewoning et al. (2012), who found a substantially faster decline in photosynthetic activity at wavelengths over 680 nm. This decrease in the quantum yield of photosynthesis at wavelengths over 680 nm is due to an imbalance in excitation of photosystem I and II, with photosystem II being under-excited at those wavelengths (Hogewoning et al. 2012).

The fact that some wavelengths may result in uneven excitation of photosystem I and II is another limitation of photosynthetic spectrum response curves. Since these curves are developed by measuring photosynthesis under narrow wavelength light, such curves do not consider possible synergistic effects of different wavelengths. McCree (1972) already recognized this limitation to his data and points out that 'it is not possible to calculate the photosynthetic efficacies of white light sources from any action spectrum or spectral quantum yield curve'. Thus, we still lack good quantitative information regarding the photosynthetic efficiency of broad spectrum light. Synergisms may explain unexpected differences in photosynthetic efficiency of different lights. For example, Zhen and van Iersel (2017) compared $\Phi_{PSII}$ and net photosynthesis of lettuce (*Lactuca sativa*) under white and red/blue LED light and found that both $\Phi_{PSII}$ and net photosynthesis were consistently higher under white light than under red/blue light, when measured at the same PPFD. In addition, $\Phi_{PSII}$ decreased more rapidly under red/blue than under white light as PPFD was increased. That resulted in increasingly large differences in net photosynthesis under red/blue versus white light with increasing PPFD (Fig. 4.7).

The synergistic effect of different wavelengths of light was first described by Emerson et al. (1957), who found that combining red and far-red light results in higher rates of photosynthesis than what would be expected based on the sum of photosynthetic rates under those two light sources by themselves (also see review by Myers 1971). Work in Emerson's laboratory subsequently led to the hypothesis that the light reactions of photosynthesis depend on two different photosystems (Hill and Bendall 1960), which was later confirmed. The exact action spectrum of photosystem I and II is still unknown, but it is clear that far-red light (>680 nm) stimulates PSI much more efficiently than photosystem II (Hogewoning et al. 2012; Laisk et al. 2014) and is needed for efficient photochemistry (Zhen and van Iersel 2017).

### 4.3.4 The Importance of Far-Red Light

The importance of far-red light for excitation of photosystem I raises questions about the optimal spectrum for sole-source LED lighting. Most LED grow lights are designed to provide most light within the wavelength range of 400–700 nm, the

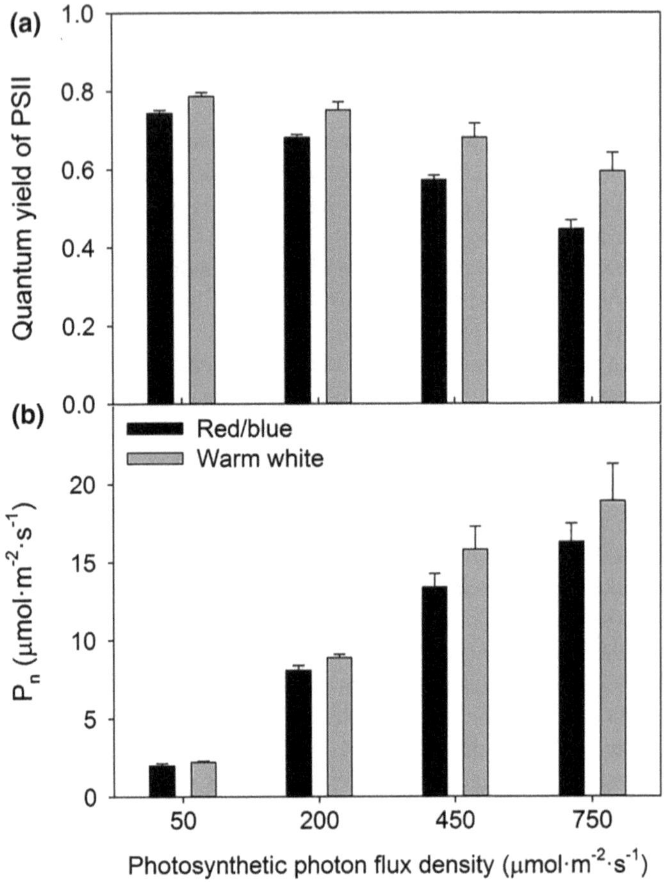


**Fig. 4.7** The quantum yield of photosystem II (*top*) and net photosynthesis ($P_n$) of lettuce (*Lactuca sativa*) under *red/blue* or *warm white* LED light provided at different PPFDs. *Warm white* consistently results in a higher quantum yield and net photosynthesis than *red/blue* light (data courtesy of Zhen and van Iersel)

part of the spectrum that is generally considered to be photosynthetically active. Grow lights made with only red and blue LEDs contain almost no far-red light, while white LEDs have a small fraction of far-red. How much far-red is present in white LED light depends on the phosphor coating of the LEDs.

To determine the effect of far-red light on $\Phi_{PSII}$ and subsequently photosynthesis, Zhen and van Iersel (2017) looked at interactions between light provided by red/blue LEDs and far-red light (peak at 735 nm). They reported that adding far-red light consistently increased net photosynthesis of lettuce exposed to red/blue light. This increase in net photosynthesis was not simply the result of increased light levels: The addition of far-red light also increased $\Phi_{PSII}$, indicating that the addition of far-red resulted in a more efficient use of the provided light in the light reactions of photosynthesis. They attributed this effect to the increased excitation of photosystem I by far-red light. This increases electron transport through photosystem I, which in turn results in a more rapid re-oxidation of the plastoquinone pool in the thylakoid membrane. This facilitates electron transfer from photosystem II to the plastoquinone pool, re-opening the reaction center of photosystem II, and allowing photosystem II to utilize excitation energy more efficiently (Zhen and van Iersel 2017). The small amount of far-red present in the spectrum of white LEDs may explain the higher $\Phi_{PSII}$ and net photosynthesis, as compared to red/blue light with practically no far-red (Fig. 4.7).

The enhancement of photosynthetic efficiency by far-red light can be important when LED lights are used as the sole source of photosynthetic lighting. Since many commercially available LED grow lights are likely deficient in far-red light, supplementing existing lights with far-red could increase photosynthetic light use efficiency, enhance crop growth and improve energy efficiency in indoor growing systems. However, this likely has little relevance to situations in which LED light is provided in the presence of sunlight since sunlight contains a large amount of far-red light (e.g., supplemental lighting in greenhouses).

### *4.3.5 Does Green Light Enhance Photosynthesis?*

Because green light is absorbed less efficiently by most plants than other wavelengths of light, green light is often considered to be used inefficiently and has received little attention in photosynthetic research. However, Kim et al. (2004) reported that adding 24% green light to light from red/blue LEDs increased lettuce biomass by 47%, even if the total PPFD was the same in both lighting treatments. They attributed the growth-stimulatory effect of green light on its ability to penetrate deeper into leaves and canopy. Because red and blue lights are absorbed efficiently by chlorophyll, most red and blue photons are absorbed within a few cell layers from the leaf surface, while green photons can penetrate further (Broderson and Vogelmann 2010). These differences in light penetration among photons of different wavelengths can have important consequences for the ability of leaves to utilize that light for photosynthesis. In an elegant study, Terashima et al. (2009) showed that the stimulatory effect of supplemental red or green light on photosynthesis depends on the PPFD of white light. They quantified leaf photosynthesis at PPFD levels from 0 to 1200 μmol $m^{-2}$ $s^{-1}$ from white light and then determined how efficiently small amounts of additional red or green light increased photosynthesis. At low light levels, red light had a greater stimulatory effect on photosynthesis than green light. This likely was the case because red light is absorbed more efficiently and has a higher $\Phi_{PSII}$ than green light. However, with high PPFDs from white light, supplemental red light increased leaf photosynthesis less than green light. With high PPFD, the cell layers close to the leaf surface already are near light saturation, many of the reaction centers are closed, and additional light will have little effect on electron transport and photosynthesis. Green light can penetrate deeper into the leaf and thus reach cells that are not yet light-saturated by the white light. As a result of the deeper penetration, green light was used more efficiently than red light, but only under high light conditions (Terashima et al. 2009).

These results, along with those of Emerson et al. (1957) and Zhen and van Iersel (2017), indicate that photosynthesis is not simply driven by PPFD, but that there are interactive effects between PPFD and light spectrum. Such effects are likely more pronounced in whole canopies than in individual leaves, because there can be major differences in spectral quality within a canopy. There is a lack of in-depth knowledge concerning the interactive effects of different wavelengths on photosynthesis and how this may depend on PPFD. Improved understanding of such interactions will help in the design of better grow lights and will allow researchers to develop better guidelines for growers.

### *4.3.6 Adaptive Control of LED Lights*

It is clear from photosynthesis models (Heuvelink and Challa 1989; Clausen et el. 2015) as well as data presented in this chapter (Figs. 4.5, 4.7, and 4.8) that light is used most efficiently when the PPFD is relatively low. To take advantage of this, supplemental light should be provided preferentially when ambient PPFD is low. van Iersel and Gianino (2017) developed a stand-alone, adaptive LED light controller that automatically does so (Fig. 4.9). This controller is designed to prevent the PPFD at the canopy level from dropping below a user-defined threshold. The PPFD at the canopy is measured using a quantum sensor, and when it drops below the threshold the duty cycle of the LEDs is automatically increased to provide enough light to reach, but not exceed, that threshold. Thus, the LED lights automatically provide more light when there is little sunlight and dim as the amount of sunlight increases. This approach to lighting control assures that supplemental light is provided when plants can use it most efficiently. Early trials with this adaptive lighting system have shown that it can reduce energy consumption by 60% with only a 10% decrease in crop biomass, as compared to using LEDs that are controlled using a timer (van Iersel and Dove 2016). The simplicity of this adaptive control approach makes it easy to implement in commercial settings, specifically in greenhouses. Benefits of this approach could be increased by also considering real-time electricity prices and providing supplemental light mainly when both ambient PPFD and electrical costs are low.

Lighting thresholds for specific crops could be determined partly based on the specific PPFD response curve for that crop (e.g., Fig. 4.5). For example, PPFD thresholds might be higher for geranium than for impatiens, because geranium can use higher PPFDs more efficiently than impatiens. The PPFD thresholds may also need to depend on the value of a crop, since relatively low $\Phi_{PSII}$ with higher electron transport rates (and growth rates) may be economical for high value, but not for low value crops.

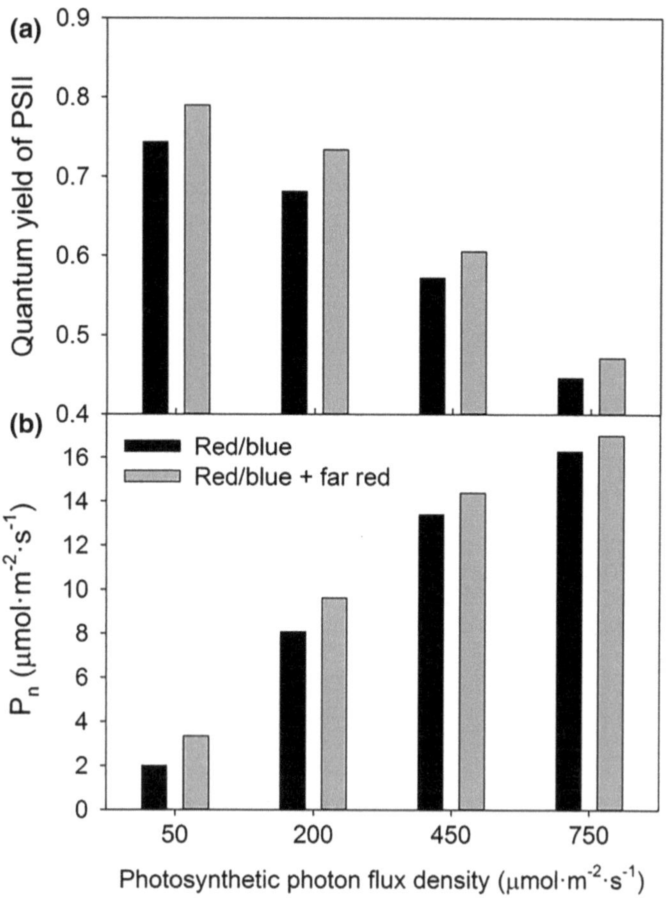


**Fig. 4.8** The quantum yield of photosystem II and net photosynthesis of lettuce (*Lactuca sativa*) under *red/blue* LED light provided at different PPFDs, with or without additional *far-red light*. The addition of *far-red light* consistently results in a higher quantum yield and net photosynthesis (data courtesy of Zhen and van Iersel)

### *4.3.7 Biofeedback Control of LED Lights*

Controlling light based on specific PPFD thresholds is relatively easy, but does not account for potential physiological changes in the crop. Rather than controlling light levels per se, it is also possible to adjust light intensity based on the physiological properties of a crop. Schapendonk et al. (2006) suggested that electron transport rate measurements can be used for lighting control. If real-time electron transport rates are determined using chlorophyll fluorescence, the PPFD provided by LED lights can be adjusted to maintain a specific electron transport rate (Fig. 4.10). The technical feasibility of this approach has been proven by van Iersel et al. (2016a, b) who

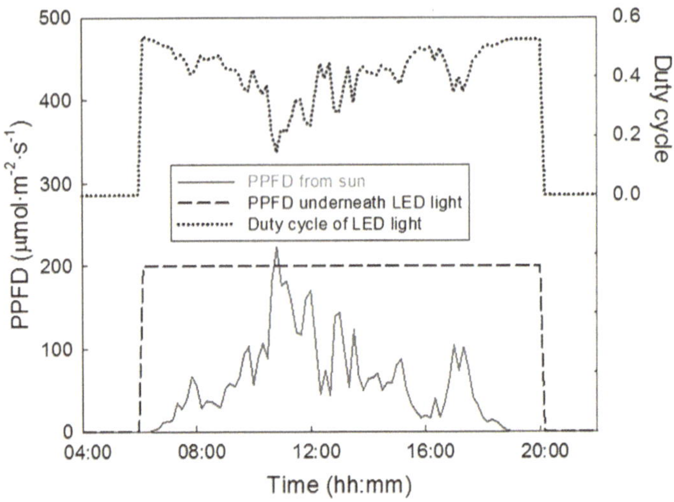


**Fig. 4.9** Performance of an adaptive LED light controller. The controller was programmed to prevent the PPFD at the canopy level from dropping below 200 μmol $m^{-2}$ $s^{-1}$. Despite large fluctuations in the PPFD provided by sunlight, the lighting controller accurately maintained the PPFD at the canopy level by constantly adjusting the duty cycle of the LED lights

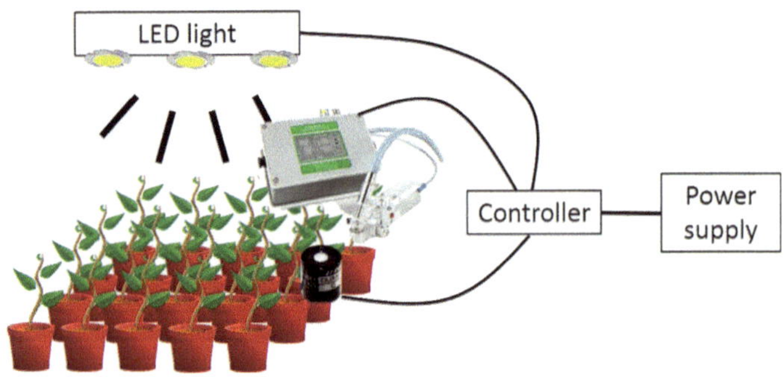


**Fig. 4.10** Diagram of a chlorophyll fluorescence-based biofeedback system to control LED lighting. A chlorophyll fluorometer and quantum sensor are used to determine the quantum yield of photosystem II and PPFD. A controller uses these data to determine the electron transport rate, compares that value to a user-defined target, and then changes the light output of the LED light, either by changing the duty cycle or current

showed that a biofeedback system can be used to maintain a range of different electron transport rates in a variety of species. Their results also showed one important issue that needs to be considered in such a biofeedback system: To determine $\Phi_{PSII}$, a short, saturating pulse of light is needed. Applying such a pulse too frequently can cause damage to photosystem II, lowering $\Phi_{PSII}$. If this happens, the leaf spot that is being measured will not be representative of the rest of the leaf, let alone the entire canopy. To minimize the risk of inducing damage to photosystem

II, it is recommended that saturating light pulses be applied at least 15 min apart (van Iersel et al. 2016a).

When a stable electron transport rate was maintained over the course of a 12-h photoperiod, the PPFD needed to be up-regulated gradually because $\Phi_{PSII}$ declined slowly throughout the photoperiod (van Iersel et al. 2016a). Such a decrease in $\Phi_{PSII}$ may be the result of an up-regulation of non-photochemical quenching (Demmig-Adams et al. 2012; Ruban 2015). However, it cannot be ruled out that it may have been partly due to damage to photosystem II caused by the saturating pulses (van Iersel et al. 2016a).

Carstensen et al. (2016) proposed a different approach for biofeedback control of lighting in controlled environments. They used a spectrophotometer to remotely sense the variable chlorophyll fluorescence emission from a plant canopy, measured at 700–780 nm. Fluorescence was induced using a blue LED, whose light output was altered in a stepwise, or sinusoidal, pattern. The resulting dynamic changes in chlorophyll fluorescence were analyzed using linear black-box models. The complexity of the resulting model appears to be indicative of the crop's light use efficiency and/or previous light-induced stress (Carstensen et al. 2016). In principle, it may be possible to use this information for feedback control of LED lights, but this has not yet been implemented. Important advantages of such a remote sensing approach over single spot measurements of chlorophyll fluorescence are (1) an entire region of the canopy can be sensed, (2) it does not require a saturating pulse of light and thus reduces the risk of photoinhibition, and (3) the sensor can be mounted remotely. Although the spectrophotometer used by Carstensen et al. (2016) is not capable of imaging, it would appear that the same general principles could be used to determine spatial variability by using an imaging sensor and appropriate filters to remotely monitor chlorophyll fluorescence.

## 4.4 Conclusions

LED lights have great potential to provide supplemental light more efficiently than traditional lights, such as HPS lamps. LED lights provide the opportunity to control both light spectrum and PPFD. To take full advantage of the opportunities provided by LEDs, we need to gain a better understanding of spectral effects on photosynthesis, as well as the physiological processes that determine light capture and photosynthetic efficiency. Interactive effects of different wavelengths of light on photosynthesis are still poorly understood and deserve more study. The controlled environment agriculture industry would benefit from smarter supplemental lighting control strategies that account for crop light use. Other factors that may need to be included in optimal lighting control strategies include real-time electricity prices and the value of the crop.

# References

Albright LD, Both AJ, Chiu AJ (2000) Controlling greenhouse light to a consistent daily integral. Trans ASAE 43:421–431

Banerjee C, Adenaeuer L (2014) Up, up and away! The economics of vetical farming. J Agric Stud 2:40–60

Björkman O, Demmig B (1987) Photon yield of $O_2$ evolution and chlorophyll fluorescence at 77 k among vascular plants of diverse origins. Planta 170:489–504

Brodersen CR, Vogelmann TC (2010) Do changes in light direction affect absorption profiles in leaves? Funct Plant Biol 37:403–412

Bugbee B (2016) Toward an optimal spectral quality for plant growth and development: the importance of radiation capture. Acta Hortic 1134:1–12

Carstensen AM, Pocock T, Bånkestad D, Wik T (2016) Remote detection of light tolerance in basil through frequency and transient analysis of light induced fluorescence. Comput Electron Agric 127:289–301

Clausen A, Maersk-Moeller HM, Corfixen Soerensen J, Joergensen BN, Kjaer KH, Ottosen CO (2015) Integrating commercial greenhouses in the smart grid with demand response based control of supplemental lighting. In: International Conference Industrial Technology Management Science (ITMS 2015), pp 199–213

Chen L, Lin CC, Yeh CW, Liu RS (2010) Light converting inorganic phosphors for white light-emitting diodes. Materials 3:2172–2195

Demmig-Adams B, Cohu C, Muller O, Adams W III (2012) Modulation of photosynthetic energy conversion efficiency in nature: from seconds to seasons. Photosyn Res 113:75–88

Demotes-Mainard S, Péron T, Corot A, Bertheloot J, Le Gourrierec J, Pelleschi-Travier S, Crespel L, Morel P, Huché-Thélier L, Boumaza R, Vian A, Guérin V, Leduc N, Sakr S (2016) Plant responses to red and far-red lights, applications in horticulture. Environ Exp Bot 121:4–21

Emerson R, Lewis CM (1943) The dependence of the quantum yield of chlorella photosynthesis on wave length of light. Amer J Bot 30:165–178

Emerson R, Chalmers R, Cederstrand C (1957) Some factors influencing the long-wave limit of photosynthesis. Proc Nat Acad Sci USA 43:133–143

Evans JR (1987) The dependence of quantum yield on wavelength and growth irradiance. Aust J Plant Physiol 14:69–79

Genty B, Briantais JM, Baker NR (1989) The relationship between the quantum yield of photosynthetic electron transport and quenching of chlorophyll fluorescence. Biochim Biophys Acta—General Subj 990:87–92

Gómez C, Mitchell CA (2016) Physiological and productivity responses of high-wire tomato as affected by supplemental light source and distribution within the canopy. J Amer Soc Hort Sci 141:196–208

Gorbe E, Calatayud A (2012) Applications of chlorophyll fluorescence imaging technique in horticultural research: a review. Scientia Hort 138:24–35

Heuvelink E, Challa H (1989) Dynamic operation of artificial lighting in greenhouses. Acta Hortic 260:401–312

Hill R, Bendall F (1960) Function of the two cytochrome components in chloroplasts: a working hypothesis. Nature 186:136–137

Hogewoning SW, Wientjes E, Douwstra P, Trouwborst G, van Ieperen W, Croce R, Harbinson J (2012) Photosynthetic quantum yield dynamics: from photosystems to leaves. Plant Cell 24:1921–1935

Hoover WH (1937) The dependence of carbon dioxide assimilation in a higher plant on wavelength of radiation. Smithson Inst. Misc. Collections 95:1–13

Horton P (2012) Optimization of light harvesting and photoprotection: molecular mechanisms and physiological consequences. Phil Trans Royal Soc B: Biol Sci 367:3455–3465

Inada K (1976) Action spectra for photosynthesis in higher plants. Plant Cell Physiol 17:355–365

Janka E, Körner O, Rosenqvist E, Ottosen CO (2015) Using the quantum yields of photosystem II and the rate of net photosynthesis to monitor high irradiance and temperature stress in chrysanthemum (*Dendranthema grandiflora*). Plant Physiol Biochem 90:14–22

Kim HH, Goins GD, Wheeler RM, Sager JC (2004) Green-light supplementation for enhanced lettuce growth under red-and blue-light-emitting diodes. HortScience 39:1617–1622

Kjaer KH, Ottosen CO, Jørgensen BN (2011) Cost-efficient light control for production of two campanula species. Scientia Hort 129:825–831

Kubota C, Chia P, Yang Z, Li Q (2012) Applications of far-red light emitting diodes in plant production under controlled environments. Acta Hortic 952:59–66

Laisk A, Oja V, Eichelmann H, Dall'Osto L (2014) Action-spectra of photosystems II and I and quantum yield of photosynthesis in leaves in state 1. Biochim Biophys Acta—Bioenerg 1837:315–325

Lawlor DW (2000) Photosynthesis. Bios Scientific Publishers, Oxford, UK

Mathieu JJ, Albright LD, Leed AR (2004) A stand-alone light integral controller. Acta Hortic 633:153–159

Maxwell K, Johnson GN (2000) Chlorophyll fluorescence—a practical guide. J Exp Bot 51:659–668

McCree KJ (1972) The action spectrum, absorptance and quantum yield of photosynthesis in crop plants. Agric Meteorol 9:191–216

Myers J (1971) Enhancement studies in photosynthesis. Ann Rev Plant Physiol 22:289–312

Nelson JA, Bugbee B (2014) Economic analysis of greenhouse lighting: light emitting diodes vs high intensity discharge fixtures. PLoS ONE 96:e99010

Ouzounis T, Rosenqvist E, Ottosen CO (2015) Spectral effects of artificial light on plant physiology and secondary metabolism: a review. HortScience 50:1128–1135

Pinto F, Damm A, Schickling A, Panigada C, Cogliati S, Müller-Linow M, Balvora A, Rascher U (2016) Sun-induced chlorophyll fluorescence from high-resolution imaging spectroscopy data to quantify spatio-temporal patterns of photosynthetic function in crop canopies. Plant, Cell Environ 39:1500–1512

Pocock T (2015) Light-emitting diodes and the modulation of specialty crops: light sensing and signaling networks in plants. HortScience 50:1281–1284

Porcar-Castell A, Tyystjärvi E, Atherton J, van der Tol C, Flexas, J, Pfündel EE, Moreno J, Frankenberg C, Berry JA (2014) Linking chlorophyll *a* fluorescence to photosynthesis for remote sensing applications: mechanisms and challenges. J Exp Bot 65:4065–4095

Ruban AV (2015) Evolution under the sun: optimizing light harvesting in photosynthesis. J Exp Bot 66:7–23

Sager JC, Smith WO, Edwards JL, Cyr KL (1988) Photosynthetic efficiency and phytochrome photoequilibria determination using spectral data. Trans ASAE 31:1882–1889

Schapendonk AHCM, Pot CS, Yin X, Schreiber U (2006) On-line optimization of intensity and configuration of supplementary lighting using fluorescence sensor technology. Acta Hortic 711:423–430

Singh D, Basu C, Meinhardt-Wollweber M, Roth B (2015) LEDs for energy efficient greenhouse lighting. Renew Sustain Energy Rev 49: 139–147

Takayama K (2015) Chlorophyll fluorescence imaging for plant health monitoring. In: Dutta Gupta S, Ibaraki Y (eds) Plant image analysis: fundamentals and applications. CRC Press, Boca Raton, pp 207–228

Terashima I, Fujita T, Inoue T, Chow WS, Oguchi R (2009) Green light drives leaf photosynthesis more efficiently than red light in strong white light: revisiting the enigmatic question of why leaves are green. Plant Cell Physiol 50:684–697

USDA-NASS (2014) 2012 Census of agriculture. United States summary and state data

van Iersel MW, Dove S (2016) Maintaining minimum light levels with LEDs results in more energy-efficient growth stimulation of begonia. In: 2016 conference program, American society for horticultural science. https://ashs.confex.com/ashs/2016/meetingapp.cgi/Paper/24412

van Iersel MW, Gianino D (2017) An adaptive control approach for LED lights can reduce the energy costs of supplemental lighting in greenhouses. HortScience 52:72–77 van Iersel MW, Mattos E, Weaver G, Ferrarezi RS, Martin MT, Haidekker M (2016a) Using chlorophyll fluorescence to control lighting in controlled environment agriculture. Acta Hortic 1134:427–433

van Iersel MW, Weaver G, Martin MT, Ferrarezi RS, Mattos E, Haidekker M (2016b) A chlorophyll fluorescence-based biofeedback system to control photosynthetic lighting in controlled environment agriculture. J Amer Soc Hort Sci 141:169–176

Wallace C, Both AJ (2016) Evaluating operating characteristics of light sources for horticultural applications. Acta Hortic 1134:435–443

Watanabe H (2011) Light-controlled plant cultivation system in Japan—Development of a vegetable factory using LEDs as a light source for plants. Acta Hortic 907:37–44

Zeidler C, Schubert D, Vrakking V (2013) Feasibility study: vertical farm EDEN. German aerospace center, institute of space systems. 117 pp

Zhen S, van Iersel MW (2017) Far-red light is needed for efficient photochemistry and photosynthesis. J Plant Physiol 209:115–122

# Chapter 5
# Economics of LED Lighting

**Bruce Bugbee**

## 5.1 Introduction

This review analyzes the state-of-the-art economics of LED lighting for plant growth. This compares the widely used, and most electrically efficient, traditional technology (1000 W high pressure sodium, HPS) with the most electrically efficient LED technology. Initial capital cost of the fixture and operating costs are included. Because LED technology is considerably more expensive per photosynthetic photon, increased fixture efficiency and decreased electric costs would need to justify this initial capital investment. However, in addition to initial cost and ongoing electric costs, there is another important difference between LED and HPS technologies. Perhaps the greatest asset of LED fixtures in their highly focused output, which can lead to significantly more efficient radiation transfer to plant leaves and thus fewer fixtures and reduced cost of electricity (Nelson and Bugbee 2014).

Readers of this book should be well aware of the fundamental difference in lighting for humans and lighting for plants, but some aspects of the differences are still not well understood. This chapter also reviews units and terminology for quantifying the efficacy of lighting for plant growth.

The effect of LED technology is often assumed to result in significantly cooler leaf temperatures than high pressure sodium technology, but most of this observation is caused by differences in the intensity of photosynthetic radiation. An energy balance model is reviewed that analyzes leaf temperature in greenhouses and sole-source indoor lighting. This model demonstrates that the thermal differences

B. Bugbee (✉)
Department of Plants Soils and Climate, Utah State University,
Logan, UT 84321, USA
e-mail: bruce.bugbee@usu.edu



S. Dutta Gupta (ed.), *Light Emitting Diodes for Agriculture*,
DOI 10.1007/978-981-10-5807-3_5

among lighting technologies are smaller than is often assumed (Nelson and Bugbee 2015). Finally, recent studies on spectral effects on photosynthesis are reviewed. These indicate that many spectral effects are primarily caused by increased leaf expansion and radiation capture, rather than by increased photosynthesis (Snowden et al. 2016).

## 5.2 Economics of LED Lighting: Initial Analysis in 2014

In a comprehensive study, Nelson and Bugbee (2014) reported photosynthetic (400–700 nm) photon efficiency and distribution pattern of multiple lighting technologies, including ten types of LED fixtures. They found that the most efficient LEDs and the most efficient double-ended HPS fixtures had nearly identical efficiencies at 1.66–1.70 μmol per joule. These fixtures were a 70% improvement over the efficiency of commonly used mogul-base HPS fixtures. The most efficient ceramic metal halide and fluorescent fixtures came in at 1.46 and 0.95 μmol per joule, respectively.

Nelson and Bugbee (2014) calculated the initial capital cost of each type of fixture per photon delivered and determined that LED fixtures cost five to ten times more than HPS fixtures. The five-year electric cost plus fixture cost per mole of photons was thus 2.3 times higher for LED fixtures, because of the high capital cost. Their analysis indicated that the long-term maintenance costs are small for both technologies. They pointed out that the unique ability of LED fixtures to focus on photons on specific areas can be used to improve the photon capture by plant canopies.

## 5.3 The Best Measure of Electrical Efficiency for Plant Growth Is μmoles per Joule

The electrical efficiency of lamps is still sometimes expressed using units for human light perception (efficacy; lumens or foot-candles out per watt in) or energy efficiency (radiant watts out per electrical watt in), but photosynthesis and plant growth are determined by moles of photons. Lighting efficiency is thus best based on photon efficiency, with units of μmoles of photosynthetic photons per joule of energy input. This is critical with LEDs where the most electrically efficient colors are deep red and blue wavelengths. A comparison of red, blue, and cool white LEDs provides a comparison of this difference (Table 5.1). The lower energy content of red photons allows more photons to be delivered per unit of input energy (energy per photon is inversely proportional to wavelength, Planck's equation). Conversely, blue LEDs have a high energy efficiency but only a relatively low photon efficiency (1.87 vs. 1.72).

**Table 5.1** Comparison of three types of LEDs and three units of measurement for efficiency

| LED color | Peak wavelength or color temperature | Photon efficiency[a] (μmol/J) | Electrical efficiency[b] (%) | Luminous efficiency[c] (lm/W) |
|---|---|---|---|---|
| Cool white | 5650 K | 1.52 | 33 | 111 |
| Red | 655 nm | 1.72 | 32 | 47 |
| Blue | 455 nm | 1.87 | 49 | 17 |

The appropriate measurement for plant growth is μmol per joule. The electrical efficiency has units of watts per watt (Reproduced with permission from Nelson and Bugbee 2014)
[a]Photon efficiency is the most appropriate measure for photosynthesis
[b]The relationship between electrical efficiency and photon efficiency is dependent on wavelength (Plank's equation $E = hc/\lambda$)
[c]Luminous efficiency is shown to demonstrate how inappropriate it is as an indicator of lighting efficiency for plants. doi:10.1371/journal.pone.0099010.t001

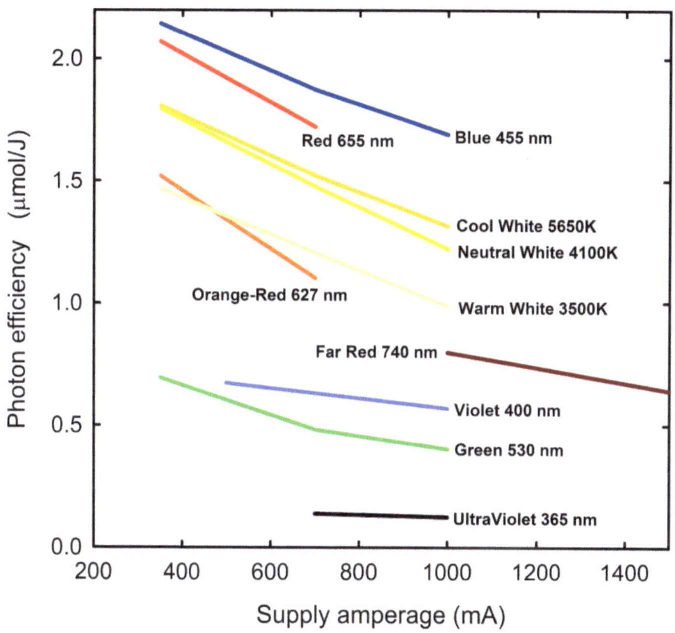


**Fig. 5.1** Effect of canopy photon capture efficiency on average annual cost over 5 years. The lowest cost of lighting is realized with 1000-W, double-ended HPS technology when all of the radiation from a beam angle of 120° can be utilized by the plant canopy (reprinted with permission from Nelson and Bugbee 2014)

## 5.4 The Value of Focused Photons from LED Fixtures

A frequently overlooked advantage of LED fixtures is their ability to focus on photons. Nelson and Bugbee (2014) pointed out that the unique ability of LED fixtures to focus on photons can be used to improve the photon capture by plant canopies. Figure 5.1 shows a typical economic crossover point for LED and HPS technologies based on bean angle of the fixture output. The lowest lighting system cost is achieved when an efficient fixture is used in a system with effective photon capture by the plants.

## 5.5 Unique Characteristics of LED Fixtures

The most electrically efficient colors of LEDs are blue, red, and cool white, respectively (Fig. 5.2), and LED fixtures thus generally come in combinations of these colors. Other monochromatic colors of LEDs are sometimes used to increase specific wavelengths in the quest to control aspects of plant growth and development (see Massa et al. 2008 for a review of unique LED applications). Ultraviolet (UV) radiation is nearly always absent in LED fixtures because UV LEDs significantly reduce fixture efficiency. There is 9% UV radiation in sunlight (percent of photosynthetic photon flux; PPF), and standard electric lights have 0.3–8% UV (percent of PPF; Nelson and Bugbee 2013). A lack of UV radiation can cause disorders in some plant species, the most common of these is intumescence (Morrow and Tibbitts 1988). This is a concern when LED fixtures are used without sunlight. LED fixtures also often have minimal far-red radiation (710–740 nm), which decreases the time to flowering in several photoperiodic species (Craig and Runkle 2013). Green light (530–580 nm) is low in many LED fixtures. Green light

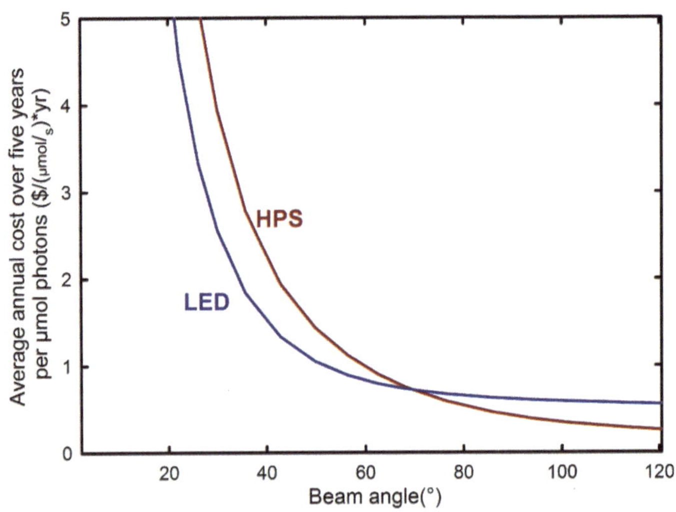


**Fig. 5.2** Effect of drive amperage and color on photon efficiency of LEDs. Data for Philips Lumileds LEDs, courtesy of Mike Bourget, Orbitec (reproduced with permission from Nelson and Bugbee 2014)

penetrates through leaves and is thus effectively transmitted to lower plant leaves (Kim et al. 2004). The lack of UV, green, and far-red wavelengths, however, is generally small when LEDs are used in greenhouses, because most of the radiation comes from the broad spectrum sunlight.

## 5.6 Advances in LED Efficacy Since 2014

Although there have been no significant fundamental advances in LED technology, manufacturers have continued to make fixtures more reliable and have further optimized the drivers and optics to increase the efficacy. In the spring of 2016, we tested a new, 600 W, white LED fixture from Fluence Bioengineering (model VYPRx PLUS). The efficacy was 2.05 μmol per joule with flat plane integration. The same fixture was tested at Rutgers University using an integrating sphere, and measured 2.02 and 2.05 μmol per joule. These values were about 20% higher than the technologies previously evaluated in our laboratory. The photon distribution from this fixture is wider than most LED fixtures, so it is difficult to take advantage of focused photons, but the distribution is still more focused than HPS fixtures. There have been several claims of high efficacy LED fixtures for plant lighting applications, but this was the first fixture we tested with an efficacy higher than 2 μmol per joule. The best previous efficacy was 1.7 μmol per joule.

In August 2016, testing of three new LED fixtures from Philips Lighting was performed at Utah State University. These fixtures had the highest efficacy that we have measured to date. All the models are of Philips GPL Top lighting. The results are as follows:

1. Deep-red/white far-red 175-W: 1.94 ± 0.07 μmol per joule;
2. Deep-red/white medium blue 200-W: 2.44 ± 0.05 μmol per joule;
3. Deep-red/blue medium blue 215-W: 2.46 ± 0.05 μmol per joule.

The values were 1.94, 2.44, and 2.46, respectively, for 3 fixtures. Two of the three fixtures (2.44 and 2.46) had the highest efficacy of any of the fixtures we have tested. The variation in efficacy mentioned above (±0.05) is the standard deviation among three replicate fixtures. One of the fixtures was also tested in an integrating sphere at Rutgers University, and the efficacy was within 2% of the values for the same fixture at Utah State University. The best efficacy of an LED fixture with white light output has been 2.05 μmol per joule (Fluence Bioengineering, see previous paragraph). All of these values are a significant increase over 1.7 μmol per joule from the technologies available in 2014.

The higher efficacy of these fixtures does not automatically mean that they are the most cost-effective plant lighting option. The Philips fixtures sell for $400 to $800 and are only 175–215 W. This is still about 5–10 times the initial capital cost of HPS technology ($400 for a 1000-W fixture). Assuming $0.10 per kWh and equal capture of photons for all types of fixtures, the time to recover the initial capital investment is 5–10 years if the fixtures are used 16 h every day (indoor cultivation), and 15–30 years if the fixtures are used with an average of 5 h a day for 365 days (supplemental lighting in a greenhouse). The variation in payback depends on the initial capital cost of the fixture.

Selection of the most cost-effective lighting technology depends on multiple factors, including (1) cost of electricity, (2) cost of fixture, (3) cost of cooling, (4) hours of operation per year, and especially, (5) the fraction of PPF captured by the plant canopy. Quantity discounts are available for most fixtures. Nelson and Bugbee (2013) developed an online calculator to facilitate a comprehensive analysis of options: http://cpl.usu.edu/htm/publications/file=15575.

The 600-W fixture from Fluence Bioengineering has a lower initial capital cost per photon of output. Although it has a lower efficacy, the payback time is similar to the Philips fixtures. Both the Philips and the Fluence Bioengineering fixtures have less focused output than the previous LED fixtures. This broader photon distribution makes it more difficult to take advantage of the narrow output of typical LED fixtures. If a user can take advantage of the more focused photon distribution from LED fixtures, the photon capture is increased and the payback time is reduced. LED efficiency is being evaluated by the present group with the advances in technology as they become available.

## 5.7 Definition of Efficacy and Efficiency

The term efficiency is typically used with ratios that have the same units in the numerator and denominator, like watts per watt. Efficacy is used when the units in the numerator and denominator are not the same, as in µmoles per joule.

When the units in the numerator and denominator are the same, a percent efficiency can be calculated, and, theoretically, the ratio is 100%. However, 100% efficiency does not make sense when the units are different. Although the efficiency of a fixture can be calculated in watts of output per watt of input, plant growth is determined by moles of photons, not by watts of energy. The most appropriate measurement is efficacy, with units of µmoles per joule. The term efficiency has often been used to refer to the ratio of µmoles per joule. This is a useful descriptive term, but it is not technically correct.

## 5.8 Thermal Effect of Electric Lighting Technologies

The use of LEDs is often assumed to result in much cooler leaf temperatures than high pressure sodium technology. The thermal properties of lighting fixtures affect plant growth, transpiration rate, and alter heating and cooling costs. Nelson and Bugbee (2015) evaluated the magnitude of this effect by measuring radiation absorbed by a leaf under four radiation sources: (1) clear sky sunlight in the field, (2) sunlight in a glass greenhouse, and indoor plants under either (3) high pressure sodium or (4) light emitting diodes. They, then, used a mechanistic energy balance model to analyze and compare leaf-to-air temperature differences. Leaf temperature was compared at equal photosynthetic photon flux. They found that the effect of plant water status and leaf evaporative cooling was much larger than the effect of radiation source. If plants are not water stressed, leaves under all four radiation sources were typically within 2 °C of air temperature. Under clear sky conditions, the cool sky temperature means that leaves are always cooler than inside a greenhouse or indoor plants—when photosynthetic photon flux, stomatal conductance, wind speed, vapor pressure deficit, and leaf size are equivalent. Leaf temperatures can increase well above air temperature as water stress increases and cooling via transpiration decreases. In a near-worst scenario of water stress and low wind, leaves can increase 6–12 °C above air temperature under any of the lighting conditions. Because LED fixtures emit much of their heat through convection rather than radiation, they result in slightly cooler leaf temperatures than leaves under HPS fixtures, but the effect of LED technology on leaf temperature is much smaller than is often assumed.

LED fixtures emit almost no near-infrared radiation (NIR; 700–3000 nm), but this radiation is poorly absorbed by leaves (Fig. 5.3). Photosynthetic (400–700 nm) and long-wave (3,000–100,000 nm) radiation absorbed are about 95%, but

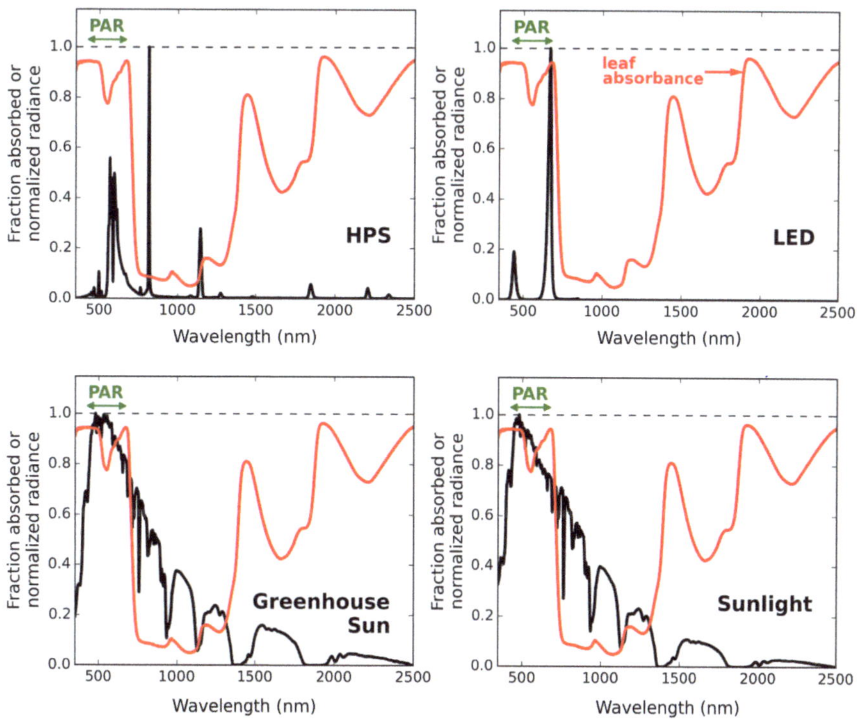


**Fig. 5.3** Radiance spectrum from four radiation sources (*black line*) and average leaf absorbance (*red line*; reprinted with permission from Nelson and Bugbee 2015)

non-photosynthetic solar NIR (700–3000 nm) is only about 20% absorbed, and thus has a smaller effect on leaf heating. Unabsorbed radiation is either transmitted or reflected.

When LED fixtures have the same electrical efficiency as HPS fixtures, they generate the same amount of heat per photosynthetic photon. LED fixtures, however, dissipate much of their heat away from the plants, while HPS fixtures dissipate more heat toward the plants.

Nelson and Bugbee (2015) found that the leaf-to-air temperature difference was always less than 2 °C except where parameters approached their extremes (Fig. 5.4). The relative order did not change, regardless of environmental conditions, with HPS > greenhouse sun > LED > clear sky sunlight.

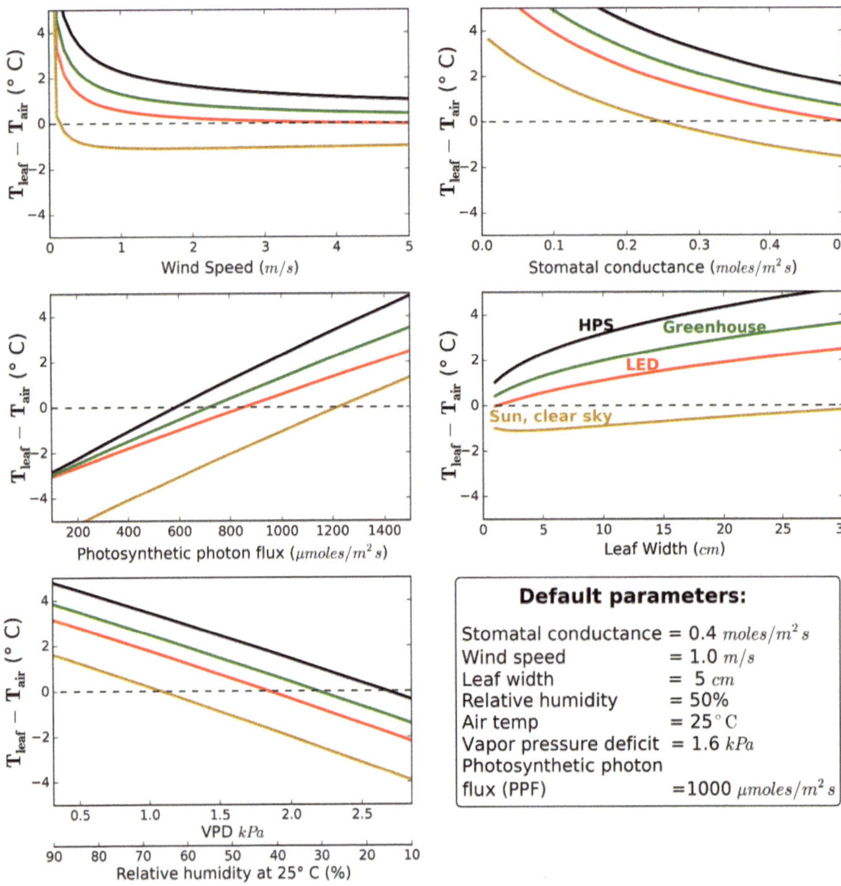


**Fig. 5.4** Calculated effects of environmental conditions on the difference between leaf temperature and air temperature under four radiation scenarios (reprinted with permission from Nelson and Bugbee 2015)

### 5.8.1 Effects of Elevated $CO_2$ on Leaf Temperature

Controlled environments often add supplemental $CO_2$, which decreases stomatal conductance 10–40% and increases leaf temperature. The analysis of Nelson and Bugbee (2015) indicates that a decrease in stomatal conductance of 30% in response to elevated $CO_2$ increased leaf temperature by 1 °C regardless of radiation source.

### 5.8.2 Effect of Light Technology on Shoot Tip Temperature

Shoot tip temperature is used to predict time to flower and plant development rates. Nelson and Bugbee (2015) found that lighting technology affects shoot tip temperature, which can alter time to flower and plant development.

### 5.8.3 Effect of Light Technology on Fruit and Flower Temperature

The analysis of Nelson and Bugbee (2015) indicated that the near-worst case analysis would be representative of flowers, fruits, and thick plant parts that have low transpiration rates, including high-value crops like tomatoes, strawberries, and *Cannabis* flowers. These thicker structures absorb more radiation than a thin leaf and have fewer stomates for evaporative cooling. Based on the analysis of Nelson and Bugbee (2015), LED technology has the potential to reduce heating of these thick, low transpiring plant parts. However, in conditions where leaves and shoot tips benefit from heating, (e.g., a greenhouse in a cool climate), HPS technology would more effectively warm the plants.

## 5.9 Spectral Effects on Single Leaf Photosynthesis

Although we have defined photosynthetic photon flux with equal weighting of all photons between 400 and 700 nm, further studies indicate that this is not strictly true. Hoover (1937) used colored filters to achieve narrow spectra and determined spectral effects on photosynthesis in 29 species (Fig. 5.1). He did not have the apparatus to determine radiation absorption, so his results were measured per incident photon. He found relatively sharp peaks in the blue and red regions and reported that differences among species were small.

Thirty-five years later, McCree (1972a, b) and Inada (1976) revisited spectral effects on photosynthesis and quantum yield. All response curves were developed from single leaves, at a low PPF, over a short-time interval (minutes). All studies include the average of more than 20 species. The studies in the 1970s confirmed the findings of Hoover (1937) and indicate only small differences among species. The differences among studies are significantly greater than differences among species within a study.

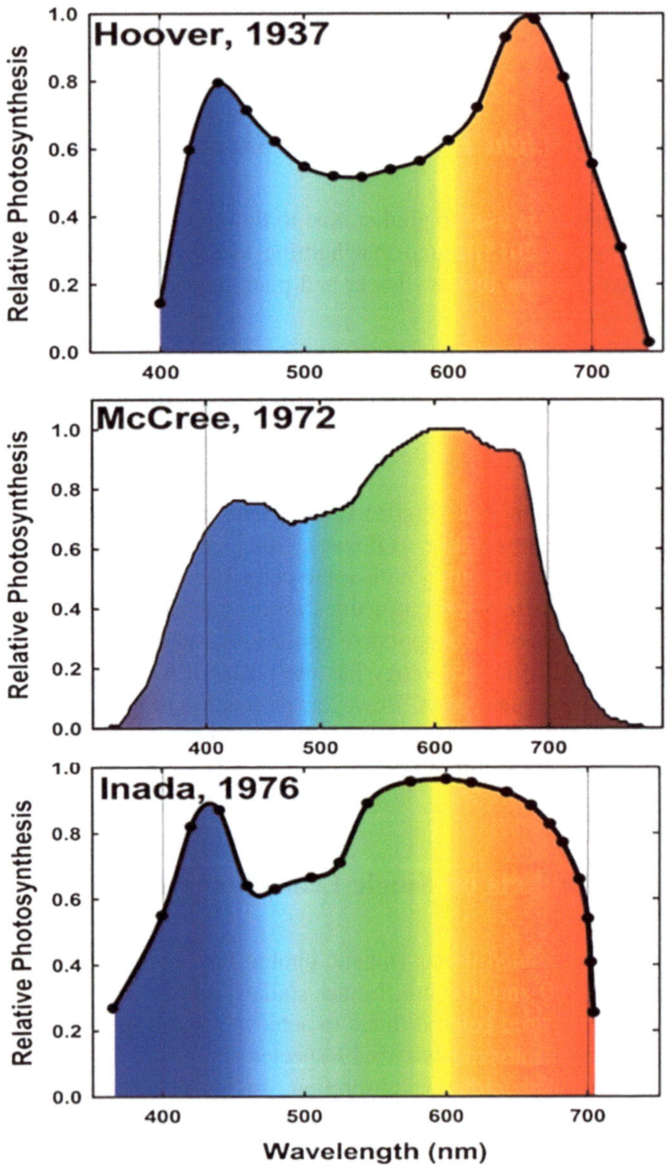


**Fig. 5.5** Spectral effects on photosynthesis from three studies: Hoover (1937), McCree (1972a, b) and Inada (1976). All curves are redrawn from the original data. *Black circles* indicate wavelengths where measurements were made. The Hoover curve is per incident photon. The McCree and Inada curves are per absorbed photon, so they reflect the quantum yield of photosynthesis. The *green light* dip in the Hoover curve would be about 15% higher if it was per incident photon (i.e., increased from 0.6 to 0.7) (reprinted with permission from Bugbee 2016)

Both McCree and Inada found that, per absorbed photon, blue/cyan photons are used less efficiently than orange/red photons, but the quantum yield increases rapidly as the color of light changes from cyan to green [between 520 and 550 nm (Fig. 5.5)].

Measurement procedures were similar among studies but the study by McCree McCree (1972a, b) has the most comprehensive discussion of principles. Differences among studies, however, indicate that the McCree curve should not be considered as a definitive reference for spectral quality and photosynthesis. More importantly, several studies over longer time intervals now indicate that it is often inappropriate to use any of these curves to predict photosynthesis in whole plants under mixed colors of light at higher PPFs.

## 5.10 Determining Whole Plant Net Photosynthesis from Crop Growth Rate and Leaf Area Index

Plant growth analysis is often used to separate crop growth rate (CGR; g dry mass per $m^2$ ground area per day) into its two component parts: net assimilation rate (NAR; grams of dry mass per $m^2$ of leaf per day) and leaf area index (LAI; leaf area index; $m^2$ of leaf per $m^2$ of ground; Hunt 1982). The equation is as follows:

$$\mathrm{CGR} = \mathrm{NAR} \times \mathrm{LAI}.$$

Crop growth rate (biomass gain) and leaf area index are not difficult to measure. The ratio of CGR to LAI yields the integrated net assimilation rate over the measurement interval (NAR = CGR/LAI). NAR is the average photosynthetic efficiency of the whole crop over time (Fitter and Hay 2012).

The NAR is related to single leaf photosynthetic rate ($P_{net}$), but there are important differences. Net photosynthesis in single leaves is typically determined by clamping a portion of a leaf in a chamber and measuring the uptake of $CO_2$ over a short-time interval (minutes). The unit of measurement is μmol of $CO_2$ per $m^2$ leaf per second. This measurement is representative of the photosynthetic rate in part of a leaf at the PPF incident on the leaf at the time of measurement. NAR integrates daily carbon gain and nighttime respiratory loss to provide a value for daily net whole plant photosynthesis.

The development of portable photosynthesis systems in the 1980s resulted in widespread use of "clamp-on" photosynthesis measurements. These systems provide a rapid indication of photosynthetic rate, and there has been great hope that these measurements would elucidate genetic and environmental effects on yield. Unfortunately, numerous studies over several decades have indicated that single leaf photosynthetic rate is poorly correlated with yield (see reviews by Evans 1993, 1998; Long et al. 2006). While it is implicit that photosynthetic efficiency is essential to growth, this is the photosynthetic efficiency of the whole crop averaged over time. The problem is that short-term measurements of $P_{net}$ on single leaves poorly predict daily photosynthesis in whole plants.

## 5.11 The Importance of Radiation Capture Efficiency

Although NAR is a more accurate predictor of environmental effects on whole plant photosynthesis than short-term measurements of $P_{net}$, it is radiation capture efficiency (fraction of radiation intercepted) that is most closely related with biomass gain (Bugbee and Salisbury 1988; Bugbee 1995; Keating and Carberry 1993; Monje and Bugbee 1998). Increases in leaf area and radiation capture are often associated with thinner leaves in which NAR is reduced (Beadle and Long 1985). In a classic study, Evans and Dunstone (1970) found that modern, high-yielding wheat cultivars had lower leaf photosynthetic rates than their wild ancestors.

Since LAI determines radiation capture and is highly correlated with canopy photosynthesis and dry mass gain (Klassen et al. 2003), several studies have sought to separate radiation capture efficiency from canopy photosynthetic efficiency (Bugbee and Monje 1992; Monje and Bugbee 1998). Improvements in radiation capture efficiency have been responsible for nearly all of the increases in yield. Increases in biomass productivity are closely related to increased leaf area, and this usually results in decreased photosynthetic rate because of increased self-shading (Evans 1993).

## 5.12 Spectral Effects on Single Leaf Photosynthetic Efficiency

In the quest to understand spectral effects on plant growth, several studies have focused on single leaf photosynthetic efficiency over short-time intervals. Numerous studies have examined the effects of increasing blue light on photosynthetic efficiency. Goins et al. (1997) and Yorio et al. (2001) demonstrated that some blue light was necessary for efficient photosynthesis. Hogewoning et al. (2010) found that increasing blue light from zero to 7% doubled the photosynthetic capacity. Terfa et al. (2013) showed that increasing blue light from 5 to 20% increased leaf thickness, which increased photosynthetic capacity. Wang et al. (2014) found that stomatal conductance and net photosynthetic rate increased with increasing blue light in cucumber. Hernández and Kubota (2015) measured a 20% increase in $P_{net}$ in cucumber as blue light fraction increased from 10 to 80%. In contrast, Ouzounis et al. (2015) found that there was no effect of blue light fraction on photosynthesis in roses, chrysanthemums and campanulas and lettuce. The results of these studies are in contrast to the spectral efficiency curves of Hoover

(1937), McCree (1972a, b), and Inada (1976), which indicate that the blue light is used less efficiently in photosynthesis. It is apparent that other interacting factors alter the effect of light quality on photosynthetic efficiency in long-term studies.

## 5.13 Effect of Fraction of Blue Light on Growth

Several studies indicate that whole plant growth (dry mass) decreases as the fraction of blue photons increases above about 5%. This has often been interpreted as an effect of increased blue light on reduced photosynthesis. This interpretation is nearly always incorrect. Photosynthetic efficiency is measured as quantum yield: moles of carbon fixed per mole of photons absorbed. Increasing blue light fraction inhibits cell division and cell expansion, and thus reduces leaf area (Dougher and Bugbee 2004). Reduced leaf area reduces photon capture. This blue-light-induced reduction in photon capture is often the primary reason for reduced growth. There is often a minimal direct spectral effect on photosynthetic efficiency. This distinction is critical when extrapolating from single leaves to whole plants and to plant communities.

## 5.14 Effect of Blue Light Fraction on Development

Plant development is, here, defined as plant size and shape. A tall plant without branches might have the same growth (dry mass) as a short highly branched plant, but they have developed differently. Although wheat, and possibly all grasses, appears to have minimal sensitivity to spectral quality (Dougher and Bugbee 2001); tomatoes are exquisitely sensitive; cucumbers, radishes and peppers have intermediate sensitivity; soybeans and lettuce have low sensitivity (Snowden et al. 2016). Blue light can alter secondary metabolism, and these compounds provide protection from biotic and abiotic challenges. Blue light can interact with radiation intensity (PPF), and responses can change with developmental stage (Cope and Bugbee 2013; Cope et al. 2014; Chen et al. 2014). The diversity of responses among species indicates that caution should be used in extrapolating from studies with *Arabidopsis* to crop plants. Similarities among groups of species, however, suggest that plants can be separated into categories by common responses.

## 5.15 Effect of Green Light Fraction on Photosynthesis and Growth

Green light can alter plant development (Folta and Maruhnich 2007; Zhang et al. 2011), although its effects may decrease as PPF increases (Wang and Folta 2013). Sun et al. (1998) found that red and blue light drive $CO_2$ fixation primarily in the upper leaf layers, while green light penetrates deeper and drives $CO_2$ fixation in the lower leaf cells. Broadersen and Vogelmann (2010) measured chlorophyll fluorescence in leaf cross sections and showed that green light penetrated much deeper than red or blue light. Accordingly, once the upper part of individual leaves and the upper canopy as a whole are saturated, a higher fraction of green light should be especially beneficial (Nishio 2000). This effect was demonstrated by Terashima et al. (2009) who reported that in a high light background, green light drives leaf $P_{net}$ more efficiently than red or blue light. Thus, whole plant $P_{net}$ could be increased by green light penetration to lower leaf cells and lower leaf layers.

Some studies have suggested that a high green light fraction can improve plant growth. Kim et al. (2004) reported that supplementing red and blue LEDs with green light (from green fluorescent lamps) increased lettuce growth by up to 48% at the same total PPF. The findings indicated that too much (51%) or too little (0%) green light caused a decrease in growth, while about 24% was optimal.

Johkan et al. (2012) grew lettuce at three PPFs using LEDs with cool white fluorescent controls. As PPF decreased and the fraction of green light increased, the lettuce plants exhibited an increased shade-avoidance response. Plants grown under cool white fluorescent lamps developed more normally and grew faster than plants grown under the LEDs. These results are in agreement with the findings of Kim et al. (2004).

In contrast to these studies, Hernández and Kubota (2015) found that the addition of 24% green light had no effect on growth (dry mass) of cucumbers. In a comprehensive study with seven species, Snowden et al. (2016) studied the effect of blue and green light fractions at PPFs of 200 and 500 $\mu$mol $m^{-2}$ $s^{-1}$. For some species, there were significant interactions between radiation quality and radiation intensity (PPF). Increasing blue light from 11 to 28%, at a PPF of 500, reduced dry mass in tomatoes, cucumbers, radishes, and peppers, but there was no significant effect on soybeans, lettuce, and wheat. At a PPF of 200, the reduction in dry mass from increasing blue light was only significant in tomatoes (Fig. 5.6).

This study used classical techniques to determine integrated net assimilation rate (photosynthetic efficiency) over the 21-day growth cycle. NAR was determined by the ratio of dry mass gain divided by leaf area. There was no evidence of decreasing in photosynthetic efficiency, in any of the seven species, with increasing blue light, but photosynthetic efficiency increased with increasing blue light in cucumbers. These results suggest that the effect of blue light on reducing leaf area and radiation interception was the underlying cause of the reduction in growth.

Snowden et al. (2016) also found that increasing blue light had a greater effect at PPF 500 than at 200 for cucumbers, radishes, and peppers, but there were no significant interactions between PPF and blue light fraction for the other four species. Green light fractions in the Snowden study varied from zero to 30%. In contrast to the significant responses to blue light, increasing green light fraction resulted in few significant differences, and there was no consistent direction of the

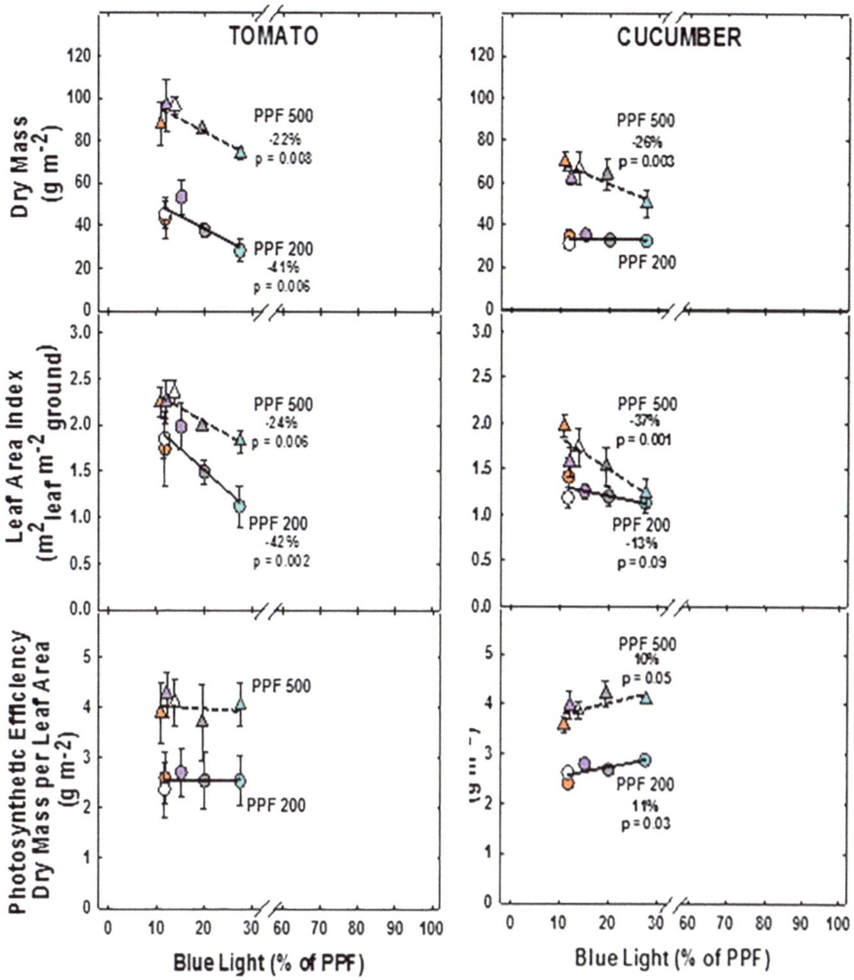


**Fig. 5.6** Effect of blue light on dry mass, leaf area index, and photosynthetic efficiency in tomatoes and cucumbers (figure developed from data in Snowden et al. 2016). Both species are highly sensitive to blue light fraction. Photosynthesis likely increased in cucumbers because of decreased self-shading at the higher blue light fractions

effect among species or PPF levels (Fig. 5.7). Overall, these results indicate significant differences in sensitivity to blue light among species. The effects of blue light were mediated by changes in leaf area, without any significant effects on photosynthesis.

Contrary to multiple reports on green light effects on growth (both increases and decreases), Snowden et al. (2016) found no consistent effect of green light among species on either growth or photosynthetic efficiency.

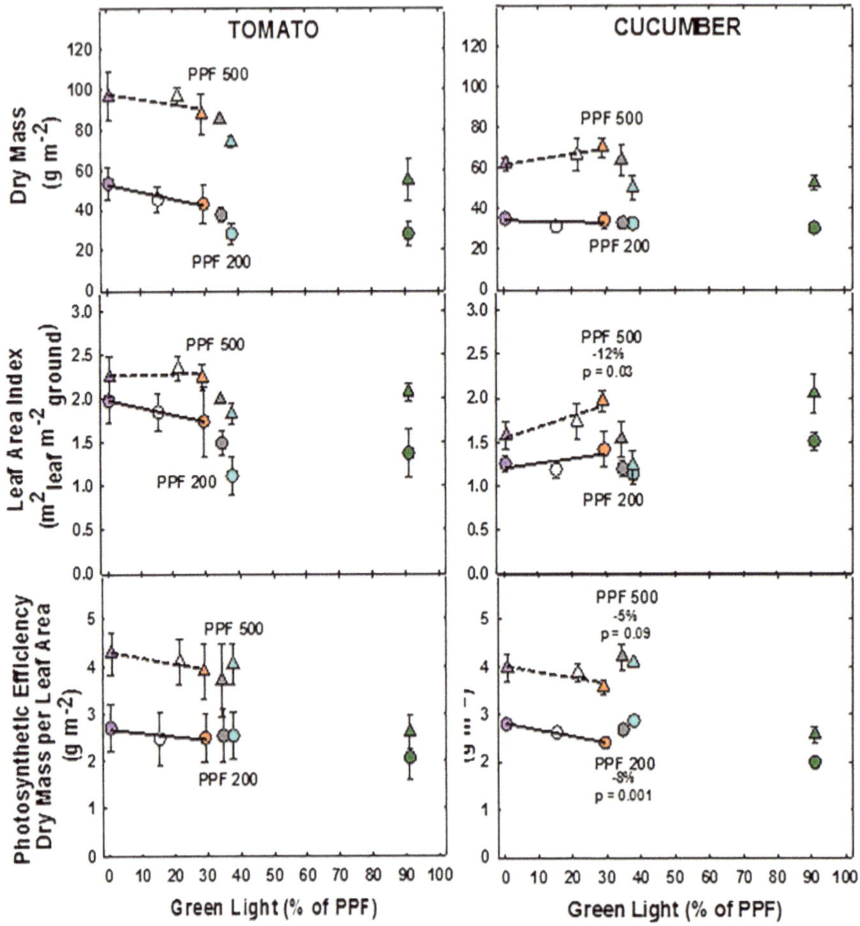


**Fig. 5.7** Effect of green light on dry mass, leaf area index, and photosynthetic efficiency in tomatoes and cucumbers (figure developed from data in Snowden et al. 2016). The *green symbols* represent light from green LEDs, which have 92% of their output between 500 and 600 nm. The regression lines connect treatments with *blue*, *green*, and *red* PPF fractions from LEDs. As the green light fraction increased, the red light decreased

## 5.16 Conclusions

The efficacy of LED fixtures has increased, and the associated time for a breakeven return on investment has decreased. An overlooked advantage of LED fixtures is their more focused light distribution. Much of the advantage of LEDs results from their small size and wattage. Because a single LED can be only one watt, it requires 1000 single LEDs to equal to power input of a single 1000 W HPS fixture. These single LEDs can be positioned to increase the efficiency of radiation transfer to plant leaves. When users can take advantage of this feature, LEDs are often the technology of choice for plant growth lighting. The effect of LED technology on leaf temperature has also been reviewed. At an equal photosynthetic photon flux, the cooling effect of LEDs is about 2 °C, which is much smaller than is often assumed. The effects of blue light on leaf expansion and radiation capture are significant, but the effects of green light are minimal. These findings are consistent with the idea that we have rediscovered the value of broad spectrum light for plant growth and development.

## References

Beadle CL, Long SP (1985) Photosynthesis—is it limiting to biomass production? Biomass 8:119–168

Broadersen CR, Vogelmann TC (2010) Do changes in light direction affect absorption profiles in leaves? Funct Plant Biol 37:403–412

Bugbee B (1995) The components of crop productivity: measuring and modeling plant metabolism. Am Soc Gravit Space Biol 8:93–104

Bugbee B (2016) Toward an optimal spectral quality for plant growth and development: the importance of radiation capture. Acta Hort 1134:1–12

Bugbee B, Monje O (1992) The limits of crop productivity: Theory and validation. BioScience 42:494–502

Bugbee B, Salisbury FB (1988) Exploring the limits of crop productivity: photosynthetic efficiency in high irradiance environments. Plant Physiol 88:869–878

Chen XL, Guo WZ, Xue XZ, Wang LC, Qiao XJ (2014) Growth and quality responses of 'Green Oak Leaf' lettuce as affected by monochromic or mixed radiation provided by fluorescent lamp (FL) and light-emitting diode (LED). Sci Hortic 172:168–175

Cope KR, Bugbee B (2013) Spectral effects of three types of white light-emitting diodes on plant growth and development: absolute versus relative amounts of blue light. HortScience 48:504–509

Cope K, Snowden MC, Bugbee B (2014) Photobiological interactions of blue light and photosynthetic photon flux: effects of monochromatic and broad-spectrum light sources. Photochem Photobiol 90:574–584

Craig DS, Runkle ES (2013) A moderate to high red to far-red light ratio from light-emitting diodes controls flowering of short-day plants. J Am Soc Hortic Sci 138:167–172

Dougher TAO, Bugbee B (2001) Differences in the response of wheat, soybean, and lettuce to reduced blue radiation. Photochem Photobiol 73:199–207

Dougher TAO, Bugbee B (2004) Long-term blue light effects on the histology of lettuce and soybean leaves and stems. J Am Soc Hort Sci 129:467–472

Evans LT (1993) Crop evolution, adaptation and yield. Cambridge University Press, Cambridge

Evans LT (1998) Greater crop production: whence and whither? In: Waterlow JC, Armstrong DG, Fowdenand L, Riley R (eds) Feeding a world population of more than eight billion people—a challenge to science. Oxford University Press, Oxford, pp 89–97

Evans LT, Dunstone RL (1970) Some physiological aspects of evolution in wheat. Aust J Biol Sci 23:725–741

Fitter A, Hay RKM (2012) Environmental Physiology of Plants, 3rd edn. Academic Press, London

Folta KM, Maruhnich SA (2007) Green light: a signal to slow down or stop. J Exp Bot 58:3099–3111

Goins G, Yorio N, Sanwo M, Brown C (1997) Photomorphogenesis, photosynthesis, and seed yield of wheat plants grown under red light-emitting diodes (LEDs) with and without supplemental blue lighting. J Exp Bot 48:1407–1413

Hernández R, Kubota C (2015) Physiological responses of cucumber seedlings under different blue and red photon flux ratios using LEDs. Environ Exp Bot 121:66–74

Hogewoning SW, Trouwborst G, Maljaars H, Poorter H, van Ieperen W, Harbinson J (2010) Blue light dose-responses of leaf photosynthesis, morphology, and chemical composition of *Cucumis sativus* grown under different combinations of red and blue light. J Exp Bot 61:3107–3117

Hoover WH (1937) The dependence of carbon dioxide assimilation in a higher plant on wave-length of radiation. Smithson Misc Collect 95:1–13

Hunt R (1982) Plant growth analysis. Institute of Terrestrial Ecology

Inada K (1976) Action spectra for photosynthesis in higher plants. Plant Cell Physiol 17:355–365

Johkan M, Shoji K, Goto F, Hahida S, Yoshihara T (2012) Effect of green light wavelength and intensity on photomorphogenesis and photosynthesis in *Lactuca sativa*. Environ Exp Bot 75:128–133

Keating K, Carberry P (1993) Resource capture and use in intercropping: solar radiation. Field Crops Res 34:273–301

Kim HH, Goins G, Wheeler R, Sager JC, Yorio NC (2004) Green-light supplementation for enhanced lettuce growth under red and blue LEDs. HortScience 39:1616–1622

Klassen SP, Ritchie G, Frantz JM, Pinnock D, Bugbee B (2003) Real time imaging of ground cover: relationships with radiation capture, canopy photosynthesis, and daily growth rate. In: Schepers J, Van Toai T (eds) Digital imaging and spectral techniques: applications to precision agriculture and crop physiology. American Society of Agronomy Special Publication No. 66, Madison, pp 3–14

Long S, Zhu X, Naidu SL, Ort D (2006) Can improvement in photosynthesis increase crop yields? Plant Cell Environ 29:315–330

Massa GD, Kim HH, Wheeler RM, Mitchell CA (2008) Plant productivity in response to LED lighting. HortScience 43:1951–1956

McCree KJ (1972a) The action spectrum absorptance and quantum yeild of photosynthesis in crop plants. Agric Meteorol 9:191–216

McCree KJ (1972b) Test of current definitions of photosynthetically active radiation against leaf photosynthesis data. Agric Meteorol 10:443–453

Monje O, Bugbee B (1998) Adaptation to high $CO_2$ concentration in an optimal environment: radiation capture, canopy quantum yield, and carbon use efficiency. Plant Cell Environ 21:315–324

Morrow RC, Tibbitts TW (1988) Evidence for involvement of phytochrome in tumor development on plants. Plant Physiol 88:1110–1114

Nelson J, Bugbee B (2013) Spectral characteristics of lamp types for plant biology. http://cpl.usu.edu/files/publications/poster/pub__6740181.pdf. Accessed 30 Nov 2016

Nelson J, Bugbee B (2014) Economic analysis of greenhouse lighting: light emitting diodes vs. high intensity discharge fixtures. PLoS ONE 9(6):e99010. doi:10.1371/journal.pone.0099010

Nelson J, Bugbee B (2015) Analysis of environmental effects on leaf temperature under sunlight, high pressure sodium and light emitting diodes. PLoS ONE 10(10):e0138930. doi:10.1371/journal.pone.0138930

Nishio JN (2000) Why are higher plants green? Evolution of the higher plant photosynthetic pigment complement. Plant Cell Environ 23:539–548

Ouzounis T, Parjikolaei BR, Fretté X, Rosenqvist E, Ottosen CO (2015) Predawn and high intensity application of supplemental blue light decreases the quantum yield of PSII and enhances the amount of phenolic acids, flavonoids, and pigments in *Lactuca sativa*. Front Plant Sci. doi:10.3389/fpls.2015.00019

Snowden MC, Cope K, Bugbee B (2016) Sensitivity of seven diverse species to blue and green light: interactions with photon flux. PLoS ONE 11(10):e0163121. doi:10.1371/journal.pone.0163121

Sun J, Nishio JN, Vogelmann TC (1998) Green light drives $CO_2$ fixation deep within leaves. Plant Cell Physiol 39:1020–1026

Terashima I, Fujita T, Inoue T, Chow WS, Oguchi R (2009) Green light drives leaf photosynthesis more efficiently than red light in strong white light: revisiting the enigmatic question of why leaves are green. Plant Cell Physiol 50:684–689

Terfa MT, Solhaug KA, Gislerød HR, Olsen JE, Torre S (2013) A high proportion of blue light increases the photosynthesis capacity and leaf formation rate of *Rosa* × *hybrida* but does not affect time to flower opening. Physiol Plant 148:146–159

Wang Y, Folta KM (2013) Contribution of green light to plant growth and development. Am J Bot 100:70–78

Wang XY, Xu XM, Cui J (2014) The importance of blue light for leaf area expansion, development of photosynthetic apparatus, and chloroplast ultrastructure of *Cucumis sativus* grown under weak light. Photosynthetica 53(2):213–222

Yorio NC, Goins G, Kagie H, Wheeler R, Sager JC (2001) Improving spinach, radish, and lettuce growth under red LEDs with blue light supplementation. HortScience 36:380–383

Zhang T, Maruhnich SA, Folta KM (2011) Green light induces shade avoidance symptoms. Plant Physiol 157:1528–1536

# Chapter 6
# An Overview of LED Lighting and Spectral Quality on Plant Photosynthesis

**Most Tahera Naznin and Mark Lefsrud**

## 6.1 Introduction

Close to half of the sun's total radiation emission reaching the earth's surface is visible light, which ranges from about 390 to 700 nm wavelengths. The sun, for example, is a broad-spectrum light source emitting photons of every wavelength continuously (with no strong emission lines) and is sensed as 'white light' by the human brain. In fact, a prism can reveal true white light which is/as an amalgamation of light including violet (400–450 nm), blue (450–520 nm), green (520–560 nm), yellow (560–600 nm), orange (600–625 nm), and red (625–700 nm; Table 6.1). Visible light is flanked on the shorter wavelength end of the spectrum by invisible ultraviolet electromagnetic radiation (10–400 nm) and on the longer wavelength end by infrared radiation (700 nm–1 mm), which constitutes roughly the other half of solar radiation incident on the earth's surface (Koning 1994; Benton 2005; Cooper 2000). These three wavelength regions of the electromagnetic spectrum are of utmost significance with respect to biological systems (Mishra 2004). Plants use light of roughly the same wavelength range as the visible spectrum for photosynthesis, from 400 to 700 nm, but instead leaves reflect a higher proportion of green than of any other color of photons, which lends them their common green color (Koning 1994).

Photosynthesis is a photobiochemical process using light energy to produce ATP and NADPH, ultimately consumed in the assembly of carbon atoms in organic molecules. Functionally, photons are harvested by protein chlorophyll-carotenoid

M.T. Naznin · M. Lefsrud (✉)
Bioresource Engineering Department, McGill University, Macdonald Stewart Building, 21111 Lakeshore, Ste-Anne-de-Bellevue, H9X 3V9 Quebec, Canada
e-mail: mark.lefsrud@mcgill.ca

*Present Address:*
M.T. Naznin
Department of Biosystems and Technology, Swedish University of Agricultural Sciences, SE 203 53 Alnarp, Sweden



S. Dutta Gupta (ed.), *Light Emitting Diodes for Agriculture*,
DOI 10.1007/978-981-10-5807-3_6

**Table 6.1** Selected properties of the ultraviolet, visible, and infrared wavelength regions of light which are of particular interest with respect to biological systems

| Color | Wavelength range (nm) | Representative wavelength (nm) | Energy (eV/photon) | Energy (kcal/mole photon) |
|---|---|---|---|---|
| Ultraviolet | <400 | 254 | 4.88 | 112.5 |
| Violet | 400–425 | 410 | 3.02 | 69.7 |
| Blue | 425–490 | 460 | 2.70 | 62.7 |
| Green | 490–560 | 520 | 2.39 | 55.0 |
| Yellow | 560–585 | 580 | 2.14 | 49.3 |
| Orange | 585–640 | 620 | 2.00 | 46.2 |
| Red | 640–740 | 680 | 1.82 | 42.1 |
| Infrared | Above 740 | 1400 | 0.88 | 20.4 |

Perception of colors which define the visible light categories is subjective and can vary from person to person. Energy values are based on the representative wavelengths (Mishra 2004)

complexes (that form the light-harvesting antenna of photosystems) and then transferred to the photosystem reaction center, where electrons are generated; these processes take place in the chloroplast (Solymosi and Keresztes 2012). If lighting is too weak, photosynthesis cannot work efficiently and etiolation symptoms appear (Solymosi and Schoefs 2010). However, excessive light generates oxygen radicals and causes photoinhibition. Both phenomena strongly limit primary productivity (Barber and Andersson 1992). Photosynthetic processes are often modified in plants grown under artificial lighting. Light-emitting diodes (LEDs) are an alternative light source and have created new opportunities for protected cultivation (Darko et al. 2014).

## 6.2 Light-Emitting Diodes (LEDs)

LEDs have low operating power, narrow bandwidth emissions, and a readily controllable spectral distribution (Brown et al. 1995, see Chap. 1 for more details). Narrow bandwidth emissions and readily controllable spectral composition are due to the nature of solid-state lighting. An LED consists of a forward-biased diode with a '*p*' and '*n*' junction, and electroluminescence is achieved by adding chemical impurities within the junction (Kasap 2001). The '*p*' junction is doped with elements (also called impurities) that have an abundance of valence electrons available for conduction, while the '*n*' junction is doped with elements that have a shortage of electrons or holes (Kasap 2001). Without externally applied voltage, an electromagnetic equilibrium is reached between the '*n*–*p*' junctions that are characterized by potential energy; however, no net current discharge occurs as the diode is in a state of equilibrium. By contrast, when external voltage is applied, equilibrium no longer exists and electrons flow from the '*n*' to the '*p*' junction, to the depletion region located between junctions (Kasap 2001). Once electrons combine with the

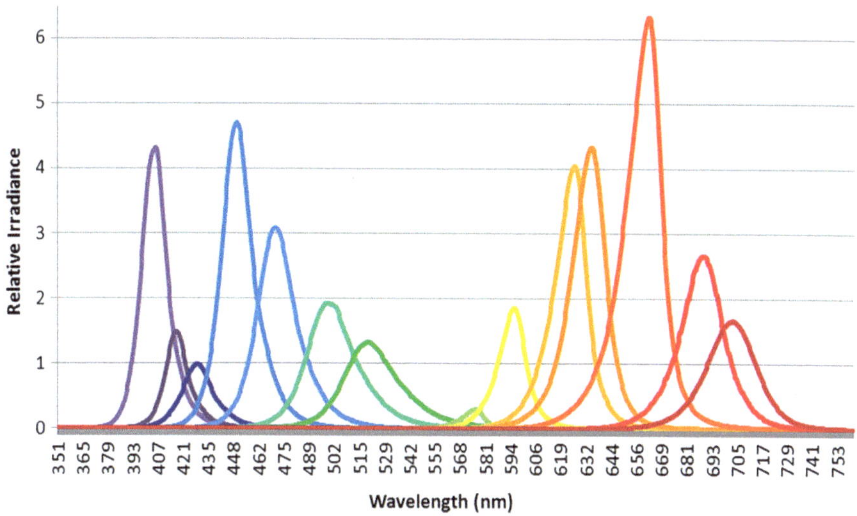


**Fig. 6.1** Spectrum of different colored LEDs. Relative irradiance versus peak wavelength of 14 different LED arrays at 1.4 A. Relative irradiance was measured from a μmol $m^{-2}$ $s^{-1}$ scale

holes in the depletion region, electrons drop from the conduction band to the valance band which results in photon emission.

The conduction band refers to the energy of free electrons that originate from the '*n*' junction, while the valence band refers to the valance energy of the holes that originate from the '*p*' junction. Photons released from LEDs correspond to the energy difference of the conduction and valence bands, also called the band gap (Kasap 2001). The band gap of an LED can be readily manipulated by altering the doping substances and the dopant concentrations. When bonds are formed within solid substrate of the LED, delocalized molecular orbitals occur (Kasap 2001). LEDs can produce light from 350 to 940 nm (Steigerwald et al. 2002), and spectral composition control is greater with LEDs than any other commercial lighting technology (Morrow 2008). The spectral composition of LEDs with peak wavelengths from 400 to 700 nm is depicted in Fig. 6.1.

## 6.3 Photosynthetic Reaction

Photosynthesis is a chemical process where the electromagnetic energy of photons is absorbed, transferred, and stored chemically in carbohydrate molecules through a complex array of oxidation/reduction reactions in photosynthetic organisms (Fig. 6.2). Photosynthetic organisms are also referred to as photoautotrophic organisms, include bacteria, algae as well as plant species and together, are the primary source of energy for all other life-forms on earth (Falkowski and Raven

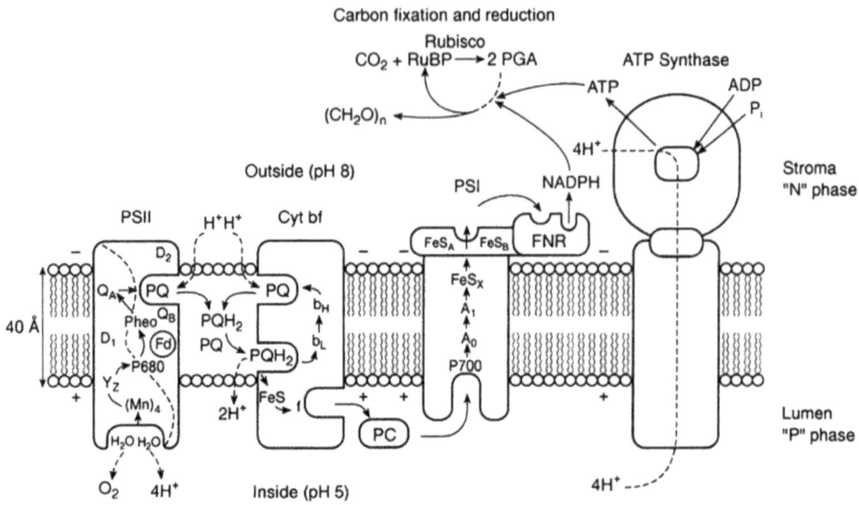


**Fig. 6.2** Schematic presentation of the photosynthetic apparatus and the chemical reactions of photosynthesis (adapted from Falkowski and Raven 2007)

2007). The photosynthetic process can be described by the following simplified equation:

$$\text{light energy} + 6CO_2 + 12H_2O \rightarrow C_6H_{12}O_6 + 6O_2 + 6H_2O$$

Photosynthesis occurs within the chloroplast, a chlorophyll bearing type of plastid organelle dedicated to energy production (Cooper 2000; Mishra 2004). These are found within the cytoplasm of mesophyll cells, mostly the palisade and spongy parenchyma cells located between the bounding epidermal layers of leaves (Mishra 2004). Within chloroplasts, the energy-generating photooxidation-reduction reactions of photosynthesis occur within the third, internal thylakoid membrane system, which forms a set of flattened thylakoid disks, often stacked in grana (Cooper 2000).

Embedded in the thylakoid membrane are five membrane protein complexes which participate in electron transport and the concomitant synthesis of the energy carrier molecules NADPH and ATP, which in turn serve to fuel the synthesis of carbohydrates. Prominent among these are the two main photosynthetic light reaction centers, membrane protein photosystem I and II complexes (PS I and PS II), named after the order of their discovery, which is counter to that of their evolution in nature. Also known as pigment systems I and II, these consist of arrays of associated chlorophyll and carotenoid antenna pigments, the molecules involved in harvesting light energy for photosynthesis, arranged in such a way as to maximize light energy capture and transfer (Cooper 2000; Mishra 2004). Chlorophyll *a* (Chl *a*) is the main pigment in photosynthesis, occurring at the light reaction centers in all photosynthetic organisms (Farabee 2007). In PS II, the reaction center Chl *a* is known as P-680 based on its excitation wavelength, while at PS I the form of Chl *a* is P-700 (Mishra 2004). Accessory antennae pigments are highly conserved in higher plants and include chlorophyll *b* (Chl *b*) and the carotenoid, β-carotene, and the carotenoids subset xanthophylls, lutein, violaxanthin, anteroxanthin, and zeaxanthin. The carotenoid and xanthophyll pigments are lipid soluble yellow, orange, and red secondary plant pigments that are uniquely synthesized in plants, algae, fungi, and bacteria (Sandmann 2001). They surround the light reaction centers where they harvest light energy and channel it, through resonance energy transfer, to Chl *a* at the reaction center. In cotton (*Gossypium hirsutum* L.), lutein is the predominant carotenoid in PS II, while β-carotene is the predominant carotenoid in PS I (Thayer and Bjorkman 1992). In the PS II complex, β-carotene is highly concentrated close to the reaction center, while lutein is present in several light-harvesting antennae components (Demmig-Adams et al. 1996). Photosynthesis is activated when sufficient photon energy excites electrons in the P-680 form of the Chl *a* pigment in PS II, ejecting the electrons from it, effectively oxidizing them. The electrons are then replaced by the photolysis of water within the thylakoid lumen, which splits it into two hydrogen ions (protons, $H^+$) and free $O^{2-}$ ions. The $O^{2-}$ ions combine to form the released diatomic $O_2$, and the protons which remain in the thylakoid lumen contribute to establishing a proton gradient across the thylakoid membrane, energizing it with a potential energy which ultimately serves in ATP synthesis and/or photoprotection (Cooper 2000; Falkowski and Raven 2007). The electron transport chain of PS II (Fig. 6.2) transfers the segregated high-energy electrons to plastoquinone (PQ) in the membrane. Plastoquinone then siphons the electrons to the second protein complex, cytochrome *bf*, where they lose energy pumping additional protons into the thylakoid lumen. Plastocyanin (PC) then transfers the depleted electrons to PS I, where photon light energy excites the P-700 Chl *a* molecule, thereby raising those same electrons back to a higher energy, excited state. When the absorption of light radiation exceeds the capacity of photosynthesis, excess excitation energy can result in the formation of triplet excited chlorophyll ($^3$Chl) and reactive singlet oxygen ($^1O_2$). Carotenoid pigments protect photosynthetic structures by quenching excited $^3$Chl to dissipate excess energy (Frank and Cogdell 1996) and binding $^1O_2$ to inhibit oxidative damage (Demmig-Adams et al. 1996). Ferrodoxin (FD) transfers these to the fourth thylakoid membrane protein complex, NADP reductase, where $NADP^+$ is reduced to NADPH in the chloroplast stroma. Finally, the fifth thylakoid membrane complex, ATP synthase, converts ADP and inorganic phosphate to ATP, using the proton motive force from the proton gradient established by the photolysis of water and the flow of electrons through the cytochrome *bf* complex to run its proton pump in reverse. In the chloroplast stroma, ATP and NADPH then fuel the fixation of atmospheric $CO_2$ and its incorporation into the three carbon sugar glyceraldehyde-3-phosphate in the Calvin cycle reactions, mediated by the enzyme RuBisCO (Ribulose-1,5-bis-phosphate carboxylase), thought to be the most abundant protein on earth (Cooper 2000; Mishra 2004; Farabee 2007; Falkowski and Raven 2007).

## 6.4 Photosynthetic Pigments

The pigments of plants have specific wavelength absorption patterns known as the absorption spectrum (Table 6.2). Chlorophyll absorbs wavelengths of light strongly in the red and blue regions, with little absorbance occurring in the green wavelengths. In acetone, Chl *a* has a peak absorbance at 430 and 663 nm, while Chl *b* peaks at 453 and 642 nm. The pigments β-carotene and lutein in acetone absorb strongly in the blue region of light with a maximum peak occurring at 454 and 448 nm, respectively (Hopkins and Huner 2004; Taiz and Zeiger 1998). These pigments have local absorption peaks with β-carotene having a second absorption peak at 477 nm, and lutein having two local peaks at 422 and 474 nm. However, peak absorption in a plant can shift up to 38 nm and is dependent on the specific environment surrounding the chloroplasts (Heber and Shuvalov 2005). The absorption of these wavelengths of light does not always directly correlate into biosynthesis of chlorophylls and carotenoids. The absorption of specific wavelengths of light required for biosynthesis is known as the action spectrum (Fig. 6.3). Wavelengths of light at 500 nm and levels greater than 700 nm result in very little biosynthesis of chlorophyll (Koski et al. 1951; Ogawa et al. 1973).

Carotenoids are secondary metabolites and vital pigments in plants which are utilized as antenna pigments to minimize the damage of photosynthetic components from the active triplet state of the chlorophyll molecule (Landrum and Bone 2001; Kopsell et al. 2009). Lutein and β-carotene are the two main carotenoids located in the antenna pigments. Ohashi-Kaneko et al. (2007) found that the carotenoid concentration was higher in spinach grown under blue fluorescent lamps than in spinach grown under white fluorescent lamps with the same PPFD (300 μmol $m^{-2}$ $s^{-1}$). Li et al. (2009) similarly showed that lutein and β-carotene concentrations in spinach were markedly increased when grown under blue fluorescent lamps with a PPFD at 300 μmol $m^{-2}$ $s^{-1}$. In contrast, Cui et al. (2009) found that the carotenoid concentration was increased in cucumber seedlings grown in plastic tunnel greenhouses and supplemented with four hours of red or yellow LED lights per day. Lefsrud et al. (2008) found that lutein and β-carotene accumulations were highest in kale under red (640 nm) LEDs and blue (440 nm) LEDs, respectively.

**Table 6.2** Local maximum absorption in acetone of plant pigments (Hopkins and Huner 2004)

| Pigments | Peak absorption in acetone (nm) | Local peak absorption in acetone (nm) |
|---|---|---|
| β-carotene | 454 | 477 |
| Chlorophyll *a* | 663 | 430 |
| Chlorophyll *b* | 642 | 453 |
| Lutein | 448 | 422, 474 |

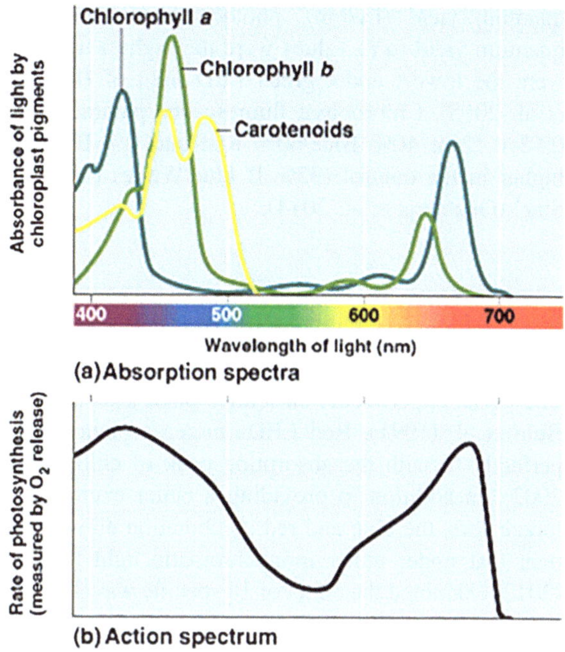


**Fig. 6.3** Absorption spectra and action spectrum of chlorophyll and antenna pigments (adapted from Cambell et al. 1999)

## 6.5 Effects of LEDs on Chlorophyll Fluorescence

Chlorophyll fluorescence is a noninvasive measurement of photosystem II (PS II) activity. The sensitivity of PS II activity to abiotic and biotic factors has made this a key technique not only for understanding the photosynthetic mechanisms but also as a broader indicator of how plants respond to environmental change (Murchie and Lawson 2013; Baker and Rosenqvist 2004). Light energy absorbed by chlorophyll molecules can (i) drive photosynthesis (photochemistry); (ii) be re-emitted as heat; or (iii) be re-emitted as light (fluorescence; Murchie and Lawson 2013). Fv/Fm represents the maximum potential quantum efficiency of photosystem (PS II), and an Fv/Fm value in the range of 0.79–0.84 is optimal for many plant species, with lowered value indicating plant stress (Maxwell and Johnson 2000). The LEDs impact on the chlorophyll fluorescence. Kim and Kim (2014) applied four levels (red, blue, red + blue, and white LED) of light quality to investigate the effects of LED light on the chlorophyll fluorescence of grafted cucumber seedlings. They observed the variable fluorescence (Fv) was the highest with red LED, while Fv significantly decreased under blue LED. Quantum yield (Fv/Fm) was the greatest with blue LED. However, Fv/Fm for scion significantly decreased under red LED. Metallo et al. (2016) found the quantum yield of PS II (Fv/Fm) was influenced by LED light in sprouting broccoli. The maximum quantum yield was observed at 5% blue/95% red but not significantly difference at 20% blue/80% red LED treatments. Another results showed that chlorophyll fluorescence parameters as the maximal quantum yield (Fv/Fm), photochemical quenching coefficient (qP), and light quantum yield (qY) values were the highest under blue LED light, and the values were the lowest under green LED light of *Houttuynia cordata* seedlings (Wang et al. 2015). Chlorophyll fluorescence parameters as Fv/Fm was in the range of 0.52–0.72 at 40% Blue/60% Red and 0% Blue/100% Red, but overall slightly higher in the control (32% B blue/White) of *Phalaenopsis* 'Vivien' and 'Purple Star' (Ouzounis et al. 2014).

## 6.6 LEDs on Plant Photosynthesis and Growth

The initial experiments on lettuce plant growth under red LEDs were reported by Bula et al. (1991). Red LEDs increase plant growth because these wavelengths perfectly fit with the absorption peak of chlorophylls and phytochrome (Schoefs 2002). In addition to providing a better excitation of the different types of photoreceptors, the blue and red combination allowed a higher photosynthetic activity than that under either monochromatic light (Opdam et al. 2005). Naznin et al. (2012) examined the effect of 14 specific wavelengths of LEDs (405, 417, 430, 450, 470, 501, 520, 575, 595, 624, 633, 662, 680, and 700 nm) on photosynthesis of tomato, lettuce, and petunia seedlings. They found photosynthesis, absorbance, quantum yield, and action spectrum peaks in the range from 417 to 450 nm and in the range from 630 to 680 nm. Stutte et al. (2009) reported that application of far-red (730 nm) with red (640 nm) increased biomass accumulation and leaf length in lettuce plant. The addition of far-red (735 nm) with red (660 nm) LEDs increased plant height and biomass accumulation in sweet pepper (*Capsicum annum* L.; Brown et al. 1995). Application of red (640 nm) LEDs as a sole source increased anthocyanin accumulation in red leaf cabbage (*Brasica olearacea* L. var. capitata; Mizuno et al. 2011).

Several experiments have been shown that the blue (400–500 nm) LEDs in combination with red LEDs effect on vegetable morphology, growth, and photosynthesis. Goins et al. (1997) found that wheat (*Triticum aestivum* L., cv. 'USU-Super Dwarf') plant can complete full life cycle under sole red LEDs but higher shoot dry mass accumulation and larger amounts of seed are produced under red LEDs supplemented with a blue light. The blue (440 and 476 nm) LEDs with red LEDs increased the chlorophyll ratio in Chinese cabbage plants (Mizuno et al. 2011 and Li et al. 2012). Naznin et al. (2016) observed higher fresh and dry mass accumulation in coriander plants cultured under different ratios of red to blue LEDs than those plants cultured under 100% red LEDs. It is well established that stomata opening is controlled by blue-light photoreceptors (Ieperen and Trouwborst 2008). This is possibly reflected in the increase of shoot dry matter with increasing levels of blue light (Nanya et al. 2012). Schwalb et al. (2014) examined the effect of different ratios of red (660 nm) and blue (435 nm) LEDs (1:10, 1:5, 1:1, 2:1, 3:1, 4:1, 5:1, 6:1, 7:1, 8:1, 9:1, 10:1, 11:1, 12:1, 13:1, 14:1, 15:1, 20:1, 25:1, 30:1, 50:1, and 100:1) on photosynthesis of lettuce and petunia seedlings with and without background broadband high-pressure sodium radiation. They found the optimum photosynthesis range occurred within the red:blue range of 5:1–15:1 except for petunia without background radiation for which the maximum occurred at 50:1.

Green (505 and 530 nm) LEDs with HPS lamps enhance better growth of cucumber (Novickovas et al. 2012). Johkan et al. (2012) reported that green LEDs with high PPF (300 μmol $m^{-2}$ $s^{-1}$) enhance the growth of lettuce plant. Seedlings grown under green, red, and blue LEDs enhance longer plant height than those grown under red (630 nm) and blue (470 nm) alone (Folta 2004). Illumination with more than 50% of green LED light causes a reduction in plant growth, whereas treatments containing up to 24% green light enhanced growth for some species (Keefe 2007). Sole green light is not sufficient for optimal plant growth because it is least absorbed by the plant, but in combination with red and blue, green light might show some important physiological effects.

## 6.7 Conclusion

In conclusion, different spectral qualities effect on plant photosynthesis, growth, and development. It has been shown that plants exhibit a high degree of physiological, morphological, and anatomical plasticity to changes in spectral quality. Study of determining the species specific optimal light spectra for maximum plant growth would be beneficial.

## References

Baker NR, Rosenqvist E (2004) Applications of chlorophyll fluorescence can improve crop production strategies: an examination of future possibilities. J Exp Bot 55(403):1607–1621

Barber J, Andersson B (1992) Too much good thing: light can be bad for photosynthesis. Trends Biochem Sci 17:61–66

Benton JJ (eds) (2005) Hydroponics: a practical guide for the soilless grower. CRC Press, Florida

Brown CS, Schuerger AC, Sager JC (1995) Growth and photomorphogenesis of pepper plants under red light-emitting diodes with supplemental blue or far-red lighting. J Am Soc Hort Sci 120:808–813

Bula RJ, Morrow RC, Tibbits TW, Barta RW, Ignatius RW, Martin TS (1991) Light emitting diodes as a radiation source for plants. HortScience 26:203–205

Cambell NA, Reece JB, Mitchell LG (eds) (1999) Biology. Wesley Longman Inc., Menlo Park, California

Cooper GM (ed) (2000) The cell: a molecular approach. Sinauer Associates, Sunderland Massachusetts, USA

Cui J, Ma ZH, Xu ZG, Zgang H, Chang TT, Liu HJ (2009) Effects of supplemental lighting with different light qualities on growth and physiological characteristics of cucumber, pepper and tomato seedlings. Acta Hortic Sin 5:663–670

Darko E, Heydarizdeh P, Schoefs B, Sabzalian MR (2014) Photosynthesis under artificial light: the shift in primary and secondary metabolism. Phil Trans R Soc B 369:1–7

Demmig-Adams B, Gilmore AM, Adams WW III (1996) In vivo functions of carotenoids in higher plants. FASEB J10:403–412
Falkowski PG, Raven RA (eds) (2007) Aquatic photosynthesis. Princeton University Press, Princeton, New Jersey
Farabee MJ (2007) On line biology book photosynthesis. Estrella Mountain Community College, Avondale, Arizona
Frank HA, Cogdell RJ (1996) Carotenoids in photosynthesis. Photochem Photobiol 63:257–264
Folta KM (2004) Green light stimulates early stem elongation, antagonizing light-mediated growth inhibition. Plant Physiol 135:1407–1416
Goins GD, Yorio NC, Sanwo MM, Brown CS (1997) Photomorphogenesis, photosynthesis and seed yield of wheat plants grown under red light-emitting diodes (LEDs) with and without supplemental blue lighting. J Exp Bot 48:1407–1413
Heber U, Shuvalov VA (2005) Photochemical reactions of chlorophyll in dehydrated photosystem II: two chlorophyll forms (680 and 700 nm). Photosyn Res 84:85–91
Hopkins WG, Huner NPA (eds) (2004) Introduction to plant physiology. John Wiley and Sons, Hoboken, New Jersey
Ieperen VW, Trouwborst G (2008) The application of LEDs as assimilation light source in greenhouse horticulture: a simulation study. Acta Hort 33:1407–141
Johkan M, Shoji K, Goto F, Hahida S, Yoshihara T (2012) Effect of green light wavelength and intensity on photomorphogenesis and photosynthesis in *Lactuca sativa*. Environ Exp Bot 75:128–133
Kasap S (2001) PN junction devices and light emitting diodes. Special custom e-book. http://www.kasap.usask.ca/samples/PNJunctionDevices.pdf. Accessed 10 Sep 2016
Keefe TJ (2007) The nature of light. http://www.ccri.edu/physics/keefe/light.htm. Accessed 5 Nov 2016
Kim YH, Kim HG (2014) Chlorophyll fluorescence characteristics of cucumber grafted seedlings graft-taken under LED illumination with different light quality and light intensity. Ag Eng 2014: topics in animal husbandry, welfare, rural buildings and greenhouses. In: International conference of agricultural engineering, Zurich, July 2014. Lecture notes in animal husbandry, welfare, rural buildings and greenhouses, vol 304. Geystiona, pp 1–8
Kopsell DA, Lefsrud M, Kopsell D (2009) Pre-harvest cultural growing conditions can influence carotenoid phytochemical concentrations in vegetable crops. Acta Hortic 841:283–294
Koning RE (1994) Light. Plant physiology information website. http://plantphys.info/plantphysiology/light.shtml
Koski VM, French CS, Smith JHC (1951) The action spectrum for the transformation of protochlorophyll to chlorophyll a in normal and albino corn seedlings. Arch Biochem Biophys 31:1–17
Landrum JT, Bone RA (2001) Lutein, zeaxanthin, and the maclar pigment. Arch Biochem Biophys 385:8–40
Lefsrud MG, Kopsell DA, Sams CE (2008) Irradiance from distinct wavelength light-emitting diodes affect secondary metabolites in Kale. HortScience 43:2243–2244
Li J, Hikosaka S, Goto E (2009) Effects of light quality and photosynthetic photon flux on growth and carotenoid pigments in spinach (*Spinacia oleracea* L.). Acta Hortic 907:105–110
Li H, Tang C, Xu Z, Liu X, Han X (2012) Effects of different light sources on the growth of non-heading chinese cabbage (*Brassica campestris* L.). J Agr Sci 4:262–273
Maxwell K, Johnson GN (2000) Chlorophyll fluorescence a practical guide. J Exp Bot 51:659–668
Metallo RM, Kopsell DA, Sams CE, Morrow RC (2016) Management of LED light quality to maximize biomass and chlorophyll fluorescence in sprouting broccoli in controlled environments. In: ASHS 2016: topics in growth chambers and controlled environments 1. American Society for Horticultural Sciences annual conference, Atlanta, August 2016. Poster notes in growth chambers and controlled environments 1. HortScience, p 213
Mishra S (2004) Photosynthesis in plants. Discovery Publishing House, New Delhi, India
Mizuno T, Amaki W, Watanabe H (2011) Effects of monochromatic light irradiation by LED on the growth and anthocyanin contents in laves of cabbage seedlings. Acta Hortic 907:179–184

Morrow RC (2008) LED Lighting in horticulture. HortScience 43:1951–1956
Murchie EH, Lawson T (2013) Chlorophyll fluorescence analysis: a guide to good practice and understanding some new applications. J Exp Bot 64(13):3983–3998
Nanya K, Ishigami Y, Hikosaka S, Goto E (2012) Effects of blue and red light on stem elongation and flowering of tomato seedlings. Acta Hortic 956:261–266
Naznin MT, Lefsrud M, Gravel V, Hao X (2016) Different ratios of red and blue LEDs light affect on coriander productivity and antioxidant properties. Acta Hortic 1134:223–229
Naznin MT, Lefsrud M, Gagne JD, Schwalb M, Bissonnette B (2012) Different wavelengths of LED light affect on plant photosynthesis. In: ASHS 2012: topics in crop physiology. American Society for Horticultural Sciences annual conference, Florida, August 2012. Lecture notes in crop physiology, vol 47(9). HortScience, p S191
Novičkovas A, Brazaitytė A, Duchovskis P, Jankauskienė J, Samuolienė G, Viršilė A, Sirtautas R, Bliznikas Z, Žukauskas A (2012) Solid-statelamps (LEDs) for the short-wavelength supplementary lighting in greenhouses: experimental results with cucumber. Acta Hortic 927:723–730
Ogawa T, Inoue Y, Kitajima M, Shibata K (1973) Action spectra for biosynthesis of chlorophylls a and b and β-carotene. Photochem Photobiol 18:229–235
Ohashi-Kaneko K, Takase M, Kon N, Fujiwara K, Kurata K (2007) Effect of light quality on growth and vegetable quality in leaf lettuce, spinach and komatsuna. Environ Control Biol 45:189–198
Opdam JG, Schoonderbeek GG, Heller EB, Gelder A (2005) Closed greenhouse: a starting point for sustainable entrepreneurship in horticulture. Acta Hortic 691:517–524
Ouzounis T, Fretté X, Ottosen CO, Rosenqvist E (2014) Spectral effects of LEDs on chlorophyll fluorescence and pigmentation in Phalaenopsis 'Vivien' and 'Purple Star'. Physiol Plant 154(2):314–327
Sandmann G (2001) Carotenoid biosynthesis and biotechnological application. Arch Biochem Biophys 385:4–12
Schoefs B (2002) Chlorophyll and carotenoid analysis in food products. Properties of the pigments and methods of analysis. Trends Food Sci Technol. doi:10.1016/S0924-2244(02)00182-6
Schwalb M, Naznin MT, Lefsrud M (2014) Determination of the effect of red and blue ratios of LED light on plant photosynthesis.In: ASHS 2014: topics in growth chambers and controlled environments 2. American Society for Horticultural Sciences annual conference, Florida, July 2014. Lecture notes in growth chambers and controlled environments 2, vol 49(9). HortScience, p S241
Steigerwald DA, Bhat JC, Collins D, Fletcher RM, Holcomb MO, Ludowise MJ, Martin PS, Rudaz SL (2002) Illumination with solid state lighting technology. IEEE J Sel Top Quant Electron 2(8):310–320
Stutte GW, Edney S, Skerritt T (2009) Photoregulation of bioprotectant content of red leaf lettuce with light-emitting diodes. HortScience 44:79–82
Solymosi K, Schoefs B (2010) Etioplast and etio-chloroplast formation under natural conditions: the dark side of chlorophyll biosynthesis in angiosperms. Photosynth Res 105(2):143–166
Solymosi K, Keresztes A (2012) Plastid structure, diversification and interconversions II. Landplants. Curr Chem Biol 6(3):187–204
Taiz L, Zeiger E (eds) (1998) Plant physiology. Sinauer Associates Inc., Sunderland, Massachusetts, USA
Thayer SS, Bjorkman O (1992) Carotenoid distribution and deepoxidation in thylakoid pigment-protein complexes from cotton leaves and bundle-sheath cells of maize. Photosynth Res 33:213–235
Wang Z, Tian J, Yu B, Yang L, Sun Y (2015) LED light spectrum affects the photosynthetic performance of *Houttuynia Cordata* seedlings. Am J Opt Photonics 3(3):38–42

# Chapter 7
# LED Lighting in Horticulture

**Akvilė Viršilė, Margit Olle and Pavelas Duchovskis**

## 7.1 Introduction

Supplemental lighting in horticulture has been used for over a century with the purpose of enhancing plant growth and development (Wallace and Both 2016). However, this technology has been borrowed from the lighting industry that was not originally designed or intended for plants (Mitchell et al. 2015). Several lamp technologies have been used for plant cultivation and research, such as incandescent, fluorescent, metal halide, and high-intensity discharge lamps. The application of each technology has been optimized for a wide range of horticultural crops for photoperiod control, changing plant morphology, and enhancing photosynthesis (Wallace and Both 2016). Today, we are in the midst of a revolution in lighting (Patisson et al. 2016). Light-emitting diodes (LEDs) are replacing conventional lamps in almost every indoor and outdoor lighting application, and the rapid technological progress of LEDs provides opportunities for advancements in horticultural lighting (Olle and Viršilė 2013).

The specific advantages of LEDs include capability to control spectral output and light intensity (photosynthetic photon flux PPF) (Mitchell et al. 2015). Light-emitting diodes emit narrow-band wavelengths from UV-C ($\sim$250 nm) to infrared ($\sim$1,000 nm) (Bourget 2008). It is the first light source that enables the selection of specific wavelengths in the lighting spectrum that match the absorbance of plant photoreceptors (Morrow 2008) and therefore impacts specific vital plant processes.

A. Viršilė (✉) · P. Duchovskis
Lithuanian Research Centre for Agriculture and Forestry,
Institute of Horticulture, 30 Kauno Str., Babtai 54333, Lithuania
e-mail: a.virsile@lsdi.lt

M. Olle
Estonian Crop Research Institute, J. Aamissepa 1, 48309 Jogeva Alevik, Estonia



S. Dutta Gupta (ed.), *Light Emitting Diodes for Agriculture*,
DOI 10.1007/978-981-10-5807-3_7

The introduction of LEDs offers the possibility to control plant growth, development, and metabolism by tailoring light parameters. However, in horticultural systems, it is difficult to attain the balance between natural plant requirements and technological options based on commercial purposes. The main objectives of the horticultural industry are the high-quality and productivity vegetable produce with minimal costs (van Ieperen 2016). Therefore, horticultural plants are often cultivated under conditions that do not match the technological aims of the grower better than natural physiological needs of the plant. In natural habitats, plants have adopted their physiology in accordance with the variations in solar spectrum and intensity. Generally, they are well-equipped to survive and flourish in a variety of environments (van Ieperen 2016) and are able to adapt to artificial lighting conditions; however, exposure to light parameters that are beyond the natural tolerance zone of the plant may also have negative results (van Ieperen 2016). Innovative LED systems add a completely new dimension to lighting control (van Ieperen 2016); however, LED functionality depends on specific photobiological, physiological, and technological knowledge for proper operation. For example, narrow-spectrum LEDs must be proportioned carefully to obtain the desired plant responses (Mitchell et al. 2012). Therefore, LED applications in horticulture are closely interconnected with the knowledge of plant photomorphogenesis. The first research results on LED application for plant lighting have been published in the early 1990s, with a considerable increase in the number of publications during the last five years. However, due to the variability of results obtained from different plant species and varieties, no general LED lighting model was yet established, and questions about LED lighting parameters, ideally for different horticultural plants in different developmental stages along with the grower's objectives, are still open.

## 7.2 The Concept of Horticultural LED Lighting and Its Emergence

Initial reports on LED applications in plant cultivation have been published in the early 1990s in the USA; they mainly focused on developing plant cultivation systems for space missions (Barta et al. 1992; Yorio et al. 2001; Massa et al. 2008) and were reviewed by Morrow (2008). At that time, only red LEDs ($\sim$660 nm) had a photosynthetic photon flux output adequate for meeting plant requirements. The results of the first experiments with lettuce (Bula et al. 1991), spinach, radish, potato (Yorio et al. 2001), and wheat (Goins et al. 1997) revealed the need of blue light for normal growth and photosynthesis. The spectrum was then enriched with the blue fluorescent lights (Bula et al. 1991; Yorio et al. 2001); however, comprehensive research on LEDs began with the development of high-power blue LEDs. Numerous studies have confirmed that the spectral combination of red and blue lights in different ratios is adequately efficient for the cultivation of various plants under greenhouse conditions (Brazaitytė et al. 2006;

Hogewoning et al. 2010; Johkan et al. 2010; Mizuno et al. 2011). The use of red and blue LEDs has been the prime selection for the producers, as these wavelengths are efficiently absorbed by the primary plant pigments (chlorophylls) (Ouzounis et al. 2015a). Moreover, the combination of red and blue lights provides the highest photon efficiency as compared to other LED colors (Nelson and Bugbee 2014). These are the main reasons for the predominance of such bicomponent LED spectra in early commercial applications.

### *7.2.1 The Concept of Light Spectral Efficiency*

A generalized lighting spectrum for photosynthesis, suggested by McCree (1971) nearly half a century ago, indicated that red and orange light photons are the most efficient, while green photons have remarkably lower efficiency as compared to red and blue lights (Bugbee 2016). Following that, one of the most discussed advantages of LEDs was the potential to combine lighting spectra by selecting only the physiologically efficient light wavelengths, avoiding wasting the energy for unproductive colors, such as green and yellow. However, photosynthetic response curves were developed from single leaves, at a low photosynthetic photon flux density, over a short time interval. In recent years, this trend has changed to a more comprehensive approach. Light efficiency is no longer assessed by single leaf response, but rather the response of the whole plant canopy, with light distribution within different canopy layers, showing the importance of a wide light spectrum for plant growth and development (Bugbee 2016; Snowden et al. 2016). In addition to primary photosynthetic chlorophyll pigments, other plant pigments, such as carotenoids and anthocyanins, are also capable of harvesting light. All these pigments have different absorption spectra, allowing the plants to absorb the wide composite light spectrum (Ouzounis et al. 2015a). However, being the main source of energy for photosynthesis, light parameters also act as signals in processes driving gene expression, physiology, morphology, and metabolism. A plant's response to the light environment is determined by the actions of distinct photoreceptors. The signaling pathways of phytochromes, cryptochromes, phototropins, and UVR8 sensors are integrated to fine-tune the developmental and photosynthetic status of the plant (Ouzounis et al. 2015a, Fig. 7.1). Understanding the individual plant responses as well as the synergy between the photoreceptors and photosynthetic signaling networks assists in the selection and timing of LED light programs for the regulation of crop growth (Pocock 2015). Selective activation of specific light-sensing pathways by customized LED luminaries allows growers to control plant productivity, quality, and production timing. Carvalho and Folta (2014) have generalized these concepts and proposed the meaning of 'environmentally modified organisms': by tailoring controllable environmental parameters (including light), it is possible to adjust plant traits within the genetic potential and produce desirable changes in plant productivity, development, or metabolism in a significantly shorter time than breeding or other genetic modifications.

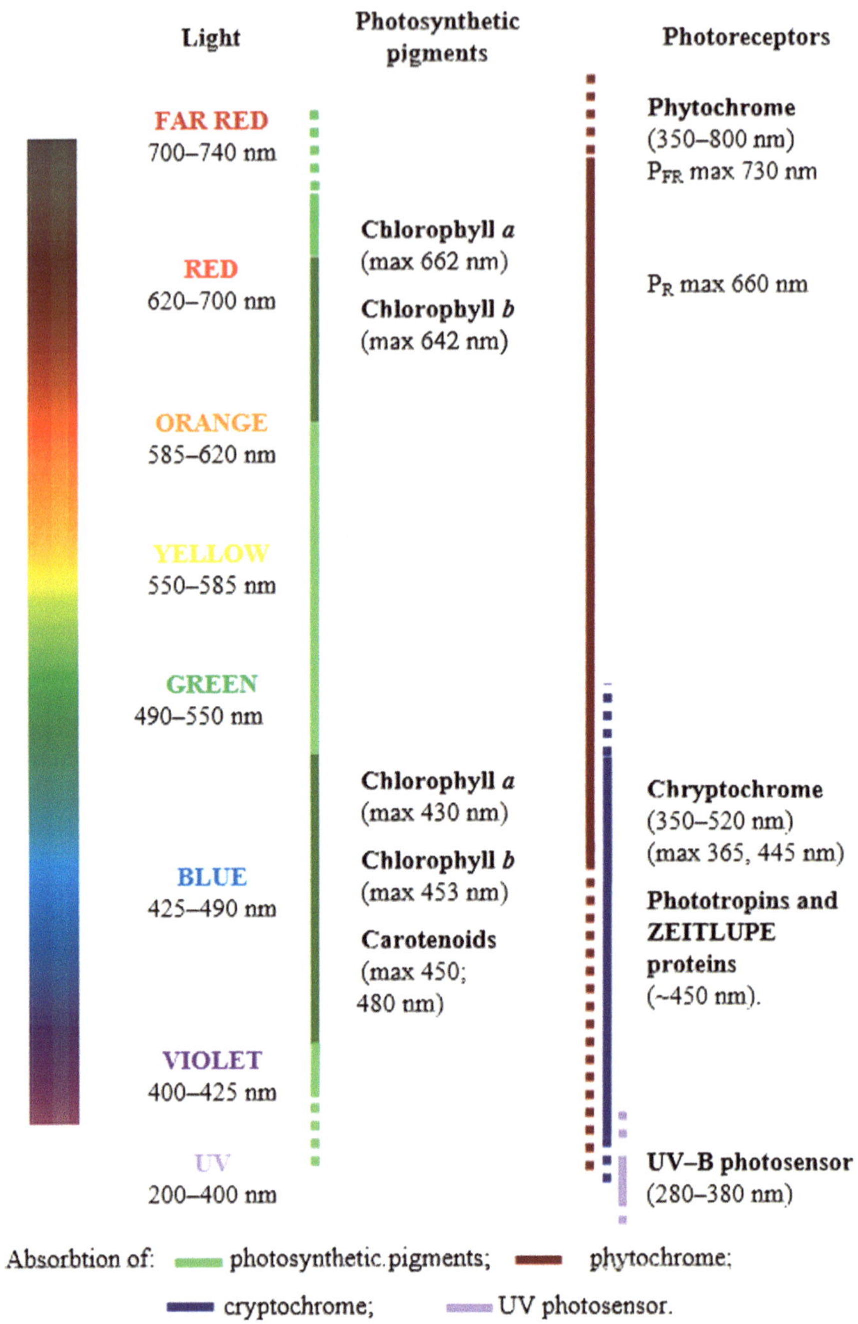


**Fig. 7.1** Light spectra and photoreceptors involved in plant growth and development

### 7.2.2 LEDs in Greenhouse and Closed Environment Horticulture

Using LED light sources provides an opportunity for directly intervening in plant growth, development, and metabolism; however, it requires extended knowledge, as the effects of light quality are complicated and the reported results are often controversial. However, the variability of research results is possibly due to different experimental conditions and plant varieties, which makes it difficult to compare these results. A global model for LED lighting parameters has not yet been developed (Nicole et al. 2016). Moreover, plant responses to monochromic light also depend on the background spectrum; the effects of separate light wavelengths differ when applied in combination with other, differential light wavelengths. Therefore, lighting strategies optimized for closed environment chambers not always produce equivalent results in greenhouse conditions. Even the low flux of natural daylight in a greenhouse, especially the variation in its spectrum, quantity, and photoperiod, has an important physiological impact on plants. Therefore, lighting conditions in greenhouses and closed environment cultivation systems should be optimized separately.

Most applications of LED lighting in greenhouses choose the combinations of red and blue wavelengths with high photon efficiency. In plant factories, the new form of protected horticulture (Kozai 2015) combinations of red/blue or red/white are also suitable (Nicole et al. 2016). Green or white light, containing substantial amounts of green wavelengths, has the positive physiological impact on plants and also are beneficial for improving the visual appearance of plants in closed environments. The combination of blue and red lights creates the purplish-gray image of plants for the human eye, therefore hindering the visual evaluation of plant health and injuries. A small flux of green light is useful to resolve this issue (Massa et al. 2008). White LEDs containing red, blue, and green wavelengths have also been tested as an attractive, human vision-friendly source of plant lighting. Cope and Bugbee (2013) evaluated the impacts of warm, neutral, and cool white LEDs (with 11, 19, and 28% of blue light, respectively) on the growth and development of radish, soybean, and wheat. They concluded that cool white LEDs could be the light source of choice because they are more electrically efficient as compared to neutral and warm white LEDs. Further, the high percentage of blue light in cool white LEDs fulfills the blue light requirements for normal plant growth and development. Dynamic lighting with the ability of shifting from cool to warm white LEDs in different growth stages also promotes plant growth. During initial stages of growth, a high percentage of blue light in the spectrum of cool white LEDs results in short, sturdy hypocotyls (Cope and Bugbee 2013). In later developmental stages, cool white LEDs could be replaced by warm while LEDs. This light spectrum would promote leaf expansion; and in final growth stages, cool white LEDs should be used again to prevent excessive stem elongation (Cope and Bugbee 2013). The dynamic lighting parameters follow the patterns of variable natural lighting and could be assigned to the innovative approach of biomimicry in technologies.

Biomimicry (bio—life and mimesis—to copy (*in Greek*)) is a growing field that aims to interpolate natural biological mechanisms and structures into a wide range of applications (Lurie-Luke 2014). The dynamic control of supplemental LED lighting intensity has been proven to reduce electricity consumption by ~20%, without affecting crop quality or productivity (Pinho et al. 2013; Schwend et al. 2016).

However, the modeling of artificial dynamic lighting parameters, even when following the principles of natural lighting, calls for comprehensive knowledge. A thorough understanding of the plant (as a part of a community in its habitat) lighting requirements during different growth stages may allow to develop a controlled spectrum LED system, which would be much more beneficial for the plants than that of white LED treatment (Singh et al. 2015). Multi-wavelength hybrid package LEDs, as smart and tunable light sources, emitting wide light spectrum (Son and Oh 2015) would be remunerative for the development of dynamic LED lighting systems, once the horticulture LED market matures.

## 7.3 LEDs Versus High-Pressure Sodium Lighting

Some examples of main light sources used in the 1990s are high-pressure sodium, high-pressure mercury, and fluorescent lamps (Olle 2015). High-pressure sodium (HPS) lamps have been the main supplemental light source in greenhouses in northern latitudes. Their prevalence is based on low costs, high photosynthetically active radiation emission, long lifetime expectancy, and high electrical efficiency. However, the main drawback of HPS lighting is related to poor quality of its spectral emission, which is predominantly in the yellow-green and infrared region of the electromagnetic spectrum, with a low blue light emission and a red-to-far-red ratio (Pinho and Halonen 2014). In recent decades, light-emitting diode lighting has been developed as a potential substitute for high-pressure sodium light. The use of LEDs in plant lighting applications provides novel opportunities for optimization of plant growth and development. This can be achieved through controlling the quantity, periodicity, and spectrum of the light provided; such an optimization can be tailored to the specific needs of each crop species and its production conditions (Pinho et al. 2007). HPS and LED technologies are not competing in horticultural lightining, they have their own niches there—it means—supplement each other. Nelson and Bugbee (2014) state that HPS fixtures are still preferred in large greenhouses with small aisles and uniformly spaced plants, where the broad, even output pattern from HPS fixtures provides uniform light distribution. In smaller greenhouses with spaced benches, the more focused pattern typically found in LED fixtures can maximize radiation transfer to plant leaves (Nelson and Bugbee 2014). However, in most cases, the choice between different light sources is based on cost analysis and photon efficacy, expressed as the conversion efficiency of electricity to photosynthetic photons, $\mu mol\ J^{-1}$ is the main index here. Nelson and Bugbee (2014) reported that most efficient LEDs and HPS fixtures produced in the USA had nearly identical efficiencies from 1.66 to 1.70 μmol $J^{-1}$. They calculated initial capital costs of fixtures per photon delivered and determined that LED fixtures cost five to ten times more than HPS fixtures. Compared to electric costs, their analysis indicated that the long-term maintenance costs were low for both technologies. However, in northern Europe, independent measurements have shown efficiencies of commercially available Dutch and Danish LED fixtures of 2.2–2.4 μmol $J^{-1}$, whereas newest HPS lamps (1,000 W) reach up to 2.1 μmol $J^{-1}$. Therefore, LEDs are fully implementable on a commercial scale (Ouzounis et al. 2015a).

In any case, LED lighting is already irreplaceable in its niche application fields, such as intermittent illumination system (Kanechi et al. 2016), inter-lighting, or close-canopy lighting in closed environment horticulture (plant factories). Controlled environment crop cultivation under LED lighting is being envisaged as the new face of agriculture in the near future (Agarwal and Dutta Gupta 2016). The application of LED technology for close-canopy lighting commonly results in significantly lower leaf temperatures than lighting with high-pressure sodium lamps. A recent analysis has shown that photon efficacy of the most efficient commercial LED fixtures was equal to that of the most efficient HPS fixtures at 1.7 μmol photosynthetic photons $J^{-1}$ of electrical input; thus, theoretically, they generate the same amount of thermal energy per photosynthetic photon. However, LEDs dissipate much of their heat away from the plane they illuminate, while HPS fixtures dissipate more heat toward the plane they illuminate (Nelson and Bugbee 2014).

Another concern limiting the prevalence of LED lighting for plant cultivation is the heterogeneous information about optimal LED lighting parameters for different plant species. The variable experimental conditions as well as the wide variety of plant species analyzed with no systematic research approach hardly allows to compare and combine current knowledge on LED parameter effects on plants. Therefore, the user of LEDs should preferably have basic photobiological knowledge to be able to perform correct and targeted lighting operations.

## 7.4 LED Lighting for Main Horticultural Crops

### *7.4.1 Microgreens*

Microgreens are a relatively new specialty crop appearing in many upscale markets and restaurants. These crops consist of vegetables and herbs consumed at initial growth stages. Research reports propose that the reaction of these specialty crops to lighting parameters is relatively different from that of the mature plants (Brazaitytė et al. 2016). However, the blue light is of primary importance in microgreen lighting. Results show significant increases in shoot tissue pigments, glucosinolates, and essential mineral elements in *Brassica* microgreens under the exposure of higher percentages of blue LED wavelengths (Kopsell and Sams 2015). Other authors showed that supplemental blue light can also be strategically used to enhance the nutritional value of microgreens (Vaštakaitė et al. 2015) as well as the mineral contents (Kopsell et al. 2014; Gerovac et al. 2016). In indoor experiments, Brazaitytė et al. (2016) have cultivated different microgreens species under the main set of red, blue, and far-red LEDs, supplemented with yellow, orange, green, and UV-A LEDs. They concluded that supplemental green (520 nm) and orange (622 nm) light induced nitrate reduction, while yellow (595) and UV-A (366, 390 nm) were more favorable for antioxidant compound accumulation, with insignificant effects on growth parameters. In contrast, Gerovac et al. (2016) found that regardless of light quality, as the lighting integral increased from 105 to 315 mmol $m^{-2}$ $s^{-1}$, hypocotyl length decreased and percent dry weight increased for kohlrabi, mizuna, and mustard microgreens (Gerovac et al. 2016). It is still difficult to determine the common lighting patterns on various microgreens, as a wide variety of vegetable species (such as beet, kale, basil, radish) with different life strategies can be cultivated as these specialty crops, which may have specific lighting requirements even in the early developmental stages.

### *7.4.2 Lettuce and Other Leafy Greens*

Lettuce and other leafy vegetables play an important role in human diet and nutrition. Their productivity and quality depend on various environmental factors, with light playing one of the main roles (Mou 2012). At northern latitudes, where natural light level in the greenhouse during autumn/winter is low, as well as in closed plant factories where artificial lighting is the sole light source, supplemental lighting is necessary for commercial crop production. Low light intensity is the limiting factor in lettuce growth and quality (Colonna et al. 2016); however, light spectral composition also has a pronounced effect. Concentrations of pigments and metabolites, such as chlorophylls, carotenoids, anthocyanins, ascorbic acid, and sugars, are affected by supplementary light sources (Li and Kubota 2009), as well as the variation in plant size, color, texture, and flavor (Carvalho and Folta 2014). Light-emitting diodes have been widely analyzed as potential light sources in green vegetable production as well as for the control of production quality (Table 7.1).

Red light is usually the basis of the lighting spectra, and sole red LED light might be sufficient for plant growth and photosynthesis. According to previous studies, ~640 nm (Lefsrud et al. 2008; Samuolienė et al. 2012c; Žukauskas et al. 2011; Samuolienė et al. 2012a) or ~660 nm (Brazaitytė et al. 2006; Mizuno et al. 2011; Li and Kubota 2009; Tarakanov et al. 2012; Wojciechowska et al. 2015; Chen et al. 2016) red LED wavelengths are most commonly used in the cultivation of lettuce and other green vegetables. As will be reviewed later in this chapter, red LED light is usually combined with blue light for efficient plant cultivation both in greenhouses and closed environment chambers; however, specific red light treatments can also be advantageous when applied for short periods over a few days before harvesting (Carvalho and Folta 2014). For example, three days of LED red

**Table 7.1** Light spectra effects on lettuce and other green leafy vegetable growth and quality

| Light colors | LED lighting conditions | Plant | Effect | References |
|---|---|---|---|---|
| Far Red 700–850 nm | 850 nm (30 μmol $m^{-2}s^{-1}$) with white LEDs (peak 449, 548 nm, 30% or B); total PPFD 135 μmol $m^{-2}$ $s^{-1}$; PP—16 h | Lettuce (*Lactuca sativa* var. *crispa*) 'Green Oak Leaf' | • Shoot fresh weight decreased by 36% compared to sole white LEDs, plants were sparse, twisted | Chen et al. (2016) |
| | 740 nm LED in combination with 660 and 455 nm. R/B = 4.5; R/FR = 7; PP 20 h, total PPFD 150 μmol $m^{-2}$ $s^{-1}$ | Lettuce (*Lactuca sativa* L.) 'Frillice Crisp' | • The addition of far-red light resulted in increased leaf area index, fresh weight, shoot height and by 17, 29, 121, and 117%, respectively<br>• Faster growth may have been the cause of the decrease in SPAD and dry weight content by 27 and 7%, respectively<br>• Far-red light stimulated the uptake of N<br>• The uptake of K, Ca, and Mg by plants under additional far-red light increased by 27, 25, and 28%, respectively, as compared to plants, illuminated with red and blue lights | Pinho et al. (2016) |
| | 735 nm with 440, 660 nm; B + R/FR ratio 0.7, 1.2, 4.1, 8.6; B:R 2:8; total PPFD 130 μmol $m^{-2}$ $s^{-1}$; PP 12 h | Lettuce (*Lactuca sativa* L.) 'Sunmang' seedlings from 16 days old | • Improved shoot and root growth; highest fresh weight at B + R/FR ratio 1, 2 | Lee et al. (2016) |
| | 735 nm in combination of 450 nm blue and 660 nm red, R:B:FR 1:1:1, total PPFD 150 μmol $m^{-2}$ $s^{-1}$, PP 12 h | Basil (*Ocimum basilicum* L.) 'Ceasar' | • Basil, cultivated under blue, red and far-red LED combination emits higher levels of most sesquiterpenoid volatiles | Carvalho et al. (2016) |
| | 734 nm (160 μmol $m^{-2}$ $s^{-1}$) supplemental for cool white fluorescent lamps (WF), total PPFD 300 μmol $m^{-2}$ $s^{-1}$, PP 16 h | Baby leaf lettuce (*Lactuca sativa* L.) 'Red Cross' | • Fresh weight, dry weight, stem length, leaf length, and leaf width increased by 28, 15, 14, 44, and 15%, respectively, as compared to sole WF<br>• Decreased chlorophyll and carotenoid concentration by 14 and 11% as compared to WF | Li and Kubota (2009) |
| | 732 nm LEDs in combination with red 660 nm light. R/FR ratios 0.7, 1.2, 4.1, 8.6, and 100% red, total PPFD 132 μmol $m^{-2}$ $s^{-1}$ | Red lettuce (*Lactuca sativa* L.) 'Sunmang' | • Fresh and dry weight the highest, when R/FR ratio was 1.2<br>• The number of leaves increased, when R/FR ratio decreased<br>• Leaves were longer and SPAD values decreased, compared to plants under fluorescent lamp lighting | Lee et al. (2015) |
| | 730 nm (20 μmol $m^{-2}$ $s^{-1}$) LEDs in combination with red 640 nm, total PPFD 320 μmol $m^{-2}$ $s^{-1}$, PP 18 h | Red leaf lettuce (*Lactuca sativa* L.) 'Outredgeous' | • Increased total biomass, leaf elongation | Stutte et al. (2009) |

(continued)

**Table 7.1** (continued)

| Light colors | LED lighting conditions | Plant | Effect | References |
|---|---|---|---|---|
| Red 620–700 nm | 660 nm LEDs (75%) in combination with blue 460 nm LEDs (25%); total PPFD ~170 μmol $m^{-2}$ $s^{-1}$ | Indian mustard (*Brassica juncea* L.) Basil (*Ocimum gratissimum* L.) | • Delayed or inhibited plant transition to flowering as compared to HPS or 460 nm + 635 nm LED combination effects | Tarakanov et al. (2012) |
| | 660 nm LEDs, PPFD 50 μmol $m^{-2}$ $s^{-1}$; PP 16 h | Cabbages (*Brasica olearacea* var. *capitata* L.) 'Kinshun' (green leaves) and 'Red Rookie' (red leaves) | • Increased anthocyanin contents and leaf pigmentation in red leaf cabbages comparing to FL, 470-, 500-, 525 nm LEDs | Mizuno et al. (2011) |
| | 660 nm (30 μmol $m^{-2}$ $s^{-1}$) with white LEDs (peak 449, 548 nm, 30% or B); total PPFD 135 μmol $m^{-2}$ $s^{-1}$; PP 16 h | Lettuce (*L. sativa* var. *crispa*) 'Green Oak Leaf' | • Increased chlorophyll and carotenoids contents | Chen et al. (2016) |
| | 660 and 430 nm in ratios 100:0, 90:10, 70:30, 50:50 supplemental for natural lighting in greenhouse in winter; PPFD 200 μmol $m^{-2}$ $s^{-1}$; PP 16 h | Lamb's lettuce (*V. locusta* L.) 'Noordhollandse' | • Reduced ascorbic acid contents | Wojciechowska et al. (2015) |
| | 658 nm (130 μmol $m^{-2}$ $s^{-1}$) with WF; total PPFD 300 μmol $m^{-2}$ $s^{-1}$; PP 16 h | Lettuce (*L. sativa* L.) 'Red Cross' | • Increased phenolic compound concentration, as compared to WF | Li and Kubota (2009) |
| | 640 nm red LEDs (253 μmol $m^{-2}$ $s^{-1}$) applied 7 days before harvesting (pre-treatment with WF and incandescent irradiance at 275 μmol $m^{-2}$ $s^{-1}$) in controlled environment | Kale (*Brassica oleracea* L.) 'Winterbor' | • Enhanced chlorophyll *a*, *b* and lutein accumulation | Lefsrud et al. (2008) |
| | 638 nm LED (~500 μmol $m^{-2}$ $s^{-1}$) supplemental for HPS (130 μmol $m^{-2}$ $s^{-1}$) lighting and natural illumination in greenhouse 3 days pre-harvest treatment | Lettuce (*Lactuca sativa* L.) 'Grand rapids' Marjoram (*Majorana hortensis* Moench.) Green onions (*Allium cepa* L.) 'Lietuvosdidieji' | • Decreased nitrate contents | Samuolienė et al. (2009) |
| | 638 nm LEDs (210 μmol $m^{-2}$ $s^{-1}$) in combination with HPS lighting (300 μmol $m^{-2}$ $s^{-1}$) and natural illumination 3 days before harvesting in greenhouse; PP 18 h | Green baby leaf lettuce (*Lactuca sativa* L.) 'Thumper' and 'Multibaby' | • Increased antioxidant properties in 'Multibaby' lettuce: higher concentration of total phenolics (28.5%), tocopherols (33.5%), antioxidant capacity (14.5%), and sugars (52.0%);<br>• Decreased concentration of ascorbic acid, as compared to untreated plants | Samuolienė et al. (2012a) |
| | 638 nm LEDs (photoregulated flux) in combination with HPS lighting (90 μmol $m^{-2}$ $s^{-1}$) and natural | White mustard (*Sinapsis alba*), Spinach (*Spinacia oleracea*) 'Giant d'hiver', Rocket (*Eruca sativa*) | • Altered antioxidant activity, increased monosaccharide, and decreased nitrate | Bliznikas et al. (2012) |

(continued)

**Table 7.1** (continued)

| Light colors | LED lighting conditions | Plant | Effect | References |
|---|---|---|---|---|
| | illumination 3 days before harvesting in greenhouse, total PPFD maintained at 300 µmol $m^{-2}$ $s^{-1}$; PP: 5 h from 5 am and 7 h from 5 12 pm | 'Rucola', Dill (*Anethum graveolens*) 'Mammouth', Parsley (*Petroselinum crispum*) 'Plain leaved', Green onions (*Allium cepa*) 'White Lisbon' | accumulation in dill and parsley. Increase in vitamin C content in mustard, spinach, rocket, dill, and green onion<br>• Decreased nitrate accumulation in dill and parsley | |
| | 638 nm or 660 nm; PPFD 300 µmol $m^{-2}$ $s^{-1}$ (3 days' pre-harvest treatment); pre-illumination with 447, 638, 665, and 731 nm, PP 16 h | Basil (*Ocimum basilicum* L.) 'Sweet Genovese' Parsley (*Petroselinum crispum*) | • Increase in free radical scavenging activity and ascorbic acid contents | Samuolienė et al. (2016) |
| Orange 585–620 nm | 600 nm amber LEDs in combination with 450 nm blue and 660 nm red, R:B:Amber 1:1:1, total PPFD 150 µmol $m^{-2}$ $s^{-1}$, PP 12 h | Basil (*Ocimum basilicum* L.) 'Ceasar' | • Basil, grown under blue, red, and amber LED combination emit higher levels of a subset of monoterpenoid volatiles | Carvalho et al. (2016) |
| Yellow 550–585 nm | | | | |
| Green 490–550 nm | 510, 520, or 530 nm LEDs (PPFD 100, 200, and 300 µmol $m^{-2}$ $s^{-1}$) | Red leaf lettuce (*Lactuca sativa* L.) 'Banchu Ref Fire' | • High-intensity (300 µmol $m^{-2}$ $s^{-1}$) green LED light promoted lettuce growth (as compared to FL)<br>• 510 nm light had the greatest effect on plant growth | Johkan et al. (2012) |
| | 530 nm in combination with red 660 and blue 460 nm, R:G:B = 4:1:1, total PPFD 200 µmol $m^{-2}$ $s^{-1}$, continuous 24 h pre-harvest lighting | Butterhead lettuce 'De Lier' | • Increased free radical scavenging activity, phenolic compounds<br>• Decreased nitrate contents | Bian et al. (2016) |
| | 530 nm LEDs (30 µmol $m^{-2}$ $s^{-1}$) supplemental for natural solar and HPS lamp (170 µmol $m^{-2}$ $s^{-1}$) illumination in greenhouse, PP 16 h | Baby leaf lettuce: red leaf 'Multired 4,' green leaf 'Multigreen 3,' and light green leaf 'Multiblond 2' | • Reduction of nitrate concentration and increase in saccharide contents in all baby leaf lettuce varieties | Samuolienė et al. (2012d) |
| | 505, 530 nm LEDs (30 µmol $m^{-2}$ $s^{-1}$) supplemental for HPS lighting (170 µmol $m^{-2}$ $s^{-1}$) and natural illumination in the greenhouse, PP 16 h | Red leaf 'Multired 4,' green leaf 'Multigreen 3,' and light green leaf 'Multiblond 2' baby leaf lettuce (*Lactuca sativa* L.) | • 535 nm green LEDs had greater positive effect on ascorbic acid, tocopherol contents, and DPPH free radical scavenging capacity<br>• 505 nm LEDs had greater effect on total phenol and anthocyanin contents | Samuolienė et al. (2012b) |
| | 518 nm in combination with red 655 nm, blue 456 nm, R:G:B = 9:1:0, 8:1:1, 7:1:2, total PPFD 173 µmol $m^{-2}$ $s^{-1}$, PP 12 h | Lettuce (*Lactuca sativa*) cultivars, red leaf 'Sunmang', and green leaf 'Grand Rapid TBR' 18-day seedlings for 4 weeks | • The substitution of blue with green LEDs in the presence of a fixed proportion of red enhanced the growth of lettuce | Son and Oh (2015) |

(continued)

**Table 7.1** (continued)

| Light colors | LED lighting conditions | Plant | Effect | References |
|---|---|---|---|---|
| | | | • Fresh weights of red leaf lettuce shoots under R8G1B1 were about 61% higher than those under R8B2 | |
| | 520 nm in combination of 450 nm blue and 660 nm red, R:B:Green 1:1:1, total PPFD 150 μmol $m^{-2}$ $s^{-1}$, 12 h | Basil (*Ocimum basilicum* L.) 'Ceasar' | • Basil, grown under blue, red and green LED combination emit higher levels of a subset of monoterpenoid volatiles | Carvalho et al. (2016) |
| | 520 nm, supplemental for the main set of 447, 638, 660, 731 nm LEDs, PPFD 300 μmol $m^{-2}$ $s^{-1}$, PP 16 h | Various microgreens | • Increased total phenolics, total carotenoids in mustard, and parsley microgreens<br>• But decreased total carotenoids in red pak choi and tatsoi microgreens | Brazaitytė et al. (2016) |
| | 518 nm in combination with red 655 nm, blue 456 nm, R:G:B = 9:1:0, 8:1:1, 7:1:2, total PPFD 173 μmol $m^{-2}$ $s^{-1}$, PP 12 h | Lettuce (*Lactuca sativa*) cultivars, red leaf 'Sunmang', and green leaf 'Grand Rapid TBR' 18-day seedlings for 4 weeks | • The substitution of blue with green LEDs in the presence of a fixed proportion of red enhanced the growth of lettuce<br>• Fresh weights of red leaf lettuce shoots under R8G1B1 were about 61% higher than those under R8B2 | Son and Oh (2015) |
| Blue 425–490 nm | Blue LEDs (476 nm, 130 μmol $m^{-2}$ $s^{-1}$) supplemental for cool white fluorescent lamps | Baby leaf lettuce (*Lactuca sativa* L.) 'Red Cross' | • Anthocyanins concentration increased by 31%<br>• Carotenoids concentration increased by 12% | Li and Kubota (2009) |
| | Blue 470 nm LEDs 50 μmol $m^{-2}$ $s^{-1}$ | Seedlings of cabbages (*Brasica olearacea* var. *capitata* L.) 'Kinshun' (green leaves), and 'Red Rookie' (red leaves) | • Promoted petiole elongation in both cabbage varieties;<br>• Higher chlorophyll contents in green leaf cabbages | Mizuno et al. (2011) |
| | Blue (468 nm) LEDs alone or in combination with red (655 nm) LEDs. Total PPFD ~100 μmol $m^{-2}$ $s^{-1}$ | Red leaf lettuce seedlings (*Lactuca sativa* L. cv. Banchu Red Fire) | • Stimulated biomass accumulation in the roots<br>• Resulted in compact lettuce seedling morphology<br>• Promoted the growth of lettuce after transplanting<br>• Greater polyphenol contents and total antioxidant status | Johkan et al. (2010) |
| | Blue 460 nm LEDs alone and in combination with red 660 nm light (11,1% of blue light), total PPFD of 80 μmol $m^{-2}$ $s^{-1}$ | Chinese cabbage (*Brassica campestris* L.) | • Higher chlorophyll concentration<br>• Blue LEDs benefit vegetative growth, while red LEDs and blue plus red LEDs support reproductive growth | Li et al. (2012) |

(continued)

**Table 7.1** (continued)

| Light colors | LED lighting conditions | Plant | Effect | References |
|---|---|---|---|---|
| | | | • Concentration of vitamin C was the greatest under blue LEDs | |
| | 460 nm LEDs in combination with 630 nm red, R:B ratio 2:1, 4:1, 8:1, 1:0, total PPFD 150 μmol $m^{-2}$ $s^{-1}$, 48 h pre-harvest continuous lighting | Lettuce (*Lactuca sativa* L.) | • Reduced nitrate contents and significantly increased soluble sugar contents, the effect most pronounced at R:B ratio 4:1 | Wanlai et al. (2013) |
| | 450–400 nm with red 630–660 nm in R:B ratios of 8:1 and 6:3, total PPFD 50 μmol $m^{-2}$ $s^{-1}$, PP 12 h | Chinese kale 'Lybao' | • Reduced shoot and plant dry weight at higher blue light rates<br>• Increased vitamin, soluble sugar, soluble protein contents, and reduced nitrate contents at higher blue light rates | Xin et al. (2015) |
| | 449 nm LEDs in combination with red 661 nm, R:B ratios of 5:1, 10:1, and 19:1, total PPFD 120 μmol $m^{-2}$ $s^{-1}$, PP 16 h | Coriander (*Coriandrum sativum*) 'Leisure' | • Highest fresh and dry mass accumulation was observed in plants under 10:1 ratios of R:B<br>• Significantly higher antioxidant accumulation was observed when R:B ratio 5:1 | Naznin et al. (2016) |
| | Sole 440 nm blue LEDs (10,6 μmol $m^{-2}$ $s^{-1}$) applied 7 days before harvesting (pre-treatment with cool white fluorescent and incandescent irradiance at 275 μmol $m^{-2}$ $s^{-1}$) | Kale plants (*Brassica oleracea* L. cv Winterbor) | • Enhanced β-carotene contents | Lefsrud et al. (2008) |
| | Blue (440 nm, 30 μmol $m^{-2}$ $s^{-1}$) LEDs in combination with red (640 nm, 270 μmol $m^{-2}$ $s^{-1}$) | Red leaf lettuce (*Lactuca sativa* L. cv. Outredgeous) | • Increased concentration of anthocyanins, higher antioxidant potential<br>• Leaf expansion | Stutte et al. (2009) |

(continued)

**Table 7.1** (continued)

| Light colors | LED lighting conditions | Plant | Effect | References |
|---|---|---|---|---|
| UV 200–400 nm | UV-A 383–426 nm LEDs in combination with red 623–673 nm and blue 427–478 nm (R:B:UV-A 52.9:7:10.1%), total PPFD 300 μmol $m^{-2}$ $s^{-1}$, PP 12 h | *Lactuca sativa* var. crispa | • Increased shoot fresh mass | Chang and Chang (2014) |
| | UV-A LEDs (373 nm, 18 ± 2 μmol $m^{-2}$ $s^{-1}$) supplemental for cool white fluorescent lamps | 'Red Cross' baby leaf lettuce (*Lactuca sativa* L.) | • Anthocyanin concentration increased by 11% | Li and Kubota (2009) |
| | 310, 325, or 340 nm UV LEDs at 0.5 W $m^{-2}$ applied 3 days before harvest were added to white fluorescent lamps | Lettuce (*Lactuca sativa L.*) 'Red fire' | • Pre-harvest UV-B light stimulated anthocyanin and other antioxidant polyphenols<br>• Anthocyanin concentration was significantly higher at 310 nm as compared to 325 and 340 nm | Goto et al. (2016) |

*PPFD* Photosynthetic photon flux density, *PP* photoperiod, *B* blue, *R* red, *FR* far red, *G* green, *FL* fluorescent, *WF* white fluorescent, *HPS* high-pressure sodium

~640 nm light supplemental to natural lighting in a greenhouse can increase lettuce carbohydrate content and antioxidant capacity and repress undesirable nitrate contents (Samuolienė et al. 2009, 2013; Žukauskas et al. 2011). The increase in overall lettuce antioxidant activity under red light is cultivar-specific and more pronounced in green leaf-type varieties than in red leaf-types (Carvalho and Folta 2014), which naturally contain higher levels of antioxidants, protecting plants from environmental exposure, including light. Pre-harvest red 640 nm LED exposure provided variable results in different leafy vegetables. For example, parsley and dill, after a three-day supplementary red light treatment in the greenhouse, showed higher accumulation of phenolic compounds, vitamin C, carbohydrates, as well as increased total antioxidant activity and reduced accumulated nitrate contents, while no nitrate reduction was observed for mustard, spinach, rocket, and green onion (Bliznikas et al. 2012). Wanlai et al. (2013) revealed that the combination of red and blue lights, as compared to sole red light, was more efficient in nitrate reduction when continuously applied 48 h before harvest. Such short-term light treatments provide new perspectives for economic pre-harvest quality management in commercial leaf vegetable production under artificial lighting.

Far-red LED light was demonstrated to be beyond the photosynthetically active region range to support suitable lettuce photosynthesis and growth (Goins et al. 2001). However, changes in red or far-red radiation and their ratios are perceived by phytochromes (Demotes-Mainard et al. 2016) and may influence photomorphogenetic processes in plants. Far-red light, when applied in combination with red (Stutte et al. 2009; Lee et al. 2015), red and blue LEDs (Lee et al. 2016), or cool white fluorescent light (Li and Kubota 2009), had pronounced effect on lettuce growth characteristics: It increased biomass and leaf length, but negatively affected chlorophyll, anthocyanin, and carotenoid concentrations. Lettuce growth promotion under supplemental far-red lighting was the result of increased leaf area and, consequently, improved light interception (Kubota et al. 2012). Far-red light, applied together with red and blue LEDs, enhanced mineral (potassium, calcium, and magnesium) uptake in hydroponically grown lettuce (Pinho et al. 2016). Regarding these effects, supplemental far-red LEDs, in particular portions, should be considered when designing artificial lighting systems for closed-type plant factories (Lee et al. 2015).

Although red light efficiently drives photosynthesis, some blue light is typically necessary to improve growth and minimize shade avoidance responses, including excessively elongated stems (Snowden et al. 2016). Blue light activates the cryptochrome system and matches chlorophyll and carotenoid absorption spectra, thus having significant effects on green vegetable morphology, growth, photosynthesis, and antioxidant system response (Olle and Viršilė 2013). Blue LEDs (440–476 nm), used alone or in combination with red LEDs, stimulated leaf area expansion and biomass accumulation in lettuce (Johkan et al. 2010; Lin et al. 2013), Chinese cabbage plants (Li et al. 2012), spinach (Matsuda et al. 2007; Ohashi-Kaneko et al. 2007), and coriander (Naznin et al. 2016). The positive effects of increased blue light fractions on growth correspond with the increased leaf chlorophyll levels and photosynthetic rates (Yorio et al. 2001; Carvalho and Folta 2014).

Studies of whole plants revealed that photosynthesis often increases with increasing fractions of blue photons (Hogewoning et al. 2010). However, the effect of blue light on plant photosynthetic productivity is primarily determined by changes in radiation capture and not by direct effects on photosynthesis (Snowden et al. 2016). Plant growth typically tends to decrease when the fraction of blue photons exceeds 5–10%. High levels of blue light in the spectrum inhibit cell division, cell expansion, and leaf area growth, what results in reduced photon capture and diminished growth (Bugbee 2016). Notwithstanding, blue light can interact with radiation intensity, and the effects of blue light fraction were greater at higher photosynthetic flux densities (Snowden et al. 2016). There is a wide range in species sensitivity to blue light, and responses can change with the developmental stage of the plant (Bugbee 2016), thus numerous and differential results for optimal red:blue light ratios for plant growth and photosynthesis in research works are reasonable and not yet exploited.

Blue LED light also was useful in improving the nutritional quality of green vegetables. It reduced nitrate contents (Xin et al. 2015; Bian et al. 2016), stimulated antioxidant status, e.g., increasing phenolic compounds (Johkan et al. 2010; Son and Oh 2013; Bian et al. 2016; Taulavuori et al. 2016), ascorbic acid (Li et al. 2012; Xin et al. 2015), carotenoids (Lefsrud et al. 2008; Li and Kubota 2009), anthocyanin contents, and leaf coloration (Stutte et al. 2009; Li and Kubota 2009; Mizuno et al. 2011), thereby affecting leaf coloration. Studies show that it is sufficient to apply end-of-production treatments, lasting a few days, with supplemental blue light to enhance pigmentation in red lettuce varieties (Owen and Lopez 2015; Nicole et al. 2016). Studies have also confirmed that the effects of blue LED lighting are species dependent and cultivar dependent and that red-colored cultivars better acclimatize to the blue lighting conditions than green cultivars (Ouzounis et al. 2015b; Taulavuori et al. 2016).

Lettuce taste may also be affected by light conditions (Carvalho and Folta 2014). In one study, the lettuce variety Grand Rapids developed an increasingly bitter taste under blue light, compared to red or red with far-red light (Carvalho and Folta 2014). Lin et al. (2013) compared the sensual properties of Boston lettuce, cultivated under red and blue or red, blue, and white LEDs, and found that the shape, crispness, and sweetness of red and blue treated plants were not acceptable for the market, but using supplemental white LEDs resulted in higher crispness and sweeter taste due to enhanced accumulation of sugars.

Green light also has valuable physiological effects (Olle and Viršilė 2013). In several studies, 510–530 nm LED light (Johkan et al. 2012; Son and Oh 2015), comparably to green fluorescent lamps, supplemental for red and blue LEDs (Kim et al. 2004), promoted lettuce growth. Son and Oh (2015) analyzed leaf morphology, transmittance, cell division rate, and leaf anatomy under treatments with green LEDs and observed enhanced growth of the two lettuce cultivars tested. Snowden et al. (2016) and Bugbee (2016) explained that green light penetrates deeper into leaves and canopies, thereby altering plant growth and development;

however, its effects may decrease with increasing photosynthetic fluxes. Folta and Maruhnich (2007) also postulated that green light is 'a signal to slow down or stop' plant growth, but they acknowledge that the role of green light is thought to be especially important in the low light conditions that typically occur below plant canopies. Green is characterized by better transmission through leaf tissue, as compared to red or blue light wavelengths (Massa et al. 2015).

Few reports also indicate the effect of green light on the nutritional value of leafy vegetables. Supplemental 530 nm green light promoted accumulation of α-carotene and anthocyanins in romaine baby leaf lettuce, cultivated in closed environment chambers under combination of red and blue LEDs (Samuolienė et al. 2013). The same authors found that 505, 530, and 535 nm green LED light, supplemental to high-pressure sodium lamp lighting in a greenhouse, reduced nitrate or increased ascorbic acid, tocopherol, and anthocyanin concentrations in different baby leaf lettuce varieties (Samuolienė et al. 2012b, d). Green and yellow LEDs, supplemental to the red and blue ones, enhanced emission levels of monoterpenoid volatiles in basil plants (Carvalho et al. 2016).

However, UV lights have the highest impact on secondary metabolites in leafy vegetables. Due to limited UV LED availability and relatively high costs (Wargent 2016), only few research results have been published so far. Chang and Chang (2014) reported an increase in shoot fresh weight of leaf lettuce exposed to UV-A light. A small flux of UV-A LED irradiation also increased anthocyanin (Li and Kubota 2009), phenolic compounds, and α-carotene, (Samuolienė et al. 2013) contents in baby leaf lettuce. Goto et al. (2016) showed that the addition of UV light 1–3 days prior to harvest effectively increases anthocyanin concentration and antioxidant capacity in red leaf lettuce.

Selected lighting conditions can be adequately introduced into lettuce production. However, significant effort will need to be put into the development of LED lighting models for different green vegetables and lettuce cultivars. In addition, there are many variables to consider when comparing research reports, such as vegetable variety, developmental stage, time during which illumination is applied, the spectral distribution of the light sources, photosynthetic photon flux, photoperiod length, as well as temperature (Carvalho and Folta 2014) and other conditions of the cultivation environment. Moreover, in a greenhouse environment, seasonality effects cannot be eliminated. Several researches performed in northern latitudes confirm that different light spectral properties were most efficient during different seasons of the year (Samuolienė et al. 2012a, b; Wojciechowska et al. 2016) as well as in the same season in different cultivation years (Wojciechowska et al. 2015).

### *7.4.3 Vegetable Transplants*

High-quality vegetable transplants are determinants for successful vegetable production under greenhouse conditions. Vigorous vegetable transplants typically have well-developed leaves and roots with short internode length and thick stems. The flower development status is also an important attribute of transplant quality (especially for tomato), as first flower clusters often develop during the propagation period (Mitchell et al. 2015). In case of grafted transplants, requirements for seedling morphology are often opposite: rootstock seedlings preferably have longer hypocotyls to ensure that the height of the graft union is well above the soil line (Chia and Kubota 2010). Therefore, control of transplant morphology by light has a diverse practical value (Table 7.2).

When LEDs were used as the sole source of lighting in closed environment chambers, plant requirements for light spectral composition were more pronounced as compared to requirements under greenhouse conditions. Red light alone was not efficient enough, and the addition of blue light wavelengths resulted in stronger, shorter tomato seedlings in different varieties (Liu et al. 2011; Nanya et al. 2012; Ouzounis et al. 2016), shorter cucumber petioles (van Ieperen et al. 2012) and hypocotyls (Hernandez and Kubota 2016), and eliminated tomato leaf curling, which had appeared under red light treatment (Ouzounis et al. 2016). However, increasing blue light input in the light spectra suppressed dry mass accumulation in tomato seedlings (Nanya et al. 2012). Hernandez et al. (2016) report that dry mass and leaf area in tomato transplants increased by up to 30–50% with increasing blue light and then decreased. In cucumber transplants (Hernandez and Kubota 2016), dry mass also decreased with increasing blue, while chlorophyll content per leaf area, net photosynthetic rate, and stomatal conductance increased with the increase of photosynthetic flux of blue light (Hogewoning et al. 2010; Hernandez and Kubota 2016). Hernandez and Kubota (2016) proposed that for cucumber cultivation under sole red and blue LED lights, 10% of blue in total photosynthetic flux is optimal, while for tomato, 30–50% are beneficial. They also reported that the addition of green light to the red and blue spectrum did not have any influence on cucumber plant responses (Hernandez and Kubota 2016). In contrast, Brazaitytė et al. (2009) found that green light accelerated cucumber, but inhibited tomato transplant growth in a closed environment chamber.

In greenhouse environments, the impact of supplemental light quality seems to be diminishing, especially when background solar irradiance provides sufficient photosynthetically active photon flux (Mitchell et al. 2015). It is likely that there is a threshold background solar daily light integral (DLI) or a relative level of supplemental DLI that requires the additional blue photon flux through supplemental lighting (Hernández and Kubota 2012; 2014a, b; Mitchell et al. 2015). However, Gomez and Mitchell (2015) evaluated the morphological responses of six tomato cultivars to different two-week LED treatments across changing solar DLIs and found that in all cultivars evaluated, hypocotyl diameter and leaf area increased upon addition of blue light to red light. The series of experiments, performed with different varieties of cucumber, tomato, and sweet pepper transplants in the greenhouse where a high-pressure sodium lamp spectra, deficient in blue light, was supplemented with green-blue LED wavelengths (530, 505, 455, 470 nm), showed different results. In a different experiment, blue and cyan (505 nm) supplemental light resulted in increased leaf area and fresh and dry weight as well as reduced hypocotyl length in cucumber and tomato seedlings (Samuolienė et al. 2012c).

**Table 7.2** Light spectra effects on greenhouse vegetable growth and photosynthesis

| Light colors | LED lighting conditions | Plant | Effect | References |
|---|---|---|---|---|
| Far Red 700–740 nm | Far red 735 nm with red 660 nm, total PPFD 300 μmol $m^{-2}$ $s^{-1}$ | Sweet pepper (*Capsicum annuum* L.) 'Hungarian Wax' | • The addition of far-red radiation resulted in taller plants with greater stem mass than red LEDs alone | Brown et al. (1995) |
| Red 620–700 nm | 662 nm red 100% or in combination with blue 456 nm 88% red and 12% blue in controlled climate room. Total PPFD 150 μmol $m^{-2}$ $s^{-1}$, PP 16 h | 9 tomato (*Solanum lycopersicum*) genotypes | • Upward or downward leaf curling was observed in all genotypes in the 100% red treatment | Ouzounis et al. (2016) |
| | 660 nm LEDs (34 μmol $m^{-2}$ $s^{-1}$) supplemental for FL (360 μmol $m^{-2}$ $s^{-1}$) | Tomato (*Lycopersicon esculentum* L. cv. Momotaro Natsumi) | • Red LEDs were effective enhancing tomato yield | Lu et al. (2012) |
| Orange 585–620 nm | 622 nm orange LEDs (30 μmol $m^{-2}$ $s^{-1}$) supplemental for the main set of 638, 447, 669, and 731 nm LEDs lighting (total PPFD ~200 μmol $m^{-2}$ $s^{-1}$) in the growth chamber | Transplants of cucumber 'Mandy' F1 | • Accelerated growth | Brazaitytė et al. (2009) |
| Yellow 550–585 nm | | | | |
| Green 490–550 nm | 523 nm green in combination with blue 473 nm and red 660 nm 20B:28G:52R%. Total PPFD 100 μmol $m^{-2}$ $s^{-1}$, PP 18 h | Cucumber (*Cucumis sativus*) 'Cumlaude' | • The addition of green light did not have significant influence in cucumber plant responses | Hernandez and Kubota (2016) |
| | 520 nm green LEDs (12 μmol $m^{-2}$ $s^{-1}$) supplemental for the main set of 638, 447, 669, and 731 nm LEDs lighting (total PPFD ~200 μmol $m^{-2}$ $s^{-1}$) in the growth chamber | Transplants of cucumber 'Mandy' F1 | • Accelerated growth | Brazaitytė et al. (2009) |
| | 505, 530 nm LEDs (15 μmol $m^{-2}$ $s^{-1}$) supplemental for HPS lighting (90 μmol $m^{-2}$ $s^{-1}$) and natural illumination in greenhouse | Transplants of cucumber 'Mirabelle' F1, Tomato 'Magnus' F1, and Sweet pepper 'Reda' | • Supplemental 505 nm LED light resulted in increased leaf area, higher fresh and dry weight, and photosynthetic pigment contents in all vegetable transplants<br>• Supplemental 530 nm light had positive effect on development and photosynthetic pigment accumulation in cucumber transplants | Samuolienė et al. (2012c) |
| | 505, 530 nm LEDs (15 μmol $m^{-2}$ $s^{-1}$) supplemental for HPS lighting (90 μmol $m^{-2}$ $s^{-1}$) and natural illumination (100–200 μmol $m^{-2}$ $s^{-1}$) in greenhouse | Transplants of cucumber 'Mandy' F1 | • Increased leaf area, higher fresh and dry weight, decreased hypocotyl elongation | Novičkovas et al. (2012) |

(continued)

**Table 7.2** (continued)

| Light colors | LED lighting conditions | Plant | Effect | References |
|---|---|---|---|---|
| Blue 425–490 nm | 450 nm blue in combination with red 660 nm LEDs in different ratios: 0.1 (B15R135 μmol $m^{-2}$ $s^{-1}$), 0.4 (B45R105), and 1.0 (B75R75), total PPFD 150 μmol $m^{-2}$ $s^{-1}$, 16 h | Tomato seedlings 'Reiyo' | • Higher B/R ratio (1.0) resulted in shorter stem length | Nanya et al. (2012) |
| | 450 nm blue LEDs (450 nm) supplemental for 638 nm red. Total PPFD 100 μmol $m^{-2}$ $s^{-1}$; blue (B) light percentage: 0B, 7B, 15B, 22B, 30B, 50B, and 100B | Cucumber plants (*Cucumis sativus* cv. Hoffmann's Giganta) | • Necessary to prevent any overt dysfunctional photosynthesis<br>• Photosynthetic capacity increased with increasing blue percentage during growth measured up to 50% blue<br>• It was associated with an increase in leaf mass per unit leaf area, nitrogen content per area, chlorophyll content per area, and stomatal conductance | Hogewoning et al. (2010) |
| | 455 nm blue in combination with red 661 nm. B:R ratios of 0B:100R%, 10B:90R%, 30B: 70R%, 50B:50R%, 75B:25R%, 100B:0R% in growth chamber. Total PPFD 100 μmol $m^{-2}$ $s^{-1}$, PP 18 h | Cucumber (*Cucumis sativus*) 'Cumlaude' | • Hypocotyl length decreased with increasing blue light up to 75% and was significantly higher under 100% blue treatment<br>• Leaf area decreased with the increase of the percentage of blue up to 75%. 100% blue resulted in significantly higher leaf area<br>• Chlorophyll content per leaf area, net photosynthetic rate, and stomatal conductance increased with the increase of blue light percentage<br>• Shoot dry and fresh mass decreased with the increase of blue light percentage. Plants under 0% blue had the lowest fresh and dry mass, when plants under 100% blue had the highest fresh mass | Hernandez and Kubota (2016) |
| | Blue 455, 470 nm LEDs (15 μmol $m^{-2}$ $s^{-1}$) supplemental for natural solar and HPS lighting (90 μmol $m^{-2}$ $s^{-1}$) in greenhouse | Transplants of cucumber hybrid 'Mirabelle' F1, tomato hybrid 'Magnus' F1, and sweet pepper 'Reda' | • Supplemental blue light resulted in increased leaf area, fresh and dry weight, and photosynthetic pigment contents in all vegetable transplants | Samuolienė et al. (2012c) |

(continued)

**Table 7.2** (continued)

| Light colors | LED lighting conditions | Plant | Effect | References |
|---|---|---|---|---|
| | Blue 455, 470 nm LEDs (15 μmol $m^{-2}$ $s^{-1}$) supplemental for natural solar and HPS (90 μmol $m^{-2}$ $s^{-1}$) lighting in greenhouse | Transplants of cucumber 'Mandy' F1 | • Supplemental 470 nm LED lighting resulted in increased leaf area, fresh and dry weight, decreased hypocotyl length<br>• 455 nm LED light caused slower growth and development of transplants<br>• Both 455 and 470 nm enhanced photosynthetic pigment contents. | Novičkovas et al. (2012) |
| | Blue 455 nm in combination with red 661 nm (total PPFD 55.5 μmol $m^{-2}$ $s^{-1}$) in different photon flux ratios (0.4 or 16% of blue) in greenhouse | Tomato seedlings 'Komeett' | • No significant differences in tomato seedling growth and morphological parameters between different red:blue ratios | Hernández and Kubota (2012) |
| | 455 nm blue LEDs (6.7–16 μmol $m^{-2}$ $s^{-1}$) supplemental for HPS (400–520 μmol $m^{-2}$ $s^{-1}$) illumination | Tomato (*Lycopersicon esculentum* 'Trust') and cucumber (*Cucumis sativus* 'Bodega') | • Supplemental blue light inside the canopy increased plant biomass, reduced internode length and fruit yield | Menard et al. (2006) |
| | 456 nm LEDs in combination with red 662 nm in controlled climate room. 100% red, or 88% red and 12% blue. Total PPFD 150 μmol $m^{-2}$ $s^{-1}$, PP 16 h | 9 tomato (*Solanum lycopersicum*) genotypes | • The combination of red and blue lights increased total dry matter in 7 of 9 genotypes<br>• Additional blue light did not affect stomatal conductance, but increased leaf chlorophyll and flavonol contents | Ouzounis et al. (2016) |
| | 455 nm blue in combination with red 661 nm in the growth chamber. 0, 10, 30, 50, 75, and 100% blue. Total PPFD 100 μmol $m^{-2}$ $s^{-1}$, PP 18 h | Cucumber (*Cucumis sativus*) 'Cumlaude' and tomato (*Solanum lycopersicum* 'Komeett') | • Cucumber hypocotyl length decreased with increasing blue up to 75%. Dry mass and leaf area decreased, when blue increased from 10 to 75%. Optimal amount of blue was 10%, which still produced seedlings with less dry mass and taller compared to FL<br>• Tomato hypocotyl length decreased with increasing blue up to 75%. Dry mass and leaf area increased with increasing blue up to 30–50% and then decreased from 50 to 100%. Optimal amount was 30–50% of blue, where morphology and growth were comparable to FL | Hernandez et al. (2016) |

(continued)

**Table 7.2** (continued)

| Light colors | LED lighting conditions | Plant | Effect | References |
| --- | --- | --- | --- | --- |
| | 450 nm blue LEDs (450 nm) supplemental for 638 nm red. Total PPFD 100 μmol $m^{-2}$ $s^{-1}$; blue (B) light percentage: 0B, 7B, 15B, 22B, 30B, 50B, and 100B | Cucumber plants (*Cucumis sativus* cv. Hoffmann's Giganta) | • Necessary to prevent any overt dysfunctional photosynthesis<br>• Photosynthetic capacity increased with increasing blue percentage during growth measured up to 50% blue<br>• It was associated with an increase in leaf mass per unit leaf area, nitrogen content per area, chlorophyll content per area, and stomatal conductance | Hogewoning et al. (2010) |
| UV 200–400 nm | | | | |

*PPFD* Photosynthetic photon flux density, *PP* photoperiod, *B* blue, *R* red, *FR* far red, *G* green, *FL* fluorescent, *WF* white fluorescent, *HPS* high-pressure sodium

Green 530 nm light had a positive effect on cucumber transplants only (Samuolienė et al. 2012c; Novičkovas et al. 2012). In the case of sweet pepper, blue and cyan light had positive effects on the variety 'Reda,' while in F1 transplants of the pepper variety 'Figaro,' supplemental blue-green LED light suppressed growth and developmental rates (Bagdonavičienė et al. 2015; Samuolienė et al. 2012c).

### *7.4.4 Greenhouse Vegetable Production*

Transplant quality might be the limiting factor for further transplant success and yield. However, supplemental lighting during the yield formation in greenhouses may also have significant effects on crop productivity and quality, even when natural lighting is not considerably deficient. Plants benefit from equal distribution of irradiation through the canopy, when the amount of light received by each leaf is between the compensation and saturation points. In the high-wire cultivation system with high plant density, most of the light can only be intercepted to the upper part of the plant canopy, regardless of natural or artificial overhead lighting (Guo et al. 2016). Inter-lighting is a recently developed supplemental lighting technique to overcome this problem. Applying a part of the supplemental light within the crop canopy can improve light distribution through the middle or lower canopy part and thus increase light use efficiency and crop yield. Due to the high bulb temperatures of HPS lamps, its use for inter-lighting was not considered. In contrast, light-emitting diodes have low heat emission, making them potentially suitable light systems for inter-lighting (Hao et al. 2012). In one study, blue/red LED inter-lighting positively impacted cucumber leaf photosynthetic characteristics in the lower leaf layers, resulting in greater leaf mass per area and dry mass allocation to leaves, but had no effect on total biomass or fruit production (Trouwborst et al. 2010). Kumar et al. (2016) reported that mini-cucumber yield was increased by 22.3 and 30.8% by the addition of one or two rows of inter-lighting LEDs compared to no inter-lighting. However, Hao et al. (2012) revealed that using inter-lighting, mini-cucumber fruit yield was increased only in the early production period and gradually diminished toward the late production period.

Gómez et al. (2013) reported significantly lower energy requirements from supplemental lighting when intra-canopy LEDs were used, compared with overhead HPS lighting, while maintaining comparable yield in two tomato cultivars. Dzakovich et al. (2015) observed that reduced supplemental lighting energy consumption by using intra-canopy LED supplemental lighting had no negative impact on tomato fruit quality. Deram et al. (2014) analyzed different red light and blue light ratios and proposed an optimal ratio of 5:1 for tomato fruit yield enhancement. Gomez and Mitchell (2016) observed that both intra-canopy lighting and top-lighting increased tomato fruit yield relative to the control, but no significant differences in yield were determined between these two supplemental light treatments. Higher crop photosynthetic activity with intra-canopy lighting did not increase fruit yield; the remaining photo-assimilates were most likely allocated to vegetative plant parts. In sweet pepper, inter-lighting also resulted in 16% higher total marketable yield, mainly due to increased fruit number and faster fruit maturation (Jokinen et al. 2012). Guo et al. (2016) reported that enhanced sweet pepper growth and fruit yield were followed by improved fruit quality, compared to the top HPS treatment, where increased fruit dry matter content and the contents of health-promoting compounds in fruits (phenolic compounds, total carotenoids) as well as higher antioxidant activity were determined.

Light condition changes in primary or secondary plant metabolite accumulation could also be associated with plant immunity, disease development, and interaction with pests (Vänninen et al. 2010; Johansen et al. 2011). However, to date, only discrete research results are explored (Schuerger and Brown 1997; Kim et al. 2013), and plant health promotion by LED lighting parameters is still a future scenario.

### 7.4.5 Ornamental Plants

There appears to be potential for LED lighting as an alternative supplemental light source as well as the sole source for propagating seedlings and cuttings of ornamental plants (Table 7.3). The selection of LED parameters depends on the plant species and on the goals of the propagator. In closed environment cultivation systems, the increasing portion of blue LED light, complementing red light, may reduce stem extension and result in more compact plants. However, this is followed by reduced biomass accumulation and leaf expansion in impatiens, petunia, and salvia (Wollaeger and Runkle 2013). Olschowski et al. (2016) reported that root and shoot development of *Calibrachoa* cuttings was highest under a wider spectrum of white LEDs or a combination of white, blue, and red LEDs, as compared to sole red and blue. Supplementation of far-red to red and blue radiation increased photosynthetic efficiency and subsequent dry mass accumulation, without excessive leaf and stem expansion in snapdragon (Park and Runkle 2016).

In greenhouses, where LEDs were applied within background natural lighting, a proper ratio of red and blue LEDs was also suitable for raising *Vinca*, *Celosia*, bedding impatiens, *Petunia*, marigold, *Salvia*, and pansy seedlings. Seedlings grown under 85:15 and 70:30 red:blue LED light were more compact, with larger stem diameter and higher chlorophyll content than plants cultivated under HPS lamps (Randall and Lopez 2014). Analogous effects were obtained when New Guinea impatiens, *Geranium* and *Petunia,* were propagated form cuttings under red:blue LEDs (Currey and Lopez 2013).

Altering a light regime is a sound and non-polluting way to control greenhouse-grown pot and bedding plants and is a promising technique of eliminating the use of chemical plant growth regulators, which are now becoming less available and more questioned by consumers (Bergstrand et al. 2016). A light regime with 620 nm light given before the period of natural light and 525 nm light given at the end of the day effectively controlled elongation in *Calibrachoa* and *Pelargonium* (Bergstrand et al. 2016). End-of-day treatment with red and blue

**Table 7.3** Light spectra effects on ornamental plants

| Light colors | LED lighting conditions | Plant | Effect | References |
|---|---|---|---|---|
| Far Red 700–740 nm | 729 nm FR light in combination with red 660 nm: R128 µmol $m^{-2}$ $s^{-1}$; R128 + FR16; R128 + FR32; R128 + FR64; R96 + FR32; R96 + FR32, R64 + FR64 with 32 µmol $m^{-2}$ $s^{-1}$ of blue 451 nm in each treatment; PP—18 h | Seedlings of snapdragon (*Antirrhinum majus*) | • Plant height and total leaf area linearly increased, when the R:FR ratio decreased<br>• Shoot dry weight was similar under the same PPFD and linearly increased, when R was constant, and FR increased<br>• Shoot dry weight per unit leaf area linearly decreased, when R was increasingly substituted by FR radiation | Park and Runkle (2016) |
| | Night interruption (4 h, 10 µmol $m^{-2}$ $s^{-1}$) with far-red 730 nm, green 530 nm, blue 450 nm, red 660 nm, or white (400–700 nm, with 28% B, 37% R, and 15% FR) LEDs. Plants grown under short-day conditions PP—10 h, white LEDs, PPFD 180 µmol $m^{-2}$ $s^{-1}$ | *Petunia hybrida* Hort. 'Easy Wave Pink' | • Night interruption with FR light in short-day conditions resulted in higher shoot length;<br>• Night interruption with FR light in short-day conditions promoted flowering. | Park et al. (2016) |
| | 4-h night interruption by R:FR (660:730 nm) LED ratios from 100% red to 100% far red or incandescent lighting under 9-h photoperiod conditions | Chrysanthemum (*Chrysanthemum* × *morifolium*), dahlia (*Dahlia hortensis*), and African marigold (*Tageteserecta*) | • In these short-day plants, a moderate-to-high R:FR (≥ 0.66) is most effective at interrupting the long night. FR light alone does not regulate flowering<br>• For all species, stem length increased quadratically when the R:FR of the night interruption increased with a maximum at R:FR of 0.66 | Craig and Runkle (2013) |
| | 4-h night interruption by R:FR (660:730 nm) LED ratios from 100% red to 100% far red or incandescent lighting under 9-h photoperiod conditions | Marigold (*Tagetes erecta* 'American Antigua Yellow'), petunia (*Petunia multiflora* 'Easy Wave White'), and snapdragon (*Antirrhinum majus* 'Liberty Classic Cherry' | • LED treatments with a moderate R:FR ratio were effective both for promoting flowering in petunia and snapdragon and for inhibiting flowering in marigold<br>• There was insignificant effect of night interruption on inflorescence or flower bud number for marigold and petunia<br>• Plant height was greatest under moderate R:FR in marigold and petunia, while snapdragon exhibited the opposite trend | Craig and Runkle (2012) |

(continued)

**Table 7.3** (continued)

| Light colors | LED lighting conditions | Plant | Effect | References |
|---|---|---|---|---|
| Red 620–700 nm | Natural light with the addition of red 660 nm LED light during the PP of 8 h | *Chrysantemum x moryfolium* 'Cyber' and *Euphorbia pulcherrima* 'Novia' | • Supplementation of natural lighting with small flux of 660 nm light significantly reduced plant height in euphorbia, but not in chrysanthemum | Bergstrand et al. (2016) |
| | Orange (O; 596 nm), red (R; 634 nm), and hyper red (HR; 664 nm) in the proportions of O-R-HR of 20–30–30, 0–80–0, 0–60–20, 0–40–40, 0–20–60, and 0–0–80, respectively, with constant amount of blue (10%; 446 nm) and green (10%; 516 nm) lights. Total PPFD 160 μmol $m^{-2}s^{-1}$, PP—18 h | Seedlings of impatiens (*Impatiens walleriana*), petunia (*Petunia hybrida*), tomato (*Solanum lycopersicum*), and marigold (*Tagetes patula*) or salvia (*Salvia splendens*) | • There were no consistent effects of lighting treatment across species on leaf area, plant height, or shoot fresh weight<br>• Orange, red, and hyper red lights generally have similar effects on plant growth at the intensities tested when background green and blue lights are provided | Wollager and Runkle (2013) |
| Orange 585–620 nm | Orange 620, white and blue 460 nm light applied for 2 h before and after 8-h photoperiod of natural light and the combination of orange light before natural photoperiod and green after; blue before natural photoperiod and orange—after | *Calibrachoa* × *hybrida* 'Callie Bright Red' and *Pelargonium* × *hortorum* 'Americana Pink Splash' | • 620 nm light given at the period of natural lighting and 535 nm light given at the end of natural period controlled elongation in calibrachoa and pelargonium | Bergstrand et al. (2016) |
| Yellow 550–585 nm | | | | |
| Green 490–550 nm | Night interruption (4 h, 10 μmol $m^{-2}$ $s^{-1}$) with green 530 nm, blue 450 nm, red 660 nm, far-red 730 nm, or white (400–700 nm, with 28% B, 37% R, and 15% FR) LEDs. Plants grown under short-day conditions PP—10 h, white LEDs, PPFD 180 μmol $m^{-2}$ $s^{-1}$ | *Petunia hybrida* Hort. 'Easy Wave Pink' | • Night interruption with green light in short-day conditions promoted flowering in petunia | Park et al. (2016) |
| Blue 425–490 nm | 470 nm blue light in different ratios with red 660 nm; R:B 100:0, 85:15, or 70:30, total PPFD 100 μmol $m^{-2}$ $s^{-1}$, PP 16 h, supplemental for natural lighting in comparison with HPS lighting | Seedlings of *Antirrhinum majus* L. 'Rocket Pink', *Catharanthus roseus* L. G. Don 'Titan Punch', *Celosia argentea* L. var. *plumosa* L. 'Fresh Look Gold', *Impatiens walleriana* Hook. f. 'Dazzler Blue Pearl', *Pelargonium hortorum* L.H. Bailey 'Bullseye Scarlet', | • Height of Catharanthus, Celosia, Impatiens, Petunia, Tagetes, Salvia, and Viola was 31, 29, 31, 55, 20, 9, and 35% shorter, respectively, for seedlings grown under the 85:15 red:blue LEDs compared to HPS lighting | Randall and Lopez (2014) |

(continued)

**Table 7.3** (continued)

| Light colors | LED lighting conditions | Plant | Effect | References |
|---|---|---|---|---|
| | | *Petunia hybrida* Vilm.-Andr. 'Plush Blue', *Salvia splendens* Sellow ex Roem. & Schult. 'Vista Red', *Tagetes patula* L. 'Bonanza Flame', and *Viola wittrockiana* Gams. 'Mammoth Big Red' | • Stem caliper of Antirrhinum, Pelargonium, and Tagetes was 16, 8, and 13% larger for seedlings, grown under 85:15 red:blue LEDs, compared to HPS lighting<br>• The quality index was similar for Antirrhinum, Catharanthus, Impatiens, Pelargonium, and Tagetes, grown under LEDs and HPS lamps. The quality index was significantly higher for Petunia, Salvia, and Viola under 85:15, 70:30, and 100:0 red:blue LEDs than under HPS lamps | |
| | 450 nm blue in combination with red 627 nm in R:B ratios of 100:0, 85:15, or 70:30, total PPFD 70 μmol $m^{-2}$ $s^{-1}$; supplemental for natural lighting in greenhouse | Cuttings of *Impatiens hawker* W. Bull 'Celebrette Frost', *Pelargonium hortorum* L. H. Bailey 'Designer Bright Red', and *Petunia hybrid* Vilm. 'Suncatcher Midnight Blue' | • There were no significant differences among impatiens and pelargonium cuttings grown under different supplemental light sources<br>• Stem length of petunia cuttings grown under 100:0 red:blue LEDs was 11% shorter, as compared to HPS;<br>• Leaf dry mass, root dry mass, root mass ratios, and root:shoot ratio of cuttings grown under 70:30 red:blue LEDs were 15, 36, 17, and 24% higher, respectively | Currey and Lopez (2013) |
| | Night interruption (4 h) of short-day photoperiod with white (W), B (462 nm), B + R (659 nm), B + FR (737 nm), B + R + FR, or R + FR light-emitting diodes (LEDs) | Chrysanthemum (*Chrysanthemum* × *morifolium*), cosmos (*Cosmos sulfureus*), two cultivars of dahlia (*Dahlia pinnata*), and marigold (*Tagetes erecta*)] and two long-day plants: dianthus (*Dianthus chinensis*) and rudbeckia (*Rudbeckia hirta*) | • In the studied crops, low-intensity blue light during the night does not influence flowering<br>• White LEDs that emit little FR light are effective at creating long days for short-day plants | Meng and Runkle (2015) |

*PPFD* Photosynthetic photon flux density, *PP* photoperiod, *B* blue, *R* red, *FR* far red, *G* green, *FL* fluorescent, *WF* white fluorescent, *HPS* high-pressure sodium

LEDs at a ratio of 80:20 inhibited shoot elongation in poinsettias (Islam et al. 2015), while supplementation of natural sunlight with a small portion of 660 nm light significantly reduced plant height in *Euphorbia*, but not in *Chrysanthemum* (Bergstrand et al. 2016).

Photoperiodic lighting (short, low-intensity lighting) is also used by commercial crop producers to inhibit flowering of short-day plants and promote flowering of long-day plants under a short natural light photoperiod (Mitchell et al. 2015). Long-day photoperiods can be imitated by day extension or night interruption. The spectral quality of photoperiodic lighting can differently affect flowering of short-day plants and long-day plants differently (Meng and Runkle 2015). Night interruption with the moderate to high red: far-red LED lights effectively inhibited flowering in short-day *Chrysanthemum* (Craig and Runkle 2013; Liao et al. 2014) and promoted flowering in long-day *Petunia* and snapdragon (Craig and Runkle 2012). Photoperiodic lighting with blue LED light provided variable results. Night interruption with blue LEDs was not perceived as long-day signal by Petunia (Park et al. 2016), *Rudbeckia*, *Chrysanthemum* (Ho et al. 2012), *Cosmos*, *Dahlia*, and marigold (Meng and Runkle 2015); however, a mixture of blue and red LEDs promoted flowering of most long-day plants tested. Meng and Runkle (2016) state that photoperiodic lighting with blue light might efficiently regulate flowering when applied in higher intensities in *Calibrachoa*, *Coreopsis*, *Petunia*, *Rudbeckia*, and snapdragon.

The LED lighting parameters in ornamental plant cultivation seem to be even more intricate than those in vegetable lighting. The distinct variety of ornamental plant species, varieties, and cultivars, as well as different cultivation and lighting practices, result in a large diversity of light spectrum effects, highlighting the need for extended lighting research in commercial ornamental plant cultivation.

## 7.5 Conclusions

Innovative LED lighting systems add a completely new dimension to horticultural plant production. With constant energy-efficiency and light distribution improvements, light-emitting diodes are a promising alternative to current supplemental lighting technologies. Yet, significant questions remain regarding how to optimize spectral quality effects on plant growth, development, mineral nutrition, and metabolism. Specific responses of plants to the LED spectrum may sometimes be predictable based on published research, and the overall plant reaction is generally difficult to foresee due to the complicated interaction of many different internal responses (Hogewoning et al. 2010). Interactions among species, light intensity, duration, and other environmental parameters hamper our ability to make broad photobiological conclusions for many whole plant physiological responses. Therefore, the field of LED lighting research seems inexhaustible. Moreover, it encourages growers to take over the role of researchers and perform small-scale R&D activities seeking to test and optimize LED lighting parameters for certain plant varieties and specific cultivation technologies.

## References


Agarwal A, Dutta Gupta S (2016) Impact of light-emitting-diodes (LEDs) and its potential on plant growth and development in controlled-environment plant production system. Curr Biotechnol 5:28–43

Bagdonavičienė A, Brazaitytė A, Viršilė A, Samuolienė G, Jankauskienė J, Sirtautas R, Sakalauskienė S, Miliauskienė J, Maročkienė N, Duchovskis P (2015) Cultivation of sweet pepper (*Capsicum annum* L.) transplants under high pressure sodium lamps supplemented by light emitting diodes of carious wavelengths. Acta Sci Pol Hortorum Cultus 14:3–14

Barta DJ, Tibbits TW, Bula RJ, Morrow RC (1992) Evaluation of light emitting diode characteristics for a space-based plant irradiation source. Adv Space Res 12:141–149

Bergstrand KJ, Asp H, Schüssler HK (2016) Growth control of ornamental and bedding plants by manipulation of photoperiod and light quality. Acta Hortic 1134:33–39

Bian ZH, Cheng RF, Yang QC, Wang J (2016) Continuous light from red, blue, and green light-emitting diodes reduces nitrate content and enhances phytochemical concentrations and antioxidant capacity in lettuce. J Amer Soc Hort Sci 141:186–195

Bliznikas Z, Žukauskas A, Samuolienė G, Viršilė A, Brazaitytė A, Jankauskienė J, Duchovskis P, Novičkovas A (2012) Effect of supplementary pre-harvest LED lighting on the antioxidant and nutritional properties of green vegetables. Acta Hortic 939:85–91

Bourget CM (2008) An introduction to light-emitting diodes. HortScience 43:1944–1946

Brazaitytė A, Ulinskaitė R, Duchovskis P, Samuolienė G, Šikšnianienė JB, Šabajevienė G, Baranauskis K, Stanienė G, Tamulaitis G, Bliznikas Z, Žukauskas A (2006) Optimization of lighting spectrum for photosynthetic system and productivity of lettuce by using light-emitting diodes. Acta Hortic 711:183–188

Brazaitytė A, Duchovskis P, Urbonavičiūtė A, Samuolienė G, Jankauskienė J, Kasiulevičiūtė-Bonakerė A, Bliznikas Z, Novičkovas A, Breivė K, Žukauskas A (2009) The effect of light-emitting diodes lighting on cucumber transplants and after-effect on yield. Zemdirbyste-Agriculture 96:102–118

Brazaitytė A, Viršilė A, Samuolienė G, Jankauskienė J, Sakalauskienė S, Sirtautas R, Novičkovas A, Dabašinskas L, Vaštakatiė V, Miliauskienė J, Duchovskis P (2016) Light quality: growth and nutritional value of microgreens under indoor and greenhouse conditions. Acta Hortic 1134:277–284

Bugbee B (2016) Towards an optimal spectral quality for plant growth and development: the importance of radiation capture. Acta Hortic 1134:1–12

Bula RJ, Morrow RC, Tibbits TW, Barta DJ, Ignatius RW, Martin TS (1991) Light emitting diodes as a radiation source for plants. HortScience 26:203–205

Carvalho SD, Folta KM (2014) Environmentally modified organisms—expanding genetic potential with light. Crit Rev in Plant Sci 33:486–508

Carvalho SD, Schwieterman ML, Abrahan CE, Colquhoun TA, Folta KM (2016) Light quality dependent changes in morphology, antioxidant capacity, and volatile production in Sweet Basil (*Ocimum basilicum*). Front Plant Sci 7:1328. doi:10.3389/fpls.2016.01328

Chang CL, Chang KP (2014) The growth response of leaf lettuce at different stages to multiple wavelength-band light-emitting diode lighting. Sci Hortic 179:78–84

Chen XL, Xu XZ, Guo WZ, Wang LC, Qiao XJ (2016) Growth and nutritional properties of lettuce affected by mixed irradiation of white and supplemental light provided by light-emitting diode. Sci Hort 200:111–118

Chia PL, Kubota C (2010) End-of-day far-red light quality and dose requirements for tomato rootstock hypocotyl elongation. HortScience 45:1501–1506

Colonna E, Rouphael Y, Barbieri G et al. (2016) Nutritional quality of ten leafy vegetables harvested at two light intensities. Food Chem 199:702–710

Cope KR, Bugbee B (2013) Spectral effects of three types of white light-emitting diodes on plant growth and development: absolute versus relative amounts of blue light. HortScience 48:504–509

Craig DS, Runkle ES (2012) Using LEDs to quantify the effect of the red to far-red ratio of night-interruption lighting on flowering of photoperiodic crops. Acta Hortic 956:179–186

Craig DS, Runkle ES (2013) A moderate to high red to far-red light ratio from light emitting diodes controls flowering of short-day plants. J Am Soc Hort Sci 138:167–172

Currey CJ, Lopez RG (2013) Cuttings of impatiens, pelargonium, and petunia propagated under light-emitting diodes and high-pressure sodium lamps have comparable growth, morphology, gas exchange, and post-transplant performance. HortScience 48:428–434

Demotes-Mainard S, Peron T, Corot A (2016) Plant responses to red and far red lights, applications in horticulture. Eviron Exp Bot 121:4–21

Deram P, Lefsrud MG, Orsat V (2014) Supplemental lighting orientation and red-to blue ratio of light-emitting diodes for greenhouse tomato production. HortScience 49:448–452

Dzakovich MP, Gomez C, Mitchell CA (2015) Tomatoes grown with light-emitting diodes or high-pressure sodium supplemental lights have similar fruit-quality attributes. HortScience 50:1498–1502

Folta KM, Carvalho SD (2015) Photoreceptors and control of horticultural plant traits. HortScience 50:1274–1280

Folta K, Maruhnich SA (2007) Green light: a signal to slow down or stop. J Exp Bot 58:3099–3111

Gerovac JR, Craver JK, Boldt JK, Lopez RG (2016) Light intensity and quality from sole-source light-emitting diodes impact growth, morphology, and nutrient content of brassica microgreens. HortScience 51:497–503

Goins GD, Yorio NC, Sanwo MM, Brown CS (1997) Photomorphogenesis, photosynthesis and seed yield of wheat plants grown under red light-emitting diodes (LEDs) with and without supplemental blue lighting. J Exp Bot 48:1407–1413

Goins GD, Ruffe LM, Cranston NA, Yorio NC, Wheeler RM, Sager JC (2001) Salad crop production under different wavelengths of red light-emitting diodes (LEDs). SAE technical paper. In: 31st international conference on environmental Systems, 9–12 July 2001, Orlando, FL, USA, pp 1–9

Gómez C, Mitchell CA (2015) Growth responses of tomato seedlings to different spectra of supplemental lighting. HortScience 50:112–118

Gómez C, Mitchell CA (2016) Physiological and productivity responses of high-wire tomato as affected by supplemental light source and distribution within the canopy. JASHS 141:196–208

Gómez C, Morrow RC, Bourget CM, Massa GD, Mitchell CA (2013) Comparison of intracanopy light-emitting diode towers and overhead high-pressure sodium lamps for supplemental lighting of greenhouse-grown tomatoes. HortTechnology 23:93–98

Goto E, Hayashi K, Furuyama S, Hikosaka S, Ishigami Y (2016) Effect of UV light on phytochemical accumulation and expression of anthocyanin biosynthesis genes in red leaf lettuce. Acta Hortic 1134:179–185

Guo X, Hao X, Khosla S, Kumar KGS, Cao R, Bennett N (2016) Effect of LED interlighting combined with overhead HPS light on fruit yield and quality of year-round sweet pepper in commercial greenhouse. Acta Hortic 1134:71–78

Hao XM, Zheng JM, Little C, Khosla S (2012) LED inter-lighting in year-round greenhouse mini-cucumber production. Acta Hortic 956:335–340

Hernandez R, Kubota C (2016) Physiological responses of cucumber seedlings under different blue and red photon flux ratios using LEDs. Environ Exp Bot 121:66–74

Hernández R, Kubota C (2012) Tomato seedling growth and morphological responses to supplemental LED lighting red:blue ratios under varied daily light integrals. Acta Hortic 956:187–194

Hernández R, Kubota C (2014a) Growth and morphological response of cucumber seedlings to supplemental red and blue photon flux ratios under varied solar daily light integrals. Sci Hort 173:92–99

Hernández R, Kubota C (2014b) LEDs supplemental lighting for vegetable transplant production: spectral evaluation and comparisons with HID technology. Acta Hortic 1037:829–835

Hernández R, Eguchi T, Kubota C (2016) Growth and morphology of vegetable seedlings under different blue and red photon flux ratios using light-emitting diodes as sole-source lighting. Acta Hortic 1134:195–200

Ho CH, Yang CM, Hsiao CL (2012) Effects of nighttime lighting with specific wavebands on flowering of chrysanthemum crop. Environ Bioinform 9:265–277

Hogewoning SW, Trouwborst G, Maljaars H, Poorter H, van Ieperen W, Harbinson J (2010) Blue light dose–responses of leaf photosynthesis, morphology, and chemical composition of cucumis sativus grown under different combinations of red and blue light. J Exp Bot 61:3107–3117

Islam HR, Gislerød HR, Torre S, Olsen JE (2015) Control of shoot elongation and hormone physiology in poinsettia by light quality provided by light emitting diodes—a mini review. Acta Hortic 1104:131–136

Johansen NS, Väninen I, Pinto DM, Nissinen AI, Shipp L (2011) In the light of new greenhouse technologies: 2. Direct effects of artificial lighting arthropods and integrated pest management in greenhouse crops. An Appl Biol 159:1–27

Johkan M, Shoji K, Goto F, Hashida S, Yoshihara T (2010) Blue light-emitting diode light irradiation of seedlings improves seedling quality and growth after transplanting in red leaf lettuce. HortScience 45:1809–1814

Johkan M, Shoji K, Goto F, Hashida S, Yoshihara T (2012) Effect of green light wavelength and intensity on photomorphogenesis and photosynthesis in *Lactuca sativa*. Environ Exp Bot 75:128–133

Jokinen K, Särkkä LE, Näkkilä J (2012) Improving sweet pepper productivity by LED inter lighting. Acta Hortic 956:59–66

Kanechi M, Maekawa A, Nishida Y, Miyashita E (2016) Effects of pulsed lighting based light-emitting diodes on the growth and photosynthesis of lettuce leaves. Acta Hortic 1134:207–214

Kim HH, Goins GD, Wheeler RM, Sager JC (2004) Green-light supplementation for enhanced lettuce growth under red- and blue-light-emitting diodes. HortScience 39:1617–1622

Kim K, Kook H, Jang J, Lee WH, Kamala-Kannan S, Chae JC, Lee KJ (2013) The effect of blue-light-emitting diodes on antioxidant properties and resistance to *Botrytis cinerea* in tomato. J Plant Pathol Microb 4:203–207

Kopsell DA, Sams CE (2015) Blue wavelengths from LED lighting increase nutritionally important metabolites in specialty crops. HortScience 50:1285–1288

Kopsell DA, Sams CE, Barickman TC (2014) Sprouting broccoli accumulate higher concentrations of nutritionally important metabolites under narrow-band light emitting diode lighting. HortScience 139:469–477

Kozai T (2015) Plant factory, an indoor vertical farming system for efficient quality food production. Academic Press, USA

Kubota C, Chia P, Yang Li Q (2012) Applications of far-red light emitting diodes in plant production under controlled environments. Acta Hortic 952:59–66

Kumar KGS, Hao X, Khosla S, Guo X, Bennett N (2016) Comparison of HPS lighting and hybrid lighting with top HPS and intra-canopy LED lighting for high-wire mini-cucumber production. Acta Hortic 1134:111–117

Lee MJ, Park SY, Oh MM (2015) Growth and cell division of lettuce plants under various ratios of red to far-red light-emitting didoes. Hortic Environ Biote 56:188–194

Lee MJ, Son KH, Oh MM (2016) Increase in biomass and bioactive compounds in lettuce under various ratios of red to far-red LED light supplemented with blue LED light. Hortic Environ Biote 57:139–147

Lefsrud MG, Kopsell DA, Sams CE (2008) Irradiance from distinct wavelength light-emitting diodes affect secondary metabolites in kale. HortScience 43:2243–2244

Li Q, Kubota C (2009) Effects of supplemental light quality on growth and phytochemicals of baby leaf lettuce. Environ Exp Bot 67:59–64

Li H, Tang C, Xu Z, Liu X, Han X (2012) Effects of different light sources on the growth of non-heading Chinese cabbage (*Brassica campestris* L.). J Agr Sci 4:262–273

Liao Y, Suzuki K, Yu W, Zhuang D, Takai Y, Ogasawara R, Shimazu T, Fukui H (2014) Night break effect of LED light with different wavelengths on floral bud differentiation of *Chrysanthemum morifolium* Ramat 'Jimba' and 'Iwa no hakusen'. Environ Control Biol 52:45–50

Lin KH, Huang MY, Huang WD, Hsu MH, Yang ZW, Yang CM (2013) The effects of red, blue, and white light-emitting diodes on the growth, development, and edible quality of hydroponically grown lettuce (*Lactuca sativa* L. var. *capitata*). Sci Hort 150:86–91

Liu XY, Chang TT, Guo SR, Xu ZG, Li J (2011) Effect of different light quality of LED on growth and photosynthetic character in cherry tomato seedling. Acta Hortic 907:325–330

Lu N, Maruo T, Johkan M, Hohjo M, Tsukakoshi S, Ito Y, Ichimura T, Shinoara Y (2012) Effects of supplemental lighting with light-emitting diodes (LEDs) on tomato yield and quality of single-truss tomato plants grown at high planting density. Environ Control Biol 50:63–74

Lurie-Luke E (2014) Product and technology innovation: what can biomimicry inspire? Biotechnol Adv 32:1494–1505

Massa GD, Kim HH, Wheeler RM, Mitchell CA (2008) Plant productivity in response to LED lighting. HortScience 43:1951–1956

Massa G, Graham T, Haire T, Flemming C II, Newsham G, Wheeler R (2015) Light-emitting diode light transmission through leaf tissue of seven different crops. HortScience 50:501–506

Matsuda R, Ohashi-Kaneko K, Fujiwara K, Kurata K (2007) Analysis of the relationship between blue-light photon flux density and the photosynthetic properties of spinach (*Spinacia olearacea* L.) leaves with regard to the acclimation of photosynthesis to growth irradiance. Soil Sci Plant Nutr 53:459–465

McCree KJ (1971) The action spectrum, absorptance and quantum yield of photosynthesis in crop plants. Agric Meteorol 9:191–216

Meng Q, Runkle ES (2015) Low intensity blue light in night-interruption lighting does not influence flowering of herbaceous ornamentals. Sci Hortic 186:230–238

Meng Q, Runkle ES (2016) Moderate-intensity blue radiation can regulate flowering, but not extension growth, of several photoperiodic ornamental crops. Environ Exp Bot. doi:10.1016/j.envexpbot.2016.10.006

Mitchell CA, Both AJ, Bourget CM, Burr JF, Kubota C, Lopez RG, Morrow RC, Runkle ES (2012) LEDs: the future of greenhouse lighting. Chron Horticult 52:6–10

Mitchell CA, Dzakovich MP, Gomez C, Lopez R, Burr JF, Hernandez R, Kubota C, Currey CJ, Meng Q, Runkle ES, Bourget CM, Morrow RC, Both AJ (2015) Light-emitting diodes in horticulture. In: Janick J (ed) Horticultural reviews, vol 43. Wiley, Hoboken

Mizuno T, Amaki W, Watanabe H (2011) Effects of monochromatic light irradiation by LED on the growth and anthocyanin contents in leaves of cabbage seedlings. Acta Hortic 907:179–184

Morrow RC (2008) LED lighting in horticulture. HortScience 43:1947–1950

Mou B (2012) Nutritional quality of lettuce. Curr Nutr Food Sci 8(3):177–187

Nanya K, Ishigami Y, Hikosaka S, Goto E (2012) Effects of blue and red light on stem elongation and flowering of tomato seedlings. Acta Hortic 956:261–266

Naznin MT, Lefsrud M, Gravel V, Hao X (2016) Different ratios of red and blue LED light effects on coriander productivity and antioxidant properties. Acta Hortic 1134:223–229

Nelson JA, Bugbee B (2014) Economic analysis of greenhouse lighting: light emitting diodes vs. high intensity discharge fixtures. PLoS ONE 9(6):e99010. doi:10.1371/journal.pone.0099010

Nicole CCS, Charalambous F, Martinakos S, van de Voort S, Li Z, Verhoog M, Krijn M (2016) Lettuce growth and quality optimization in a plant factory. Acta Hortic 1134:231–238

Novičkovas A, Brazaitytė A, Duchovskis P, Jankauskienė J, Samuolienė G, Viršilė A, Sirtautas R, Bliznikas Z, Žukauskas A (2012) Solid-state lamps (LEDs) for the short-wavelength supplementary lighting in greenhouses: experimental results with cucumber. Acta Hortic 927:723–730

Ohashi-Kaneko K, Takase M, Kon N, Fujiwara K, Kurata K (2007) Effect of light quality on growth and vegetable quality in leaf lettuce, spinach and komatsuna. Environ Control Biol 45:189–198

Olle M (2015) Methods to avoid calcium deficiency on greenhouse grown leafy crops. Lap Lambert Academic Publishing, Germany

Olle M, Viršilė A (2013) The effects of light-emitting diode lighting on greenhouse plant growth and quality. Agr Food Sci 22:223–234

Olschowski S, Geiger EM, Herrmann JV, Sander G, Grüneberg H (2016) Effects of red, blue, and white LED irradiation on root and shoot development of *Calibrachoa* cuttings in comparison to high pressure sodium lamps. Acta Hortic 1134:245–250

Ouzounis T, Razi Parjikolaei B, Fretté X, Rosenqvist E, Ottosen CO (2015a) Predawn and high intensity application of supplemental blue light decreases the quantum yield of PSII and enhances the amount of phenolic acids, flavonoids, and pigments in *Lactuca sativa*. Front Plant Sci 6:19. doi:10.3389/fpls.2015.00019

Ouzounis T, Rosenqvist E, Ottosen K (2015b) Spectral effects of artificial light on plant physiology and secondary metabolism: a review. HortScience 50:1128–1135

Ouzounis T, Heuvelink E, Ji Y, Schouten HJ, Visser RGF, Marcelis LFM (2016) Blue and red LED lighting effects on plant biomass, stomatal conductance, and metabolite content in nine tomato genotypes. Acta Hortic 1134:251–258

Owen WG, Lopez R (2015) End-of-production supplemental lighting with red and blue light-emitting diodes (LEDs) influences red pigmentation of four lettuce varieties. HortScience 50:676–684

Park Y, Runkle ES (2016) Investigation the merit of including far-red light radiation in the production of ornamental seedlings grown under sole-source lighting. Acta Hortic 1134:259–265

Park GY, Muneer S, Soundararajan P, Manivnnan A, Jeong BR (2016) Light quality during night interruption affects morphogenesis and flowering in *Petunia hybrida*, a qualitative long-day plant. Hortic Environ Biote 57:371–377

Patisson PM, Tsao JY, Krames MR (2016) Light emitting diode technology status and directions: opportunities for horticultural lighting. Acta Hortic 1134:413–425

Pinho P, Halonen L (2014) Agricultural and horticultural lighting. In: Karileck R, Sun CC, Zissis G, Ma R (eds) Handbook of advanced lighting technology. Springer International Publishing, Switzerland

Pinho P, Lukkala R, Särkka L, Teri E, Tahvonen R, Halonen L (2007) Evaluation of lettuce growth under multi-spectral-component supplemental solid state lighting in greenhouse environment. IREE 2:854–680

Pinho P, Hytönen T, Flantanen M, Elomaa P, Halonen L (2013) Dynamic control of supplemental lighting intensity in a greenhouse environment. Lighting Res Technol 45:295–304

Pinho P, Jokinen K, Halonen L (2016) The influence of the LED light spectrum on the growth and nutrient uptake of hydroponically grown lettuce. Lighting Res Technol. doi:10.1177/1477153516642269

Pocock T (2015) Light-emitting diodes and the modulation of specialty crops: light sensing and signaling networks in plants. HortScience 50:1281–1284

Randall WC, Lopez RG (2014) Comparison of supplemental lighting from high pressure sodium lamps and light emitting-diodes during bedding plant seedling production. HortScience 49:589–595

Samuolienė G, Urbonavičiūtė A, Duchovskis P, Bliznikas Z, Vitta P, Žukauskas A (2009) Decrease in nitrate concentration in leafy vegetables under a solid-state illuminator. HortScience 44:1857–1860

Samuolienė G, Brazaitytė A, Duchovskis P, Viršilė A, Jankauskienė J, Sirtautas R, Novičkovas A, Skalauskienė S, Sakalauskaitė J (2012a) Cultivation of vegetable transplants using solid-state lamps for the short-wavelength supplementary lighting in greenhouses. Acta Hortic 952:885–892

Samuolienė G, Brazaitytė A, Sirtautas R, Novičkovas A, Duchovskis P (2012b) The effect of supplementary LED lighting on the antioxidant and nutritional properties of lettuce. Acta Hortic 952:835–841

Samuolienė G, Sirtautas R, Brazaitytė A, Viršilė A, Duchovskis P (2012c) Supplementary red-LED lighting and the changes in phytochemical content of two baby leaf lettuce varieties during three seasons. J Food Agric Environ 10:7001–7706
Samuolienė G, Sirtautas R, Brazaitytė A, Duchovskis P (2012d) LED lighting and seasonality effects antioxidant properties of baby leaf lettuce. Food Chem 134:1494–1499
Samuolienė G, Brazaitytė A, Sirtautas R, Viršilė A, Sakalauskaitė J, Sakalauskienė S, Duchovskis P (2013) LED illumination affects bioactive compounds in romaine baby leaf lettuce. J Sci Food Agric 93:3286–3291
Samuolienė G, Brazaitytė A, ViršilėA Jankauskienė J, Sakalauskienė S, Duchovskis P (2016) Red light-dose or wavelength-dependent photoresponse of antioxidants in herb microgreens. PLoS ONE. doi:10.1371/journal.pone.0163405
Schuerger AC, Brown CS (1997) Spectral quality affects disease development of three pathogens on hydroponically grown plants. HortScience 32:96–100
Schwend T, Beck M, Prucker D, Peisl S, Memper H (2016) Test of a PAR sensor-based, dynamic regulation of LED lighting in greenhouse cultivation of Helianthus annuus. Eur J Hortic Sci 81:152–156
Singh D, Basu C, Meinhardt-Wollweber M, Roth B (2015) LEDs for energy efficient greenhouse lighting. Renew Sust Energ Rev 49:139–147
Snowden MC, Cope KR, Bugbee B (2016) Sensitivity of seven diverse species to blue and green light: interactions with photon flux. PLoS ONE 11(10):e0163121. doi:10.1371/journal.pone.0163121
Son KH, Oh MM (2013) Leaf shape, growth, and antioxidant phenolic compounds of two lettuce cultivars grown under various combinations of blue and red light-emitting diodes. HortScience 48:988–995
Son KH, Oh MM (2015) Growth, photosynthetic and antioxidant parameters of two lettuce cultivars as affected by red, green, and blue light-emitting diodes. Hortic Environ Biote 56:639–653
Stutte GW, Edney S, Skerritt T (2009) Photoregulation of bioprotectant content of red leaf lettuce with light-emitting diodes. HortScience 44:79–82
Tarakanov I, Yakovleva O, Konovalova I, Paliutina G, Anisimov A (2012) Light-emitting diodes: on the way to combinatorial lighting technologies for basic research and crop production. Acta Hortic 956:171–178
Taulavuori K, Hyöky V, Oksanen L, Taulavuori E, Julkunen-Tiitto R (2016) Species-specific differences in synthesis of flavonoids and phenolic acids under increasing periods of enhanced blue light. Environ Exp Bot 121:145–150
Trouwborst G, Oosterkamp J, Hogewoning SW, Harbinson J, van Ieperen W (2010) The responses of light interception, photosynthesis and fruit yield of cucumber to LED-lighting within the canopy. Physiol Plantarum 138:289–300
van Ieperen W (2016) Plant growth control by light spectrum: fact or fiction? Acta Hortic 1134:19–24
van Ieperen W, Savvides A, Fanourakis D (2012) Red and blue light effects during growth on hydraulic and stomatal conductance in leaves of young cucumber plants. Acta Hortic 956:223–230
Vänninen I, Pinto DM, Nissinen AI, Johansen NS, Shipp L (2010) In the light of new greenhouse technologies: 1. Plant-mediates effects of artificial lighting on arthropods and tritrophic interactions. An Appl Biol 157:393–414
Vaštakaitė V, Viršilė A, Brazaitytė A, Samuoliene G, Jankauskiene J, Sirtautas R, Novičkovas A, Dabašinskas L, Sakalauskienė S, Miliauskienė J, Duchovskis (2015) The effect of blue light dosage on growth and antioxidant properties of microgreens. Sodininkystė ir daržininkystė 34 (1–2):25–35
Wallace C, Both AJ (2016) Evaluating operating characteristics of light sources for horticultural applications. Acta Hortic 1134:435–443
Wanlai Z, Wenke L, Qichang Y (2013) Reducing nitrate content in lettuce by pre-harvest continuous light delivered by red and blue light emitting diodes. J Plant Nutr 36:491–490

Wargent JJ (2016) UV LEDs in horticulture: from biology to application. Acta Hortic 1134:25–32

Wojciechowska R, Długosz-Grochowska O, Kołton A, Župnik M (2015) Effects of LED supplemental lighting on yield and some quality parameters of lamb's lettuce grown in two winter cycles. Sci Hortic 187:80–86

Wojciechowska R, Kołton A, Długosz-Grochowska O, Knop E (2016) Nitrate content in *Valerianella locusta* L. plants is affected by supplemental LED lighting. Sci Hortic 211:179–186

Wollaeger HM, Runkle ES (2013) Growth responses of ornamental annual seedlings under different wavelengths of red light provided by light-emitting diodes. HortScience 48:1478–1483

Xin J, Liu H, Song S, Chen R, Sun G (2015) Growth and quality of Chinese kale grown under different LEDs. Agric Sci Technol 16:68–69

Yorio NC, Goins GD, Kagie HR, Wheeler RM, Sager JC (2001) Improving spinach, radish and lettuce growth under red light emitting didoes (LEDs) with blue light supplementation. HortScience 36:380–383

Žukauskas A, Bliznikas Z, Breivė K, Novičkovas A, Samuolienė G, Urbonavičiūtė A, Brazaitytė A, Jankauskienė J, Duchovskis P (2011) Effect of supplementary pre-harvest LED lighting on the antioxidant properties of lettuce cultivars. Acta Hortic 907:87–90

# Chapter 8
# Light-Emitting Diodes (LEDs) for Improved Nutritional Quality

**Giedrė Samuolienė, Aušra Brazaitytė and Viktorija Vaštakaitė**

## 8.1 Introduction

Plants are an important source of nutrition for humans. Among various environmental factors, such as temperature, moisture and fertilisers, light is one of the key factors in plant production. Light and photosynthesis, which is the main vital process in plants, are intimately connected. However, only 4.6–6.0% of the total incident solar radiation energy is utilised for plant photosynthesis (Long et al. 2006). It is now understood that light quality (the spectrum) and quantity [photosynthetically active photon flux density (PPFD)] regulate phytochemical composition and content, which affects nutritional and/or postharvest quality in many plant species, especially leafy vegetables (Li and Kubota 2009; Stutte et al. 2009; Johkan et al. 2010; Costa et al. 2013; Kopsell and Sams 2013; Samuolienė et al. 2013a, b; Braidot et al. 2014; Chen et al. 2014; Kopsell et al. 2014).

Plants contain a wide variety of highly sensitive photoreceptors that perceive even minor changes in light quality and, accordingly, modulate the photosynthetic or photomorphogenetic responses. A series of studies have reported the specific wavelength-mediated responses of particular receptors and described their physiological functions through light-sensing systems. UV-B light is perceived using the UVR8 photoreceptor, which is linked to a specific molecular signalling pathway and leads to UV-B acclimation (Tilbrook et al. 2013). UV-A and blue light photoreceptors are known as cryptochromes, which modulate a number of biophysical and biochemical changes, resulting in conformational changes to propagate light

G. Samuolienė (✉) · A. Brazaitytė · V. Vaštakaitė
Lithuanian Research Centre for Agriculture and Forestry, Institute of Horticulture, Kaunas str. 30, 54333 Babtai, Kaunas distr., Lithuania
e-mail: g.samuoliene@lsdi.lt



S. Dutta Gupta (ed.), *Light Emitting Diodes for Agriculture*,
DOI 10.1007/978-981-10-5807-3_8

signals (Lin 2002; Liu et al. 2011; Wenke and Qichang 2012). According to Folta and Maruhnich (2007), green light-mediated responses affect plant processes via cryptochrome-dependent and cryptochrome-independent means. Generally, the effects of green light oppose those of red- and blue-directed wavebands and oftentimes are mediated by cryptochrome/phytochrome light-sensing systems (Wang and Folta 2013). Blue and red lights absorbed by chlorophyll are responsible for photosynthesis and metabolism of primary metabolites. Red and far-red lights and its ratio are detected by phytochrome, in which the conversion from the inactive to active form is of great importance for developmental and biochemical processes occurring in plants (Carvalho et al. 2011b).

Lighting systems, such as high-intensity discharge lighting (high-pressure sodium (HPS), metal halide and xenon lamps as well as fluorescent and incandescent lamps) are characterised by broad spectral power distribution, with limited control over the emissions of UV or infrared radiation (Morrow 2008; Mitchell et al. 2012; for more details, see Chap. 1). Such lighting is usually used as artificial lighting in greenhouses, growth rooms or plant growth chambers. However, only supplemental assimilation lighting is considered to be the most effective for plant welfare. Several studies have reported improved quality of horticultural crops in various aspects. For example, higher sugar and ascorbic acid concentrations in tomato were found under supplemental light (Dorais and Gosselin 2002). Higher light intensity especially that used in northern regions, improved fruit set, average fruit weight and yield of sweet pepper (Heuvelink et al. 2006). On the other hand, self-shading conditions that decrease light and inner zones of canopies can receive 4 times less light (Li and Yang 2015). All of these requirements can be combined and solved using light-emitting diode (LED) lamps, which have unique advantages over existing horticultural lighting. LED lamps have the ability to control spectral composition and light intensity and offer the opportunity to select the most favourable light spectra for photosynthetic and photomorphogenetic responses (Morrow 2008; Samuolienė et al. 2012b; Koga et al. 2013; Brazaitytė et al. 2015a). Moreover, vertical distribution of LED light in the canopy can be used (Heuvelink et al. 2006). In recent years, closed artificial lighting growth chambers or indoor plant factory technologies have become popular across the globe, as it is believed that these can improve the economy while solving issues of plant quality and productivity as well as agricultural sustainability in densely populated areas. However, there is an urgent need to determine the techno-economical feasibilities of the production systems as well as to develop innovative and energy-saving strategies for the operation of these systems. LEDs are an integral part of such systems due to their economic efficiency. The most important findings of past decades concerning LED illumination, which involve changes in internal quality attributes of greenhouse vegetables subjected to light quality and quantity, are discussed in this chapter.

## 8.2 Phenolic Compounds

Nowadays, there is much interest in phenolic compounds as naturally occurring secondary metabolites in horticultural plants. Phenolic components are being found as complex mixtures in all vegetables and fruits, but the quantities of compounds vary among parts of plants. Due to the ability to scavenge free radicals, flavonoids and phenolic acids are important contributing factors to antioxidant activity. The activity of antioxidants is determined as the reactivity to being hydrogen- or electron-donating agents as well as the ability to stabilise the unpaired electron and interact with other antioxidants (Rice-Evans et al. 1997). Phenolic compounds are categorised into simple phenols, flavonoids, phenolic acids (hydroxybenzoic acids and hydroxycinnamic acids), lignins and tannins by the number of constitutive carbon atoms in conjugation with the structure of the basic phenolic skeleton. These substances are responsible for colour, edible flavour, odour and antioxidant properties (Khanam et al. 2012).

The flavonoids constitute a large group of compounds, occurring as glycones, although the most common forms are glycoside derivatives in plants (Agati et al. 2012; Khanam et al. 2012). Flavonoids are mostly located in the wall and the vacuole of epidermal cells; in external surface organs (such as trichomes); in the chloroplast envelope; and in the leaf interior, both in the palisade and spongy mesophyll cells, and their distribution depends on sunlight irradiance to which the plants are subjected (Agati et al. 2013). The flavonols are the most ubiquitous flavonoids in food, as they are present in almost all vegetables. Khanam et al. (2012) reported that isoquercetin (quercetin-3-glucoside) and rutin (quercetin-3-rutinoside) are the most common flavonols found in leafy vegetables. Quercetin is found in collard greens, mustard, kale, okra, sweet potato greens, purple hull peas, and purslane. Other flavonols, such as kaempferol, are present in broccoli. The flavones are much less common than other flavonoids in the human diet. The main sources of flavones are herbs, such as parsley and celery. They exist mainly as glycosides of luteolin and apigenin. The isoflavones are present in several legumes, but soybeans have been identified as the principal dietary source for humans (Vauzour et al. 2012).

Several studies revealed that flavonoids have antioxidant functions in higher plants that are challenged with a range of environmental stresses (Havaux and Kloppstech 2001; Lillo et al. 2008; Agati et al. 2012). Stress conditions inactivate antioxidant enzymes while up-regulating the biosynthesis of flavonols. Conversely, an increase in the antioxidant enzyme activity upon UV-B radiation is negatively correlated with flavonol production. The biosynthesis of antioxidative flavonoids is enhanced by excess light (excitation energy) caused by the interaction of high light intensity with other environmental conditions. The excess light is stressful to plants and can reduce the activity of antioxidants in chloroplasts while up-regulating the biosynthesis of flavonoids, even in the absence of UV irradiance (Agati et al. 2012). Flavonols may protect plants more from long-term visible light-induced oxidative damage in comparison with xanthophylls (Havaux and Kloppstech 2001). The biosynthesis of quercetin glycosides and kaempferol increases under low or high light intensity (Lillo et al. 2008) and does not depend on solar wavelength proportions in common spectra (Agati et al. 2012). In addition, quercetin-3-O- and luteolin-7-O-glycosides accumulate in response to UV-B (Agati et al. 2011). The quercetin derivatives may protect chloroplasts from the visible light-induced generation of $^1O_2$ (Agati et al. 2007).

The anthocyanins (anthocyanin glycosides) are water-soluble flavonoids, which are found in some vegetables, fruits or berries, and give them a pink, red, blue or purple colour (Rice-Evans et al. 1996; Heim et al. 2002; Balasundram et al. 2006; Vauzour et al. 2012). There are six anthocyanidins, which occur in plants the most frequently: pelargonidin, cyanidin, peonidin, delphinidin, petunidin and malvidin. The anthocyanins can be graded as antioxidants, as well as other phenolics, due to ability to donate hydrogen to highly reactive radicals and prevent further radical formation (Rice-Evans et al. 1996; Heim et al. 2002; Balasundram et al. 2006). Most of the researches on anthocyanins have focussed on their photoinduction by wavelengths in the UV, visible and far-red regions. The anthocyanin synthesis can be triggered by the dark, and high level of UV-B radiation, probably via DNA damage. Although photoinduction of anthocyanins has been demonstrated in both the laboratory and the field, the actual photoreceptors responsible have not been clearly identified. In early research, Lindoo and Caldwell (1978) theorised that far-red and UV-B radiation (and therefore their photoreceptors) acted independently on anthocyanin formation. Mohr et al. (1984) stated that the specific induction wavelength of anthocyanins varies among species. The other studies suggested that anthocyanin synthesis is induced by the UV-B photoreceptor, or some combination of it with phytochrome and cryptochrome. However, the anthocyanin induction can be development stage dependent and affected by environmental conditions such as temperature (Chalker-Scott 1999).

The phenolic acids, as well as flavonoids, are widely available in the plant kingdom and are produced from phenylalanine and tyrosine via the shikimic acid pathway. These substances constitute about one-third of dietary phenols, which may be presented in free and bound forms. Khanam et al. (2012) revealed that hydroxybenzoic acids are the most abundant compounds in leafy vegetables. Salicylic acid is the most prominent individual hydroxybenzoic acid in komatsuna, followed by pak choi and red amaranth. Vanillic acid is the second-most abundant hydroxybenzoic acid, and the highest concentrations in red and green amaranths were determined. Syringic acid and gallic acid are also common in leafy vegetables, with the highest amounts in komatsuna and green amaranth. Ellagic acid was detected only in salad spinach. The most common hydroxycinnamic acids in leafy vegetables are *p*-coumaric acid, ferulic acid and *m*-coumaric acid. Caffeic acid and chlorogenic acid also have been observed in leafy vegetables, with the highest concentrations in green amaranth. High amounts of sinapic acid were determined in mizuna, pak choi and komatsuna (Khanam et al. 2012). Caffeic acid, chicoric acid and chlorogenic acid are the main phenolic compounds in lettuce (Romani et al. 2002). Hydroxycinnamic acid derivatives have a molar extinction coefficient in the 290–320-nm spectral region and are much more effective than flavonoids, which have a molar extinction coefficient beyond 350 nm, in absorbing the shortest solar wavelengths (Agati et al. 2012, 2013). It was shown that hydroxycinnamic acids accumulate in shade-adapted plants but are absent in leaves exposed to full sunlight (Tattini et al. 2000; Agati et al. 2002). The existence of flavonoids, hydroxybenzoic acid and hydroxycinnamic acid in daily diets may enhance cellular antioxidant defences and prolong healthy life (Carvalho et al. 2011a).

A measure of total antioxidant capacity helps to understand the functional properties of vegetables and fruits (Shalaby and Shanab 2013). Several radical scavenging capacity assays are widely used for rapid screening and evaluation of novel antioxidant preparations using 2,2-diphenyl-1-picrylhydrazyl (DPPH) radicals. The stable DPPH free radical, which has an unpaired valence electron at one atom of the nitrogen bridge, is still highly utilised in hydrophilic and lipophilic antioxidant research due to its simple reaction systems, which involve only the direct reaction between the radical and the antioxidant, and has no other interference, such as enzyme inhibition or presence of multiple radicals, although they are not physiologically relevant (Cheng et al. 2006; Sharma and Bhat 2009).

The protective effects of flavonoids in biological systems are ascribed to their capacity to transfer electron free radicals, chelate metal catalysts, activate antioxidant enzymes, reduce $\alpha$-tocopherol radicals and inhibit oxidases (Rice-Evans et al. 1996; Heim et al. 2002; Balasundram et al. 2006). It is known that high intake of flavonoids helps to prevent heart diseases, cancer and neurodegradation (Vazour et al. 2012).

Among artificial lighting systems, LEDs present the maximum photosynthetically active radiation efficiency (80–100%; Darko et al. 2014) which is used for formation of various metabolic pathways, such as those involved in the synthesis of phenolic compounds. Lee et al. (2016) revealed that the total phenolic contents (TP) decreased with an increasing far-red PPFD in a ratio with red light in red-leaf lettuce. It has also been reported that the effect of red light alone on TP is species-dependent (Lee et al. 2014).

Red light led to increased TP in common buckwheat but, in contrast, decreased TP in tartary buckwheat sprouts. Deep red light alone also acted positively in the accumulation of TP in basil but negatively on parsley microgreens (Samuolienė et al. 2016), and it had no effect on synthesis in Chinese kale sprouts (Qian et al. 2016) nor in *Brassica* microgreens (Brazaitytė et al. 2016a). The optimal deep red ratio with blue light of 90 and 10% was determined from growing Lamb's lettuce during a 2-year study (Wojciechowska et al. 2015). Red alone, as well as deep red, affected the synthesis of TP in various green vegetables or sprouts. Samuolienė et al. (2011b) demonstrated that red light stimulated TP synthesis in lentil, wheat and radish seedlings in comparison with darkness. Depending upon plant species, red light increased (Brazaitytė et al. 2016b), decreased (Brazaitytė et al. 2016a; Samuolienė et al. 2016) or had no effect (Brazaitytė et al. 2016a) on TP content in microgreens. There is no doubt that red light can affect the synthesis of TP content in combination with other light sources. Samuolienė et al. (2016) revealed that red LEDs in combination with other (blue, red and far-red) or in combination with HPS lights had positive effects on TP in basil microgreens during 3 days of treatment. In addition, a similar trend in other microgreen species (Samuolienė et al. 2012a;

Brazaitytė et al. 2016b), romaine baby leaf lettuce (Samuolienė et al. 2012b), red-leaf and light-green-leaf lettuce (Samuolienė et al. 2011b) or green-leaf lettuce (Žukauskas et al. 2011) was determined. Other reports stated that light in orange–yellow regions is also involved in accumulation of TP. During orange or yellow treatment, TP increased in romaine baby leaf lettuce (Samuolienė et al. 2013b), mustard and beet microgreens (Brazaitytė et al. 2016b), leafy radish sprouts (Urbonavičiūtė et al. 2009a, c; Samuolienė et al. 2011b), wheat leaves (Urbonavičiūtė et al. 2009b), tomato fruit and apple peels (Kokalj et al. 2016). Some of these results are similar to treatments with green LEDs. However, green light that is supplemental to HPS light had negative effects on TP in romaine baby leaf lettuce (Samuolienė et al. 2013b). Blue light alone increased TP in Chinese kale sprouts up to 69% in comparison with darkness (Qian et al. 2016). It can be assumed that photoinduction of blue light receptors (cryptochromes) is directly linked to production of TP. Furthermore, an increased blue ratio in combination with deep red led to higher accumulation of TP in lettuce in comparison with deep red alone or small amounts of blue ratio treatments (Son and Oh 2013). Due to common photoreceptors with blue light, UV-A also plays an important role in polyphenol synthesis in parsley (Brazaitytė et al. 2016b) and basil (Vaštakaitė et al. 2015a) microgreens, romaine baby leaf lettuce (Samuolienė et al. 2013b) and leafy radish (Urbonavičiūtė et al. 2009a) (Table 8.1).

The increased PPFD level of far-red in combination with red light reduced total anthocyanins (TA) in red-leaf baby and other lettuce (Li and Kubota 2009; Stutte et al. 2009). However, phytochrome-dependent synthesis of TA (Li and Kubota 2009) varies among various horticultural plant species. Brazaitytė et al. (2016a) reported that deep red lighting alone increased the amount of TA in tatsoi microgreens, but, in the same growing conditions, deep red light alone decreased the amount in mustard and had no effect on red pak choi microgreens. In addition, deep red light alone increased TA in other green vegetables, such as Chinese kale sprouts and red-leaf cabbage (Mizuno et al. 2011; Qian et al. 2016). A review of several studies revealed that red light combined with HPS lamps had more positive effects on anthocyanin synthesis in various green vegetable tissues in comparison with only red light (Samuolienė et al. 2011a, 2012a; Brazaitytė et al. 2013, 2016a). An optimal PPFD level of deep red and red in combination with far-red and blue light for stimulation of anthocyanin synthesis in *Brassica* microgreens at 300 μmol $m^{-2}$ $s^{-1}$ was determined (Samuolienė et al. 2013a). The effects of green light on TA in green vegetables depend on green light wavelengths and plant species. A greater increase of TA was achieved by combining different green light with other wavelengths of LEDs or HPS lamps. Green light in combination with HPS led to significantly increased contents of TA in romaine baby leaf (Samuolienė et al. 2013b) and red-leaf lettuce (Samuolienė et al. 2012b). Moreover, there is evidence that the synthesis of anthocyanins depends not only on plant variety but also on seasonality. Samuolienė et al. (2012b) reported that the highest amount of TA under supplemental green light to HPS was in red-leaf lettuce in November. In contrast, the lowest amount of TA under the same light and seasonal conditions was measured in green-leaf lettuce in comparison with HPS lighting without green. Several

**Table 8.1** The summarised effects of LED light spectral composition on plant metabolites

| LED light treatment | Plant | Increase | Decrease | No impact | References |
|---|---|---|---|---|---|
| 850 nm with W (449, 548 nm); PPF 135 µmol $m^{-2}$ $s^{-1}$; control—W; PP—16 h | Lettuce (*Lactuca sativa* var. *crispa*, 'Green Oak Leaf') | AA | TC | | Chen et al. (2016) |
| 735 nm with 440, 660 nm; PPFD 0.7, 1.2, 4.1, 8.6 µmol $m^{-2}$ $s^{-1}$, B:R 2:8; control—FL or 440, 660 nm LEDs (2:8); PPFD 130 ± 4 µmol $m^{-2}$ $s^{-1}$; PP—12 h | Lettuce (*Lactuca sativa* L., 'Sunmang') | TP | TP | | Lee et al. (2016) |
| 734 nm with WF (PPFD 160.4 µmol $m^{-2}$ $s^{-1}$); total PPFD 300 µmol $m^{-2}$ $s^{-1}$; control—WF; PP—16 h | Lettuce (*L. sativa* L., 'Red Cross') | | TA, XA, BC | | Li and Kubota (2009) |
| 730 nm with 640 nm; total PPFD ~320 µmol $m^{-2}$ $s^{-1}$; control—FL; PP—18 h | Lettuce (*L. sativa* L., 'Outredgeous') | | | TA | Stutte et al. (2009) |
| R, PPFD 4.75 µmol $m^{-2}$ $s^{-1}$ (unified into 100 bulbs); control—B, PPFD 12.41 µmol $m^{-2}$ $s^{-1}$; R and B, PPFD 9.19 µmol $m^{-2}$ $s^{-1}$ (unified into 100 bulbs) | Common buckwheat (*Fagopyrum esculentum*, 'Kitawase') | TP | | | Lee et al. (2014) |
| | Tartary buckwheat (*F. tataricum*, 'Hokkai T8') | | | TP | |
| R with B in ratio 9:1, PPFD 600 µmol $m^{-2}$ $s^{-1}$; control-WF; PP—12, 16, 20, 24 h | Lettuce (*L. sativa* L., 'Dasusheng') | AA | | | Shen et al. (2014) |

(continued)

**Table 8.1** (continued)

| LED light treatment | Plant | Increase | Decrease | No impact | References |
|---|---|---|---|---|---|
| 665 nm; PPFD 300 μmol $m^{-2}$ $s^{-1}$ (during 3-days treatment); control—447, 638, 665, 731 nm; PPFD 300 μmol $m^{-2}$ $s^{-1}$; PP—16 h | Basil (*Ocimum basilicum* L., 'Sweet Genovese') | TP, DPPH | BC, LU, AA | | Samuolienė et al. (2016) |
| | Parsley (*Petroselinum crispum*) | DPPH, BC, AA | | | |
| | Mustard (*Brassica juncea* L., 'Red Lion') | | TA, LU | TP, C | Brazaitytė et al. (2016b) |
| | Red pak choi (*Brassica rapa* var. *chinensis*, 'Rubi F1'), | BC, LU | | TP, TA | |
| | Tatsoi (*Brassica rapa* var. *rosularis*) | TA, LU | BC | TP, DPPH | |
| 665 nm with 447, 638, 731 nm; PPFD 204.9 μmol $m^{-2}$ $s^{-1}$ (during 3-days treatment); total PPFD 300 μmol $m^{-2}$ $s^{-1}$; control—447, 638, 665, 731 nm; PP—16 h | Basil (*O. basilicum* L., 'Sweet Genovese') | DPPH, AA | LU | BC | Samuolienė et al. (2016) |
| | Parsley (*P. crispum*) | DPPH, AA | TP, LU | | |
| 660 nm; PPFD 30 μmol $m^{-2}$ $s^{-1}$; control—darkness; PP—16 h | Chinese kale (*Brassica oleracea* var. *alboglabra* Bailey, 'DFZC') | TA | | TP | Qian et al. (2016) |
| 660 nm; PPFD 54.2 μmol $m^{-2}$ $s^{-1}$; control—natural light | Barley (*Hordeum vulgare* L., 'Nichinohoshi') | γ-T | AA | T | Koga et al. (2013) |
| 660 nm, PPFD 50 μmol $m^{-2}$ $s^{-1}$; comparing with FL, 470, 500, 525 nm; PP—16 h | Cabbage (*Brassica oleracea* var. *capitata* L., 'Red Rookie', 'Kinshun') | TA | | | Mizuno et al. (2011) |

(continued)

**Table 8.1** (continued)

| LED light treatment | Plant | Increase | Decrease | No impact | References |
|---|---|---|---|---|---|
| 660 nm, PPFD 80 μmol $m^{-2}$ $s^{-1}$; control—FL; PP—12 h | Non-heading Chinese cabbage (*Brassica campestris* L., cultivar 605) | AA | TC | | Li et al. (2012) |
| 660, 460 nm (8:1); PPFD 80 μmol $m^{-2}$ $s^{-1}$; control—FL; PP—12 h | | AA | | TC | |
| 660 nm; PPFD 50 μmol $m^{-2}$ $s^{-1}$; control—W LEDs; PP—16 h | Tartary buckwheat (*F. tataricum* Gaertn., 'Hokkai T8') | ZEA | TC, BC, LU | | Tuan et al. (2013) |
| 660 ± 22 nm, PPFD 100 μmol $m^{-2}$ $s^{-1}$; control—FL; PP—16 h | Red-leaf lettuce (*L. sativa* L. cv. 'Banchu Red Fire') | | TP, TA, TC at 17 DAS | TP, TA, TC at 45 DAS | Johkan et al. (2010) |
| 655 ± 2, 467 ± 21 nm, 100 μmol $m^{-2}$ $s^{-1}$; control—FL; PP—16 h | | TP, TA, TC at 17 DAS | | TP, TA, TC at 45 DAS | |
| 660 nm with W (449, 548 nm); PPF 135 μmol $m^{-2}s^{-1}$; control—W; PP—16 h | Lettuce (*L. sativa* var. *crispa* 'Green Oak Leaf') | TC, AA | | | Chen et al. (2016) |
| 660, 440 nm in ratios 100, 90:10, 70:30, 50:50; PPFD 200 μmol $m^{-2}$ $s^{-1}$; control—HPS; PP—16 h | Lamb's lettuce (*Valerianella locusta* L., 'Nordhollandse') | PA (90:10, 50:50—winter, autumn), PA (70:30—autumn)<br>Flavonoid glycosides (70:30, 50:50—autumn)<br>Free flavonoid (90:10 B—winter) | | | Długosz-Grochowska et al. (2016) |
| 660, 430 nm in ratios 90:10, 70:30, 50:50 with HPS; total PPFD 200 μmol $m^{-2}$ $s^{-1}$; control—HPS, 660, 430 nm in ratio 1:1; PP—16 h | Lamb's lettuce (*V. locusta* L., 'Nordhollandse') | TP (90:10 in 1st year), TP (90:10, 50:50 in 2nd year), DDPH | DDPH (HPS, 100% 660 nm in 2nd year) | DPPH (90:10, 70:30, 50:50 in 2nd year) | Wojciechowska et al. (2015) |
| 660, 466 nm (5:2); PPFD 140, 200, 285 μmol $m^{-2}$ $s^{-1}$; control—natural light or negative; PP—14 and 20 h | Tomato (*Solanum lycopersicum* Mill.) | AA | | | Verkerke et al. (2015) |

(continued)

**Table 8.1** (continued)

| LED light treatment | Plant | Increase | Decrease | No impact | References |
|---|---|---|---|---|---|
| 660, 445 nm (9:1); PPFD 52.3 μmol $m^{-2}$ $s^{-1}$; control—natural light | Barley (*H. vulgare* L., 'Nichinohoshi') | | T, AA | | Koga et al. (2013) |
| 660, 454 nm; PFD 210 μmol $m^{-2}$ $s^{-1}$; control—FL; PP—16 h | Boston lettuce (*L. sativa* L. var. *capitata*) | | | TC | Lin et al. (2013) |
| 660, 454 nm, W (500–600 nm); PFD 210 μmol $m^{-2}$ $s^{-1}$; control—FL; PP—16 h | | | | TC | |
| 658 ± 12 nm, PFD 150 μmol $m^{-2}$ $s^{-1}$; control—dysprosium lamp (CK); PP—12 h | Non-heading Chinese cabbage (*B. campestris* L. 'Te'aiqing') | | TC | | Fan et al. (2013) |
| 658 ± 12, 460 ± 11 (6:1), PFD 150 μmol $m^{-2}$ $s^{-1}$; control—dysprosium lamp (CK); PP—12 h | | | | TC | |
| 658 nm with WF (PPFD 177 μmol $m^{-2}$ $s^{-1}$); total PPFD 307.4 μmol $m^{-2}$ $s^{-1}$; control—WF; PP—16 h | Lettuce (*L. sativa* L., 'Red Cross') | TP | | XA, BC | Li and Kubota (2009) |
| 650 nm, 1200 lx; control—W 380–760 nm; PP—16 h | Non-heading red cabbage (*Brassica oleracea*) seedlings | | TC | | Matioc-Precup and Cachiţă-Cosma (2013) |
| 650, 470 nm (7:1), PPFD 170 μmol $m^{-2}$ $s^{-1}$ or 391 μmol $m^{-2}$ $s^{-1}$; control—HPS; PP—24 h | Chinese cabbage (*Brassica chinensis* L., 'Vesnyanka') | | | TC, AA | Avercheva et al. (2014) |
| 640, 430 nm (7:3); PPFD 137 μmol $m^{-2}$ $s^{-1}$; control—darkness; PP—16 h | Tartary buckwheat (*Fagopyrum tataricum* Gaerth., 'Hokkai T10') | Rutin, chlorogenic acid | | Quarcetin | Seo et al. (2015) |

(continued)

**Table 8.1** (continued)

| LED light treatment | Plant | Increase | Decrease | No impact | References |
|---|---|---|---|---|---|
| 640 nm (86%), 460 nm (14%); PFD 200 µmol $m^{-2}$ $s^{-1}$; control—FL | Lettuce (*L. sativa*, L. cv. 'Grand Rapids') | | | TC | Urbonavičiūtė et al. (2007) |
| 638–667 nm; PPFD 0, 12.5, 25, 50, 100 µmol $m^{-2}$ $s^{-1}$; PP—18 h | Alfalfa (*Medicago sativa* L.), red radish (*Raphanus sativus* L. var. *sativus*) | | TP | | Kwack et al. (2015) |
| 638 nm; PPFD 300 µmol $m^{-2}$ $s^{-1}$ (during 3-days treatment); control—B 447 nm, R 638 nm, R 665 nm and Fr 731 nm, PP—16 h | Basil (*O. basilicum* L., 'Sweet Genovese') | TP, DPPH, α-T | BC, LU | | Samuolienė et al. (2016) |
| | Parsley (*P. crispum*) | BC, α-T, AA | TP, LU | DPPH | |
| | Mustard (*B. juncea* L., 'Red Lion') | DPPH, C | TA, LU | TP | Brazaitytė et al. (2016b) |
| | Red pak choi (*B. rapa* var. *chinensis*, 'Rubi F1'), | BC, LU | TP | TA | |
| | Tatsoi (*B. rapa* var. *rosularis*) | TP, TA | | DPPH | |
| 638 nm, PPFD 200 µmol $m^{-2}$ $s^{-1}$; control—HPS or 445, 638, 669, 731 nm; PP—18 h | Wheat (*Triticum aestivum* L.) | AA | TP, DPPH | | Urbonavičiūtė et al. (2009c) |
| 638 nm with 447, 665, 731 nm, PPFD 171.3 µmol $m^{-2}$ $s^{-1}$ (during 3-days treatment); total PPFD 300 µmol $m^{-2}$ $s^{-1}$; control—447, 638, 665, 731 nm; PP—16 h | Basil (*O. basilicum* L., 'Sweet Genovese') | TP, DPPH, α-T, AA | | | Samuolienė et al. (2016) |
| | Parsley (*P. crispum*) | DPPH | TP, α-T, AA | | |

(continued)

**Table 8.1** (continued)

| LED light treatment | Plant | Increase | Decrease | No impact | References |
|---|---|---|---|---|---|
| 638 nm with HPS, PPFD 210 μmol $m^{-2}$ $s^{-1}$, total PPFD 300 μmol $m^{-2}$ $s^{-1}$ (3 days before harvest); control—HPS, PP—16 h (18 h for perilla) | Basil (*O. basilicum* L., 'Sweet Genovese'), | TP, α-T | | | Samuolienė et al. (2016) |
| | Parsley (*P. crispum*) | TP | | | |
| | Various microgreens | TP (except mustard, beet); TA; DPPH (except beet) | | | Brazaitytė et al. (2016a) |
| | Lettuce (*L. sativa* L., 'Thumper') | T | BC | TP, TA, DPPH, AC | Samuolienė et al. (2013a) |
| | Lettuce (*L. sativa* L., 'Thumper') | TP (in November, March), DPPH (in November) | DPPH (in March) | | Samuolienė et al. (2012b) |
| | Lettuce (*L. sativa* L., 'Multibaby') | TP (in November), DPPH (in November, March) | TP (in March) | | |
| | Various microgreens | TP; TA; DPPH; AA (amaranth, pea, kale, broccoli, mustard) | TP (amaranth); TA (borage, mustard, beet); DPPH (beet); AA (basil, borage) | TA (basil); DPPH (pea, broccoli, amaranth) | Samuolienė et al. (2012c) |
| | Lettuce (*L. sativa* L., 'Multired 4') | TP, AA | TA | | Samuolienė et al. (2011b) |
| | Lettuce (*L. sativa* L., 'Multigreen 3') | TA | | AA | |
| | Lettuce (*L. sativa* L., 'Multiblond 2') | TP | | AA | |
| | Perilla (*Perilla frutescens* var. *crispa*) | TA | | DPPH | Brazaitytė et al. (2013) |

(continued)

**Table 8.1** (continued)

| LED light treatment | Plant | Increase | Decrease | No impact | References |
|---|---|---|---|---|---|
| 638 nm with HPS; PPFD 170 μmol $m^{-2}$ $s^{-1}$, total PPFD 300 μmol $m^{-2}$ $s^{-1}$ (3 days before harvest); control—HPS, PP—From 5 till 10 am and from 5 till 12 pm | Various leafy vegetables | AA | | DDPH, TP, α-T (spinach) | Bliznikas et al. (2012) |
| | Lettuce (*L. sativa* L., 'Lolo Bionda') | DPPH, AC | | | Žukauskas et al. (2011) |
| | Lettuce (*L. sativa* L., 'Grand Rapids') | TP, DPPH, AC | | | |
| | Lettuce (*L. sativa* L., 'Lollo Rossa') | TP, DPPH | C | | |
| 638 nm; PPFD 100 μmol $m^{-2}$ $s^{-1}$; control—darkness; PP—12 h | Wheat (*Triticum aestivum* L.), radish (*Raphanus sativus* L.), lentil (*Lens esculentum* Moench.) | TP | | | Samuolienė et al. (2011a) |
| 638 nm; PPFD 200 μmol $m^{-2}$ $s^{-1}$; control—HPS | Radish (*R. sativa* L., 'Tamina') | DPPH | TPC | | Urbonavičiūtė et al. (2009b) |
| 638, 660, 455, 735 nm PPFD 545, 440, 330, 220, 110 μmol $m^{-2}$ $s^{-1}$; control—638, 660, 455, 735 nm, PPFD 220 μmol $m^{-2}$ $s^{-1}$; PP—16 h | Kolhrabi (*Brassica oleracea* var. *gongylodes*, 'Delicacy Purple') | TP, DPPH (545–440 μmol); TA (440 μmol) | TP, DPPH (110 μmol) | | Samuolienė et al. (2013b) |
| | Mustard (*B. juncea* L., 'Red Lion') | DPPH (545–440 μmol); α-T (545 μmol) | TP, DPPH (110 μmol); TA (220 μmol) | TP | |
| | Red pak choi (*B. rapa* var. *chinensis*), 'Rubi F1') | DPPH, AA (545–440 μmol); TP (440–330 μmol); TA (330 μmol) | DPPH, TP (110 μmol); α-T (545 μmol) | | |
| | Tatsoi (*B. rapa* var. *rosularis*) | DPPH, TP (545–440 μmol); TA (330 μmol); AA | DPPH, TP (110 μmol); α-T (545 μmol) | | |
| | Borage (*Borago officinalis*) | TP (440–545 μmol); DPPH (330–545 μmol) | TP (110–330 μmol); DPPH (110–220 μmol) | | Viršilė and Sirtautas (2013) |

(continued)

**Table 8.1** (continued)

| LED light treatment | Plant | Increase | Decrease | No impact | References |
|---|---|---|---|---|---|
| 638 nm with 669, 455, 731 nm PPFD 154 μmol $m^{-2}$ $s^{-1}$; total PPFD 200 μmol $m^{-2}$ $s^{-1}$; control—HPS | Radish (*R. sativa* L., 'Tamina') | TP, DPPH | | | Urbonavičiūtė et al. (2009b) |
| 625–630 nm; 128 lx; 96 h continuous radiation; control—dark | Pea (*Pisum sativum* L.) | BC | | | Wu et al. (2007) |
| 630 nm; PPF 133 μmol $m^{-2}$ $s^{-1}$; control—FL; PP—14 h | Lettuce (*L. sativa* var. *crispa*, 'Green Oak Leaf') | | | TC, AA | Chen et al. (2014) |
| 630 nm with FL; PPF 133 μmol $m^{-2}$ $s^{-1}$; control—FL; PP—14 h | | TC | | AA | |
| 630 nm with B 460 (1:1), PPF 133 μmol $m^{-2}$ $s^{-1}$; control—FL; PP—14 h | | | | TC, AA | |
| 627, 447 nm (95%/5%); PFD 250 μmol $m^{-2}$ $s^{-1}$; control—fluorescent/incandescent light; PP—16 h | Broccoli (*Brassica oleracea* var. *italica*) | TC, BC, LU | VIO | ZEA, NEO, ANT | Kopsell et al. (2014) |
| 627, 447 nm (80%/20%); PFD 250 μmol $m^{-2}$ $s^{-1}$; control—fluorescent/incandescent light; PP—16 h | | TC, BC, LU | NEO, VIO | ZEA, ANT | |
| 627, 447 nm (90%/10%, 80%/20%, 60%/40%); PFD 250 μmol $m^{-2}$ $s^{-1}$; control—fluorescent/incandescent light; PP—16 h | Chinese kale(*Brassica oleracea* var. *alboglabra* 'Green Lance') | TC, BC, LU, ZEA, NEO, ANT, VIO | | | Kopsell et al. (2016) |
| 622 nm with 455, 638, 670, 735 nm; PPFD ~175 μmol $m^{-2}$ $s^{-1}$; control—447, 638, 670, 735 nm; PP—16 h | Lettuce (*L. sativa* L., 'Thumper') | TP | | DPPH, AC, BC | Samuolienė et al. (2013a) |
| 622 nm with 447, 638, 665, 731 nm; PPFD 300 μmol $m^{-2}$ $s^{-1}$; control—447, 638, 665, 731 nm; PP—16 h | Various microgreens | TP (mustard, beet); TC, AC, BC, LU, NEO (mustard); VIO (tatsoi) | VIO (mustard); TC, AC, BC, LU, NEO, VIO (red pak choi, tatsoi) | | Brazaitytė et al. (2015a, b, c, 2016a) |

(continued)

**Table 8.1** (continued)

| LED light treatment | Plant | Increase | Decrease | No impact | References |
|---|---|---|---|---|---|
| 596 nm with W (449, 548 nm); PPF 135 μmol $m^{-2}$ $s^{-1}$; control—W; PP—16 h | Lettuce (*L. sativa* var. *crispa* 'Green Oak Leaf') | | TC | AA | Chen et al. (2016) |
| Flashing 596 nm with HPS; PPFD 70 μmol $m^{-2}$ $s^{-1}$; control—HPS; PP—18 h | Wheatgrass (*Triticum aestivum* L., 'Širvinta') | AA | | TP, DPPH, AA | Urbonavičiūtė et al. (2009a) |
| | Barley grass (*Hordeum vulgare* L., 'Luokė') | | | TP, DPPH | |
| | Radish (*R. sativus* L., 'Tamina') | TP, DPPH | AA | | |
| 595 nm with 455, 638, 670, 735 nm; PPFD 175 μmol $m^{-2}$ $s^{-1}$; control—455, 638, 670, 735 nm; PP—16 h | Lettuce (*L. sativa* L., 'Thumper') | | | DPPH, AC, BC | Samuolienė et al. (2013a) |
| 595 nm with 455, 638, 669, 735 nm; PPFD 100 μmol $m^{-2}$ $s^{-1}$; control—darkness; PP—12 h | Wheat (*T. aestivum* L.), radish (*R. sativus* L.), lentil (*L. esculentum* Moench.) | DPPH, α-T (wheat); TP (radish) | | | Samuolienė et al. (2011a) |
| 595 nm with 447, 638, 665, 731 nm; PPFD 300 μmol $m^{-2}$ $s^{-1}$; control—447, 638, 665, 731 nm; PP—16 h | Various microgreens | TP (mustard, beet, tatsoi, parsley); TC, AC, BC, LU, NEO (mustard); TC, VIO (tatsoi) | VIO (mustard); TC, BC, LU, NEO (red pak choi, tatsoi) | AC, VIO (red pak choi) | Brazaitytė et al. (2015a, b, c, 2016a) |
| 595 nm with 455, 638, 669, 731 nm; total PPFD 200 μmol $m^{-2}$ $s^{-1}$; control—HPS | Radish (*R. sativa* L., 'Tamina') | TP, DPPH | | | Urbonavičiūtė et al. (2009b) |
| Amber (Y) 595 nm with 455, 638, 669, 731 nm; total PPFD 200 μmol $m^{-2}$ $s^{-1}$; control—HPS; PP—18 h | Wheat (*Triticum aestivum* L., 'Širvinta 1', 'Ada', 'Tauras', 'Milda', 'Alma') | TP; DPPH ('Širvinta 1'); AA ('Širvinta 1', 'Ada') | | | Urbonavičiūtė et al. (2009c) |

(continued)

**Table 8.1** (continued)

| LED light treatment | Plant | Increase | Decrease | No impact | References |
|---|---|---|---|---|---|
| 590 nm; PFD 150 μmol $m^{-2}$ $s^{-1}$; control—dysprosium lamp (CK); PP—12 h | Non-heading Chinese cabbage (*B. campestris* L. 'Te'aiqing') | | TC | | Fan et al. (2013) |
| 590 nm; 1200 lx; control—W (380–760 nm); PP—16 h | Non-heading red cabbage (*Brassica oleracea*) | | TC | | Matioc-Precup and Cachiţă-Cosma (2013) |
| 590 nm; PPFD 8.3 μmol $m^{-2}$ $s^{-1}$; control—darkness | Apple (*Malus domestica*, 'Granny Smith') | TP, DPPH, AA | α-T, γ-T, δ-T | | Kokalj et al. (2016) |
| | Tomato (*Solanum lycopersicum* L.) | TP, AA | α-T, γ-T, δ-T | | |
| | Red bell pepper (*Capsicum annuum*) | DPPH, α-T, γ-T, AA | | | |
| 535 nm with HPS, PPFD—200 μmol $m^{-2}$ $s^{-1}$ (during three seasons); control—HPS, PP—16 h | Lettuce (*L. sativa* L., 'Multired 4') | TP, DPPH, AA (January); T (November) | DPPH (November) | | Samuolienė et al. (2012a) |
| | Lettuce (*L. sativa* L., 'Multigreen 3') | TP, DPPH, AA (January); T (November) | TA (November) | | |
| | Lettuce (*L. sativa* L., 'Multiblond 2') | T (November) | | | |
| 530 nm with HPS; PPFD 115 μmol $m^{-2}$ $s^{-1}$; control—HPS; PP—16 h | Lettuce (*L. sativa* L., 'Thumper') | TA, DPPH | TP, AC, BC | | Samuolienė et al. (2013a) |
| 530 nm with 640, 440 nm, PPFD 300 μmol $m^{-2}$ $s^{-1}$; control—FL; PP—18 h | Lettuce (*L. sativa* L., 'Outredgeous') | | | TA | Stutte et al. (2009) |

(continued)

**Table 8.1** (continued)

| LED light treatment | Plant | Increase | Decrease | No impact | References |
|---|---|---|---|---|---|
| 530 nm with 627, 447 nm (10%/85% 5%), PFD 250 μmol $m^{-2}$ $s^{-1}$; control—fluorescent/incandescent light; PP—16 h | Broccoli (*B. oleracea* var. *italica*) | LU, BC | NEO | TC, ZEA, VIO, ANT | Kopsell et al. (2014) |
| 530 nm (10%) with 627, 447 nm (10%/70%20%), PFD 250 μmol $m^{-2}$ $s^{-1}$; control—fluorescent/incandescent light; PP—16 h | | BC, LU, TC | NEO, VIO | ZEA, ANT | |
| 526 nm with WF (PPFD 166.9 μmol $m^{-2}$ $s^{-1}$); total PPFD 306 μmol $m^{-2}$ $s^{-1}$; control—WF; PP—16 h | Lettuce (*L. sativa* L., 'Red Cross') | TA | | XA, BC | Li and Kubota (2009) |
| 522 nm with W (449, 548 nm); PPF 135 μmol $m^{-2}$ $s^{-1}$; control—W; PP—16 h | Lettuce (*L. sativa* var. *crispa* 'Green Oak Leaf') | AA | TC | | Chen et al. (2016) |
| 520 nm with 447, 638, 638, 731 nm; PPFD 300 μmol $m^{-2}$ $s^{-1}$; control—447, 638, 638, 731 nm; PP—16 h | Various microgreens | TP (mustard, beet, parsley); TC, AC, BC, LU, NEO (mustard); VIO (in tatsoi); NEO (red pak choi) | VIO (mustard); TC, AC, BC, LU (red pak choi, tatsoi) | | Brazaitytė et al. (2015a, b, c, 2016a) |
| 520 nm; PFD 150 μmol $m^{-2}$ $s^{-1}$; control—dysprosium lamp (CK); PP—12 h | Non-heading Chinese cabbage (*B. campestris* L. 'Te'aiqing') | | TC | | Fan et al. (2013) |
| 510 nm with 455, 638, 670, 735 nm; PPFD ~175 μmol $m^{-2}$ $s^{-1}$; control—455, 638, 670, 735 nm; PP—16 h | Lettuce (*L. sativa* L., 'Thumper') | TA, AC | | DPPH, BC | Samuolienė et al. (2013a) |
| 510 nm with 455, 638, 669, 731 nm, PPFD ~200 μmol $m^{-2}$ $s^{-1}$; control—darkness; PP—12 h | Wheat (*T. aestivum* L.) | TP, DDPH, α-T | | | Samuolienė et al. (2011a) |
| | Radish (*R. sativus* L.), | α-T, AA | DPPH | | |
| | Lentil (*L. esculentum* Moench.) | TP, DDPH, α-T, AA | | | |

(continued)

**Table 8.1** (continued)

| LED light treatment | Plant | Increase | Decrease | No impact | References |
|---|---|---|---|---|---|
| 510 nm with 455, 638, 669, 731 nm, PPFD 200 μmol $m^{-2}$ $s^{-1}$; control—HPS | Radish (*R. sativa* L., 'Tamina') | TP, DPPH | | | Urbonavičiūtė et al. (2009b) |
| 510 nm with 455, 638, 669, 731 nm, PPFD 200 μmol $m^{-2}$ $s^{-1}$; control—HPS; PP—18 h | Wheat (*T. aestivum* L., 'Širvinta 1', 'Ada', 'Tauras', 'Milda', 'Alma') | TP ('Širvinta 1', 'Ada', 'Milda'); DPPH ('Tauras', 'Ada', 'Milda'); AA ('Širvinta 1', 'Ada') | | | Urbonavičiūtė et al. (2009c) |
| 505 nm with HPS; PPFD 200 μmol $m^{-2}$ $s^{-1}$ (during three seasons); control—HPS, PP—16 h | Lettuce (*L. sativa* L., 'Multired 4') | TA (November); T (November) | DPPH (November); TA (January) | | Samuolienė et al. (2012a) |
| | Lettuce (*L. sativa* L., 'Multigreen 3') | T (November) | TA (November) | | |
| | Lettuce (*L. sativa* L., 'Multiblond 2') | T (November) | | | |
| 505 nm with HPS, PPFD ~115 μmol $m^{-2}$ $s^{-1}$; control—HPS; PP—16 h | Lettuce (*L. sativa* L., 'Thumper') | AC | TP | BC | Samuolienė et al. (2013a) |
| 500, 640 nm (10%/90%); PFD 200 μmol $m^{-2}$ $s^{-1}$; control—FL | Lettuce (*L. sativa*, L. cv. 'Grand Rapids') | | | TC | Urbonavičiūtė et al. (2007) |
| B with HPS, PPFD 45 μmol $m^{-2}$ $s^{-1}$ (from 6 to 8 am; from 9 pm to 8 am; from 5 to 7 pm), 80 μmol $m^{-2}$ $s^{-1}$ (from 5 to 7 pm); control—HPS and daylight, PPFD level of 90 ± 10 μmol $m^{-2}$ $s^{-1}$ and 6.1 mol $m^{-2}$ $d^{-1}$; PP—HPS from 9 pm–8 am to 12 pm—5 pm | Lettuce (*L. sativa* L., 'Batavia', 'Lollo Rossa') | PA, chlorogenic acid, caffeic acid ('Lollo Rossa') | | PA ('Batavia') | Ouzounis et al. (2015) |
| 476 nm with FL (PPFD 166.9 μmol $m^{-2}$ $s^{-1}$); total PPFD 306 μmol $m^{-2}$ $s^{-1}$; control—FL; PP—16 h | Lettuce (*L. sativa* L., 'Red Cross') | TA, XA, BC | | | Li and Kubota (2009) |

(continued)

**Table 8.1** (continued)

| LED light treatment | Plant | Increase | Decrease | No impact | References |
|---|---|---|---|---|---|
| 470 nm, PPFD 30 μmol $m^{-2}$ $s^{-1}$; control—darkness; PP—16 h | Chinese kale (*B. oleracea* var. *alboglabra* Bailey, 'DFZC') | TP, TA | | | Qian et al. (2016) |
| 470 nm with HPS; PPFD ~115 μmol $m^{-2}$ $s^{-1}$; control—HPS; PP—16 h | Lettuce (*L. sativa* L., 'Thumper') | | TP, AC, BC | | Samuolienė et al. (2013a) |
| 470 nm with HPS; PPFD 200 μmol $m^{-2}$ $s^{-1}$ (during three seasons); control—HPS; PP—16 h | Lettuce (*L. sativa* L., 'Multired 4') | TA, T (November) | DPPH (November); TA (January) | | Samuolienė et al. (2012a) |
| | Lettuce (*L. sativa* L., 'Multigreen 3') | TP, DPPH (January); T (November) | TA (November) | | |
| | Lettuce (*L. sativa* L., 'Multiblond 2') | T (November) | | | |
| 470 nm; PPFD 41 μmol $m^{-2}$ $s^{-1}$; control—470, 627 nm (12%/88%); PP—24 h | Broccoli (*B. oleracea* var. *italica*) | BC, VIO | | LU, ZEA, NEO, ANT | Kopsell and Sams (2013) |
| 470 nm; PPFD 50 μmol $m^{-2}$ $s^{-1}$; control—WL; PP—16 h | Tartary buckwheat (*F. tataricum* Gaertn., 'Hokkai T8') | ZEA | TC, BC, LU | | Tuan et al. (2013) |
| B 470 nm continuous light; PPFD 50 and 100 μmol $m^{-2}$ $s^{-1}$ PPFD; control—dark | Satsuma mandarin (*Citrus unshiu* Marc.), Valencia orange (*Citrus sinensis* Osbeck)—juice sacs | TC, AC, BC, LU, *β*-cryptoxanthin, all-trans-violaxanthin | | 9-cis-violaxanthin (100 μmol) | Zhang et al. (2015a, b) |

(continued)

**Table 8.1** (continued)

| LED light treatment | Plant | Increase | Decrease | No impact | References |
|---|---|---|---|---|---|
| 468 nm; PPF 100 µmol $m^{-2}$ $s^{-1}$; control—FL; PP—16 h | Lettuce (*L. sativa* L. cv. 'Banchu Red Fire') | TC at 17 DAS | | TC at 45 DAS | Johkan et al. (2010) |
| 465–470 nm; 112.29 lx; 96 h continuous radiation; control—dark | Pea (*P. sativum* L.) | BC | | | Wu et al. (2007) |
| 465 nm; 1200 lx; control—W (380–760 nm); PP—16 h | Non-heading red cabbage (*B. oleracea*) | | TC | | Matioc-Precup and Cachiţă-Cosma, (2013) |
| 460 nm; PPFD 80 µmol $m^{-2}$ $s^{-1}$; control—FL; PP—12 h | Non-heading Chinese cabbage (*B. campestris* L., cultivar 605) | TC, AA | | | Li et al. (2012) |
| 460 nm, PFD 150 µmol $m^{-2}$ $s^{-1}$; control—dysprosium lamp (CK); PP—12 h | Non-heading Chinese cabbage (*B. campestris* L. 'Te'aiqing') | | TC | | Fan et al. (2013) |
| 460 nm; PPF 133 µmol $m^{-2}$ $s^{-1}$; control—FL; PP—14 h | Lettuce (*L. sativa* var. *crispa* 'Green Oak Leaf') | | | TC, AA | Chen et al. (2014) |
| 460 nm with FL, PPF 133 µmol $m^{-2}$ $s^{-1}$; control—FL; PP—14 h | | TC | AA | | |
| 456, 665 nm (0B/100R, 13B/87R, 26B/74R, 35B/65R, 47B/53R, 59B/41R); PPFD 171 µmol $m^{-2}$ $s^{-1}$; control—FL and HPS | Lettuce (*L. sativa* L., 'Sunmang', 'Grand Rapid TBR') | TP due to increased 456 nm ratio | | TP (0B/100R, 13B/87R) | Son and Oh (2013) |
| 455 nm with HPS; PPFD ~115 µmol $m^{-2}$ $s^{-1}$; control—HPS; PP—16 h | Lettuce (*L. sativa* L., 'Thumper') | | TP, AC, BC | | Samuolienė et al. (2013a) |

(continued)

**Table 8.1** (continued)

| LED light treatment | Plant | Increase | Decrease | No impact | References |
|---|---|---|---|---|---|
| 455 nm with HPS; PPFD 200 μmol $m^{-2}$ $s^{-1}$ (during three seasons); control—HPS; PP—16 h | Lettuce (*L. sativa* L., 'Multired 4') | TP, DPPH (January); TA, T (November); | DPPH (November); TA (January) | | Samuolienė et al. (2012a) |
| | Lettuce (*L. sativa* L., 'Multigreen 3') | T (November) | TA (November) | | |
| | Lettuce (*L. sativa* L., 'Multiblond 2') | TP (January); T (November) | | | |
| 450 nm with W (449, 548 nm); PPF 135 μmol $m^{-2}$ $s^{-1}$; control—W; PP—16 h | Lettuce (*L. sativa* var. *crispa* 'Green Oak Leaf') | AA, TC | | | Chen et al. (2016) |
| 447 nm with 638, 665, 731 nm; PPFD 0, 25, 50, 75, 100 μmol $m^{-2}$ $s^{-1}$ (0, 8, 16, 25, 33%, respectively), total PPFD 302.5 μmol $m^{-2}$ $s^{-1}$; results compared to a trial mean; PP—16 h | Red pak choi (*B. rapa* var. *chinensis*, 'Rubi' $F_1$) | TP (25 μmol); TA (0–25 μmol) | DPPH (0 and 100 μmol) | | Vaštakaitė et al. (2015b) |
| | Tatsoi (*B. rapa* var. *rosularis*) | TP (100 μmol); TA (75 μmol); DPPH (0 and 100 μmol) | DPPH (25–75 μmol) | | |
| | Basil (*O. basilicum* L., 'Sweet Genovese') | TP (100 μmol), TA | DPPH (25–75 μmol) | | |
| 440, 640 nm; PPFD ~300 μmol $m^{-2}$ $s^{-1}$; control—FL; PP—18 h | Lettuce (*L. sativa* L., 'Outredgeous') | TA | | | Stutte et al. (2009) |
| 430 nm, PPFD 177 μmol $m^{-2}$ $s^{-1}$; control—darkness; PP—16 h | Tartary buckwheat (*F. tataricum* Gaerth., 'Hokkai T10') | TA | | Quarcetin | Seo et al. (2015) |
| 390 nm with HPS; PPFD ~125 μmol $m^{-2}$ $s^{-1}$; control—HPS; PP—16 h | Various microgreens | TP, DPPH (parsley); TA (mustard, red pak choi, tatsoi, beet) | | | Brazaitytė et al. (2016a) |

(continued)

**Table 8.1** (continued)

| LED light treatment | Plant | Increase | Decrease | No impact | References |
|---|---|---|---|---|---|
| 390 nm with HPS; PPFD ~125 μmol $m^{-2}$ $s^{-1}$; control—HPS; PP—16 h | Basil (*O. basilicum* L., 'Sweet Genovese', 'Dark Opal') | TP ('Sweet Genovese') | TP, TA ('Dark Opal') | TA ('Sweet Genovese') | Vaštakaitė et al. (2015a) |
| 380 nm 455, 638, 670, 735 nm; PPFD ~175 μmol $m^{-2}$ $s^{-1}$; control—455, 638, 670, 735 nm; PP—16 h | Lettuce (*L. sativa* L., 'Thumper') | TP, AC | | DPPH, BC | Samuolienė et al. (2013a) |
| 385 nm with 455, 638, 669, 731 nm; PPFD 200 μmol $m^{-2}$ $s^{-1}$; control—HPS | Radish (*R. sativa* L., 'Tamina') | DPPH | | TP | Urbonavičiūtė et al. (2009b) |
| 385 nm with 455, 638, 669, 731 nm; PPFD 200 μmol $m^{-2}$ $s^{-1}$; control—HPS; PP—18 h | Wheat (*Triticum aestivum* L., 'Širvinta 1', 'Ada', 'Tauras', 'Milda', 'Alma') | AA ('Širvinta 1', 'Ada') | TP, DPPH | | Urbonavičiūtė et al. (2009c) |
| 373 nm with FL (PPFD 20.9 μmol $m^{-2}$ $s^{-1}$); total PPFD 325.5 μmol $m^{-2}$ $s^{-1}$; control—FL; PP—16 h | Lettuce (*L. sativa* L., 'Red Cross') | TA | | XA, BC, AA | Li and Kubota (2009) |
| 366, 390, 402 nm (PPFD 6.2 or 12.4 μmol $m^{-2}$ $s^{-1}$) with 447, 638, 665, 735 nm; total PPFD 300 μmol $m^{-2}$ $s^{-1}$; control—447, 638, 665, 735 nm; PP—16 h | Basil (*O. basilicum* L., 'Sweet Genovese') | DPPH | α-T (6.2 μmol); AA (366 nm, 12.4 μmol) | TA (6.2 μmol) | Brazaitytė et al. (2015a, b, c) |
| | Beet (*Beta vulgaris* L., 'Bulls Blood') | DPPH; TP (402 nm, 12.4 μmol); TA (366 and 390 nm, 12.4 μmol); AA, α-T (366 nm, 12.4 μmol) | α-T (6.2 μmol) | TP, TA (6.2 μmol) | |
| | Red pak choi (*B. rapa* var. *chinensis*, 'Rubi $F_1$') | DPPH; TP (366 nm, 6.2 and 12.4 μmol); TA (366- and 390 nm, 12.4 μmol); AA, α-T (366 nm, 12.4 μmol) | α-T (6.2 μmol) | TA (6.2 μmol) | |

(continued)

**Table 8.1** (continued)

| LED light treatment | Plant | Increase | Decrease | No impact | References |
|---|---|---|---|---|---|
| 365, 640 nm (8%/92%); PFD 200 μmol $m^{-2}$ $s^{-1}$; control—FL | Lettuce (*L. sativa* L. cv. 'Grand Rapids') | | | TC | Urbonavičiūtė et al. (2007) |
| 365 nm; PPFD 32.2 μmol $m^{-2}$ $s^{-1}$; control—natural light; PP—3 h during day or night | Pea (*P. sativum* L., 'Shenchun') | TA (day/night 6/3 h) | TA (night 6 h) | TP, flavonoids, AA | Wenke and Qichang (2012) |
| 365 nm with 680, 460 nm (80R/20B), PPFD 360 μmol $m^{-2}$ $s^{-1}$; control—natural sunlight, PPFD 289 μmol $m^{-2}$ $s^{-1}$; PP—14 h, followed by 2-h illumination with 365 nm | Perilla (*P. Frutescens* var. *purpurea* Makino, 'Akajiso') | PA, caffeic acid, rosmarinic acid, 7-o-glucoside | | | Iwai et al. (2010) |

*A* Amber, *B* blue, *Fr* far-red, *G* green, *R* red, *O* orange, *W* white
*PPFD* Photosynthetic photon flux density
*PP* Photoperiod
*FL* Fluorescent lamps, *WF* white fluorescent lamps, *W* white LEDs, *HPS* high-pressure sodium lamps
*DAS* Days after sowing
*AA* Ascorbic acid, *T* tocopherols, *α-T* α-tocopherol, *γ-T* γ-tocopherol, *TP* total phenolics, *TA* total anthocyanins, *DPPH* DPPH radical scavenging activity, *PA* phenolic acids, *TC* total carotenoids, *LU* lutein, *AC* β-carotene, *BC* β-carotene, *NEO* neoxanthin, *VIO* violaxanthin, *ZEA* zeaxanthin, *ANT* antheraxanthin, *XA* xanthophylls

reports revealed that the biosynthesis of phenolic compounds is linked to induction of the blue light photoreceptor cryptochrome (Iwai et al. 2010). Seo et al. (2015) found that blue light alone increased the content of TA in tartary buckwheat sprouts in comparison with darkness or other light sources. In addition, blue light in combination with high-intensity discharge lamps (fluorescent or HPS) or in combination with other wavelength LEDs increased TA accumulation in baby leaf lettuce (Li and Kubota 2009), red-leaf lettuce (Stutte et al. 2009), Chinese kale sprouts (Qian et al. 2016), tatsoi (Vaštakaitė et al. 2015b) and other various plants. The photoreceptors perceiving blue light react the same to UV-A, and some positive effects on anthocyanin synthesis in various microgreens (Brazaitytė et al. 2015b, 2016b), baby leaf lettuce (Li and Kubota 2009) and pea seedlings (Wenke and Qichang 2012) were also determined (Table 8.1).

The effects of various LED light treatments on individual phenolic compounds were also observed. Deep red light in the amount of 50–70% in combination with blue light increased the concentration of flavonoid glycosides and, in the amount of 90%, free flavonoids in Lamb's lettuce (Długosz-Grochowska et al. 2016). The deep red and blue ratio (7:3) increased the flavonol rutin, but the ratio had no impact on quercetin synthesis in tartary buckwheat sprouts (Seo et al. 2015; Table 8.1).

Several studies revealed that synthesis of phenolic acids (PA) can be initiated by wide spectra of LED wavelengths. The enrichment of far-red light to spectra can increase chlorogenic and caffeic acid levels in lettuce (Lee et al. 2016); however, this phenomenon depends on the far-red ratio in the lighting spectra. Długosz-Grochowska et al. (2016) reported that deep red in combination with blue (ratios 90R/10B, 50R/50B and 70R/30B) during different growing seasons initiated accumulation of PA in Lamb's lettuce tissues. Tartary buckwheat sprouts grown under similar ratios of 70% red and 30% blue showed ability to synthesise chlorogenic acid compared to dark conditions (Seo et al. 2015). It is known that blue light is closely related to metabolic pathways of PA. Ouzounis et al. (2015) demonstrated that blue light alone increased the concentration of chlorogenic acid in red-leaf lettuce, but no significant impact on PA accumulation in green-leaf lettuce was determined. In addition to these comprehensive studies, Iwai et al. (2010) also reported that UV-A led to increased contents of PA, such as caffeic or rosmarinic acids, in red perilla leaves (Table 8.1).

Both positive and negative effects on DPPH radical scavenging activity were determined in studies related to antioxidant properties of various horticultural plants. The antioxidant activity is typically related to secondary metabolites that have antioxidant potential, such as phenols, and strongly depends on lighting conditions, as well. The DPPH radical activity usually correlates with TP, but not in all studies was this observed. Brazaitytė et al. (2016a) demonstrated that deep red decreased DPPH in tatsoi microgreens. As mentioned above, the same decrease of TP in *Brassica* microgreens was measured. In addition, under deep red alone or compared to other LED lights, an increase of DPPH in basil and parsley microgreens was determined (Samuolienė et al. 2016). Samuolienė et al. (2012b) demonstrated that antioxidant activity varied between romaine and curly baby leaf lettuce grown under supplemental short-term red lighting in three growing seasons, mostly following similar trends as TP. A highly increased antioxidant activity at PPFD levels of 440–545 μmol $m^{-2}$ $s^{-1}$ in various species of microgreens was determined, in contrast to 110–220 μmol $m^{-2}$ $s^{-1}$ PPFD (Samuolienė et al. 2013a; Viršilė and Sirtautas 2013; Table 8.1).

## 8.3 Carotenoids

Carotenoids are lipophilic isoprenoid and are broadly presented in fruits and vegetables (Botella-Pavia and Rodriguez-Concepcion 2006; Kopsell and Kopsell 2006; Maiani et al. 2009; Cuttriss et al. 2011; Flores-Perez and Rodriguez-Concepcion 2012). According to various sources in the literature, 500–700 carotenoids have been known in nature. People regularly uses 40 of them (Stahl and Sies 2005; Kopsell and Kopsell 2006). Generally, carotenoids have such functions as free radical scavenging, improving the immune response, repressing cancer development and defending eye tissues. α-Carotene (AC), β-carotene (BC) and β-cryptoxanthin, lycopene are mostly related to decrease of cardiovascular diseases. Zeaxanthin (ZEA) and lutein (LU) protect eyes from light-induced damage. (Botella-Pavia and Rodriguez-Concepcion 2006; Kopsell and Kopsell 2006). In plants, carotenoids are light-harvesting pigments in chloroplasts and protect plants from photo-oxidative damage (Botella-Pavia and Rodriguez-Concepcion 2006; Lefsrud et al. 2007; Cuttriss et al. 2011). Generally, carotenoids are characterised by such functions as free radical scavenging, enhancing the immune response, suppressing cancer development and protecting eye tissues, but individual carotenoids differ in their protective roles. α-Carotene (AC), β-carotene (BC) and β-cryptoxanthin, which are provitamin A carotenoids, are mostly associated with cardiovascular disease reduction. Zeaxanthin (ZEA) and lutein (LU) are components of the macular pigment in the eye and protect the macula from light-induced damage. Lycopene prevents cardiovascular diseases and prostate cancer (Botella-Pavia and Rodriguez-Concepcion 2006; Kopsell and Kopsell 2006). In plants, carotenoids are light-harvesting pigments in chloroplasts and are important in the protection of plants against photo-oxidative damage (Botella-Pavia and Rodriguez-Concepcion 2006; Lefsrud et al. 2007; Cuttriss et al. 2011). Generally, carotenoids protect plants from photo-oxidative damage through thermal dissipation by means of the xanthophyll cycle (converting violaxanthin (VIO) to ZEA) (Stange and Flores 2012). Chlorophyll molecules, in addition to their participation in photosynthesis, are the precursors of tocopherols (Zhang et al. 2015a, b), which are also distinguished by antioxidant properties. The changes of mentioned compounds accumulation depend on the environmental conditions during growth and show different results for different plant species (Kopsell and Kopsell 2006). Since carotenoids are closely related to photosynthesis, the most important factors influencing carotenoid content changes are light quality and quantity. Properly chosen light spectra and intensity, which, with current technology of LEDs, allows the possibility to use specific light wavelengths in the range from ultraviolet to infrared, can lead to higher carotenoid content in vegetables grown in greenhouses and indoors (Tamulaitis et al. 2005; Morrow 2008). Many scientific experiments regarding effects of light spectra on carotenoid content changes associated with blue, red and their combinations as well as supplementation of other lamp spectra using LED illumination of these wavelengths have been performed. It is known that chlorophylls and carotenoids have high light absorption at 400–500 and at 630–680 nm (Lin et al. 2013). In addition, the absorption peaks of various carotenoids differ: LU absorbs at 448 nm, BC at 454 nm and, generally, xanthophylls (XA) at 446 nm (Lefsrud et al. 2008; Li and Kubota 2009). However, some researchers determined two peaks of maximum LU and BC in kale at 440 and 640 nm, which closely conform to the action spectrum previously reported for wheat (Ogawa et al. 1973; Lefsrud et al. 2008). Although the literature data show that carotenoid concentrations increased under blue light, treatments by Lefsrud et al. (2008) showed that red light and its ratio with blue light is important for changes in carotenoid accumulation. The relatively low number of treatments concerning red and blue LED lighting and their ratio showed that impact of these light spectra on carotenoid content was contradictory and depended on plant species. Some authors reported that the concentrations of TC were enhanced in seedlings of non-heading Chinese cabbage and red-leaf lettuce under blue LEDs compared to FL, red–blue and red LED lighting (Johkan et al. 2010; Li et al. 2012). Monochromatic blue light also affected composition of carotenoids. Short-duration exposure to blue light increased BC and VIO levels in broccoli microgreens compared to red–blue LEDs (Kopsell and Sams 2013), increased BC in pea seedlings compared to darkness (Wu et al. 2007) and increased ZEA in sprouts of tartary buckwheat compared to white LEDs. Furthermore, Zhang et al. (2015a, b) reported that blue LED light was effective at inducing accumulation of carotenoids, such as BC, β-cryptoxanthin, all-*trans*-violaxanthin, AC and LU, in the juice sacs of Satsuma mandarin and Valencia orange. However, blue LED treatment did not affect concentrations of LU, ZEA and neoxanthin (NEO) in broccoli microgreen tissues compared to red–blue LEDs (Kopsell and Sams 2013) nor did it affect TC in red- and green-leaf lettuce compared to FL (Johkan et al. 2010; Chen et al. 2014). Other authors reported that monochromatic blue LEDs caused decreases of BC, LU and TC in sprouts of tartary buckwheat and non-heading red cabbage compared to white LEDs (Matioc-Precup and Cachiţă-Cosma 2013; Tuan et al. 2013) as well as a decrease of TC in Chinese cabbage compared to plants grown under a dysprosium lamp (Fan et al. 2013; Table 8.1).

Some experiments regarding effects of monochromatic red LED light on carotenoid content and composition showed different responses by various plants. Wu et al. (2007) showed that leaves of pea seedlings irradiated with red LED light for 96 h presented a significant increase in BC concentration compared to blue and white LED illumination, but no significant difference of BC content was observed in stems of such seedlings. Tuan et al. (2013) reported an increase of ZEA in sprouts of tartary buckwheat under red LEDs, but a decreased content of BC, LU and TC was found. Monochromatic red LED illumination also resulted in lower carotenoid content in non-heading red cabbage, Chinese cabbage and red-leaf lettuce, and no significant differences were observed in TC in leaves of green- and red-leaf lettuce (Johkan et al. 2010; Li et al. 2012; Fan et al. 2013; Matioc-Precup and Cachiţă-Cosma 2013). Short-term red only LED pre-harvest illumination, when plants are not be able to absorb the light of other parts of the spectrum, resulted in an increase of TC in *Brassica* microgreens, with the exceptions of LU in mustard and BC in tatsoi under red light (Brazaitytė et al. 2016a). Such illumination increased BC and decreased LU in parsley microgreens, but these carotenoids decreased in basil microgreens (Samuolienė et al. 2016; Table 8.1).

Literature data show that mixed red–blue LEDs have no effect on TC in lettuce (Urbonavičiūtė et al. 2007; Johkan et al. 2010; Lin et al. 2013; Chen et al. 2014) and Chinese cabbage seedlings (Avercheva et al. 2009; Li et al. 2012; Fan et al. 2013) compared to FL or HPS. However, some authors reported that carotenoid contents in lettuce and Chinese cabbage seedlings were higher compared to monochromatic red and blue LED illumination (Fan et al. 2013; Chen et al. 2014). Kopsell et al. (2014, 2016) reported a slight importance of red and blue light ratios on carotenoid content. Increasing the percentage of blue light among the LED light treatments did not result in higher carotenoid accumulation in kale and broccoli microgreen tissues. However, red–blue LED light treatments resulted in much higher carotenoid concentrations compared to the fluorescent/incandescent light treatment.

Most studies using red and blue LEDs and their combination were performed without supplemental broad-spectrum irradiation (Johkan et al. 2010; Li et al. 2012; Kopsell and Sams 2013; Chen et al. 2014; Kopsell et al. 2014). Although red and blue LEDs have a great influence on photosynthesis, plants in nature are adapted to utilise a wide spectrum of light to control various physiological processes. Several studies involved mixed lights of FL or HPS lamps with LEDs, but only a few were related to effects on carotenoid contents in leafy vegetables. Chen et al. (2014) reported that carotenoid contents of green-leaf lettuce irradiated with FL plus red or blue LEDs were significantly increased compared to FL. Furthermore, they detected more carotenoid contents under blue than under red light. However, among mixed light treatments, carotenoid accumulations appeared to be the highest under FL and red LEDs, followed by FL lamps and blue LEDs and then red–blue LED illumination; the authors also stated that the remaining portion of the spectrum besides blue and red, which are known as the absorption spectra of carotenoids, might also have effects on the induction of pigment enhancement. However, Li and Cubota (2009) reported contradictory results with red-leaf baby lettuce. Carotenoid concentrations (xanthophylls and BC) increased by 6–8% with supplemental blue LED light, but no effects with supplemental red LED light were found. The highest carotenoid content was obtained under a white–red LED illumination treatment and was 49% higher compared to white LEDs; on the other hand, there were no significant differences between white–blue and white–red LEDs (Chen et al. 2016). Lin et al. (2013) stated that white–red–blue LEDs had no effect on carotenoid content in Boston lettuce compared to red–blue LED and FL light. Blue LED light supplemental to HPS lamps resulted in a decrease of AC and BC in romaine baby leaf lettuce (Samuolienė et al. 2013b). Short-term high-PPFD red LED light supplemental to HPS lamps caused a significant decrease in BC content in such lettuce and had no significant effect on α-carotene accumulation; increased AC content in green-leaf lettuce ('Lolo Bionda' and 'Grand Rapids') but reversed effects were found (Žukauskas et al. 2011; Samuolienė et al. 2013b). Furthermore, such treatment led to increased LU and BC in basil microgreens but had no effect on parsley (Samuolienė et al. 2016; Table 8.1).

The majority of overview-type studies examined only a few selected light qualities, and there are some reports examining the effects of different monochromic light, such as green, yellow or orange, which have various functions in driving physiological process despite the fact that chlorophylls and carotenoids have low light absorption at 530–610-nm light. However, plant responses to monochromic light and their interaction with other spectral compositions showed contradictory results (Li and Kubota 2009; Lin et al. 2013; Chen et al. 2016). In the literature, the most examined are green light effects on plants. Green light penetrates into plant canopies better than red and blue light, and green light may promote growth and increased content of bioactive compounds in various plants (Kim et al. 2004; Bouly et al. 2007; Folta and Maruhnich 2007; Johkan et al. 2010; Samuolienė et al. 2013b). However, monochromatic green LED light caused a decrease of TC content in Chinese and red cabbage compared to other light sources (Fan et al. 2013; Matioc-Precup and Cachiţă-Cosma 2013). Green LED light in combination with other light sources, such as white LED and blue–red–far-red LED illumination, decreased TC content in lettuce, red pak choi and tatsoi microgreens but increased it in mustard (Brazaitytė et al. 2015a). Furthermore, green light influences changes in carotenoid composition. Supplemental green light decreased VIO content in mustard but increased contents of other carotenoids. In contrast, this illumination increased VIO content in tatsoi. In red pak choi, supplemental green light caused an increase of xanthophyll-cycle carotenoids, such as VIO and NEO (Brazaitytė et al. 2015a). Monochromatic green light caused some decreases of LU and BC in kale (Lefsrud et al. 2008); in combination with fluorescence lamps, this light had no effect on xanthophylls and BC in red baby leaf lettuce (Li and Kubota 2009), and in combination with HPS lamps and with blue–red–far-red illumination, this light increased BC in green baby leaf lettuce (Samuolienė et al. 2013b; Table 8.1).

Dougher and Bugbee (2001) showed that yellow light from 580- to 600-nm suppressed chlorophyll or chloroplast formation in lettuce, thus inhibiting lettuce growth. According to literature data, monochromatic yellow LED light results in a decrease of TC in Chinese and red cabbage compared to that using a dysprosium lamp and red–blue or natural light (Fan et al. 2013; Matioc-Precup and Cachiţă-Cosma 2013). Chen et al. (2016) reported that carotenoid contents in lettuce were significantly inhibited by yellow light supplemental to white LEDs. Similar results were obtained in red pak choi microgreens under yellow LED light supplemental to standard illumination of blue, red and far-red LEDs (Brazaitytė et al. 2015a). However, such illumination increased TC in mustard and tatsoi microgreens.

Furthermore, yellow light supplemental to blue, red and far-red illumination had an influence on carotenoid composition, with such effects depending on microgreen species. All studied carotenoids increased in mustard, but only VIO increased in tatsoi. However, almost all studied carotenoid contents decreased in red pak choi (Brazaitytė et al. 2015a). Supplemental yellow light had no effect on AC and BC content in romaine lettuce leaves (Samuolienė et al. 2013b). Similar effects on above-mentioned carotenoids were determined in lettuce and *Brassica* microgreens under orange light supplemental to blue, red and far-red illumination (Samuolienė et al. 2013b; Brazaitytė et al. 2015a; Table 8.1).

Far-red light reverses the status of phytochrome and is important with respect to changes in gene expression, plant architecture and reproductive responses (Yeh and Chung 2009). Monochromatic and supplemental far-red LED to FL, to white LEDs and to blue–red–far-red LEDs decreased contents of various carotenoids in kale, baby lettuce, lettuce and mustard microgreens (Lefsrud et al. 2008; Li and Kubota 2009; Brazaitytė et al. 2015a). However, red pak choi and tatsoi microgreens accumulated higher contents of various carotenoids under blue–red–far-red LEDs (Brazaitytė et al. 2015a).

Supplemental UV-A radiation to FL had no effect on xanthophylls and BC in red baby leaf lettuce (Li and Kubota 2009); in combination with blue–red–far-red LED illumination, UV-A increased AC in green baby lettuce (Samuolienė et al. 2013b) as well as LU and BC in *Brassica* microgreens (Brazaitytė et al. 2015c).

Irradiance dosage is not only important in regulating photosynthetic processes, but it also plays an important role in the metabolism of secondary plant compounds, such as carotenoids. However, limited research is available on the impact of irradiance dosage concerning LED illumination in a controlled environment. Applications of 300–400 μmol $m^{-2}$ $s^{-1}$ irradiance levels from LED illumination resulted in notable increases of carotenoid concentrations in *Brassica* microgreens and Chinese cabbage (Avercheva et al. 2014; Brazaitytė et al. 2015a). However, some data concerning irradiance level treatments using FL and incandescent lamps showed that the carotenoid concentration increased from 125 to 300 μmol $m^{-2}$ $s^{-1}$ in kale, spinach and mustard, but above these irradiance levels, the carotenoid contents started to decrease (Lefsrud et al. 2006; Kopsell et al. 2012). Increased carotenoid accumulation is very important for reducing stress caused by high irradiance, but such irradiance also could decrease carotenoid concentrations due to photodegradation of the pigment molecules under such irradiance (Demmig-Adams et al. 1996; Lefsrud et al. 2006; Lee et al. 2007a, b; Table 8.1).

An overview of data demonstrated effective manipulation of carotenoid contents in plants by various spectra and irradiance levels, but effects were species-dependent. However, only a few studies examined changes in individual carotenoids, such as AC, BC, LU and ZEA, in different light conditions. These carotenoids are important for human health, and investigations into increasing their content using manipulation of lighting could be carried out more in the future.

## 8.4 Tocopherols

Tocopherols are a mixture of four (α-, β-, γ- and δ-) homologues. These lipophilic antioxidants play an important role in efficient scavenging of singlet oxygen, protecting lipid damage against photo-oxidative stress (Yadav et al. 2013). Tocopherols also play an important role in adaptation to low temperature and are essential for preventing non-enzymatic lipid oxidation during seed germination and dormancy (DellaPenna and Maeda 2007). Structural analyses show that molecules having vitamin E antioxidant activity include four tocopherol and four tocotrienol homologues (Schneider 2005). Tocopherols are synthesised in plastids, where chlorophyll synthase catalyses the final step in chlorophyll biosynthesis, followed by the prenylation of homogentisate with phytyl diphosphate, catalysed by homogentisate phytyl transferase (Zhang et al. 2015a, b). Vitamin E is one of the most light-sensitive vitamins. α-Tocopherol (α-T) has strong vitamin E activity in human cells, while β-, γ- and δ-tocopherol (β-, γ- and δ-T, respectively) are distinguished by strong antioxidant activity in plants. In addition, the composition of tocopherols depends on tissue type, as α-T is predominant in leaves (Abbasi et al. 2007), while γ-T is typically present in dicot seeds (Grusak and DellaPenna 1999). The antioxidant activity of tocopherols is determined by the amount of methyl groups attached to the phenolic ring of the polar head structure (Voll and Abbasi 2007). Therefore, α-T is the most efficient, and its single molecule can neutralise up to 120 singlet oxygen molecules (Wu and Tang 2004). One of the most important functions of tocopherols in biological membranes is that they act as recyclable chain reaction terminators of polyunsaturated fatty acid free radicals generated by lipid oxidation (Schneider 2005). On the other hand, tocopherols can be distinguished by non-antioxidant functions, such as modulation of membrane fluidity, stabilisation of membrane structure, participation in photosystem II protection, protection of membranes against deleterious effects and inhibition of cell proliferation (Voll and Abbasi 2007). Moreover, plants are almost the lone (aside from cyanobacteria) source of tocopherols.

The contents of particular nutrients are mainly determined genetically, and their metabolism might be controlled or changed, manipulating various environmental factors and horticultural growth strategies. There are literature data regarding improvement of nutritional quality by plant breeding (Mou 2009) or biotechnology (Lee et al. 2007a, b) but not much about the response of secondary metabolites to light spectra or intensity. Tocopherols are not directly associated with light reactions, but tocopherols quite easily accept light-driven manipulations, as their metabolic pathways are related to photosynthetic pigments (Stange and Flores 2012). The action of light spectral components is usually complex and often is reported with mixed results. Liu et al. (2008) noticed that combinations of antioxidant properties among tocopherols, ascorbic acid, lycopene and β-carotene are capable of producing synergistic antioxidant effects and may result in enhanced antioxidant effectiveness of natural antioxidants. The total vitamin E content in red and blue LED-irradiated barley decreased to 65%, and no difference was found in red LED-irradiated plants compared to natural light (Koga et al. 2013). It was significant that γ-T in red LED-irradiated barley was 50% greater compared to other light treatments, and this was explained by the fact that red radiation suppressed the activity of γ-T methyltransferase, which is a synthetic enzyme catalysing the conversion from γ-T to α-T. Kokalj et al. (2016) found that yellow LED light had positive effects on α-T accumulation in apple and bell pepper fruit. Our studies generally showed that the photoresponse of tocopherols is controlled by light dose and spectrum reactions and moreover is species-dependent. Increased α-T concentrations were observed under lower (110–220 μmol $m^{-2}$ $s^{-1}$) LED light intensities in mustard, red pak choi, tatsoi and kohlrabi microgreens compared to 545 μmol $m^{-2}$ $s^{-1}$ (Samuolienė et al. 2013a), and increased red light dosage conditioned an increase of α-T in basil but not in parsley (Samuolienė et al. 2016). Greater accumulation of α-T was found under higher UV-A irradiation in basil, beet and pak choi (Brazaitytė et al. 2015b). The inconsistent results were obtained with supplemental green and blue LEDs with HPS lighting. Generally, more positive effects on tocopherol accumulation were observed by supplemental green light compared to supplemental blue light and depended on lettuce variety and season (Samuolienė et al. 2012c; Table 8.1). Koga et al. (2013) suggested that the visible decrease of tocopherols might be due to the suppression by blue light radiation of the activity of homogentisate phytyl transferase, an enzyme that controls the total amount of tocopherols. Although tocopherols do not directly participate in light reactions, it is clear that there is strong interaction between photoreceptor activation and antioxidant response via an enzymatic pathway.

## 8.5 Ascorbic Acid

Ascorbic acid (AA) is a compound that inhibits the action of reactive oxygen species, preventing oxidative damage of cells (Pastori et al. 2003; Du et al. 2012). AA acts as a key component in the expression of photonic energy dissipation mechanisms, such as the xanthophyll cycle (Yabuta et al. 2007). As it is known, the uptake of phytochemicals through the consumption of plant tissues is more effective than intake through artificial supplements. Therefore, fruits with higher AA contents might be of great importance in a healthy diet. Moreover, there is evidence that AA participates in photoprotective defence against reactive oxygen species (Darko et al. 2014). The effects of LED light quality, intensity and photoperiod on nutrient accumulation in various vegetables were reviewed recently (Bian et al. 2015; D'Souza et al. 2015). In general, research has shown that various LED light treatments result in the accumulation of AA in various seeds (Samuolienė et al. 2011b; Wenke and Qichang 2012), different varieties of microgreens (Samuolienė et al. 2011a, 2012a, 2016; Bliznikas et al. 2012; Brazaitytė et al. 2015a, b, c), baby leaf lettuce (Li and Kubota 2009), different varieties of lettuce (Shen et al. 2014), tomato (Verkerke et al. 2015; Kokalj et al. 2016), Chinese cabbage (Avercheva et al. 2014) and other plants. For better understanding of the photophysiological and biological responses to different parameters of light, improvement strategies, through photosynthetic and photomorphogenetic light receptors can be developed (Franklin and Whitelam 2004; Hogewoning et al. 2010). Mou (2009) stated that nutrient contents of vegetables are first determined by genetic differences but can be modified by environmental influences or horticultural type and/or by interaction of all these components. Moreover, the minor constituents (such as vitamins) are found in the microgramme range in plants; thus, the most feasible quantitative changes might be done. According to Grusak (2002), at the genetic level, minimal diversion of precursors and only limited modifications in a plant's ability to store or sequester the target phytochemical is needed. However, the exact physiological mechanisms of how nutritional content, especially of metabolites that are not direct receptors of light, is enhanced through light are not fully understood.

Smirnoff et al. (2013) suggested that ascorbate accumulation is controlled by a complex interplay between cryptochrome, photosynthesis and end product repression. The first committed step in ascorbate biosynthesis from GDP-mannose is catalysed by GDP-L-galactose phosphorylase, encoded by *VTC2* and *VTC5*. Authors found that VTC2 and VTC5 reporter protein expression was blue light intensity-dependent and was rapidly repressed by exogenous ascorbate and its precursor, L-galactose. The reduction of AA in plant tissues was observed with respect to declining light intensity in spinach, tomato, lettuce, sweet pepper and strawberry (Gruda 2005). *Arabidopsis thaliana* leaves accumulated ascorbate in a light dose-dependent manner; this response was consistent with AA functions in photoprotection of photosynthesis (Smirnoff et al. 2013). Gautier et al. (2009) found that tomato fruit irradiance had an impact on AA metabolism, whereas leaf irradiance had an impact on photosynthesis and sugar transport to the fruits. Thus, the mentioned findings are in agreement with those of Rosales et al. (2011), who stated that accumulation of ascorbate might be enhanced by light quantity through the stimulation of secondary metabolism. Increased red light doses resulted in an accumulation of AA in basil but had no effect on parsley microgreens (Samuolienė et al. 2016). Verkerke et al. (2015) found that vitamin C concentrations of different tomato cultivars increased with increasing intensity of red and blue LEDs (140, 200 and 285 μmol $m^{-2}$ $s^{-1}$); moreover, the oxygen radical absorbance capacity also increased. The opposite results were found in red pak choi and tatsoi microgreens, in which the lowest (110 μmol $m^{-2}$ $s^{-1}$) PPFD level resulted in about 3.8 times higher AA concentrations than under 220 μmol $m^{-2}$ $s^{-1}$ LED light; however, the AA accumulations in mustard and kohlrabi were not significantly affected (Samuolienė et al. 2013b; Table 8.1). Despite ascorbate not being involved in the photoprotection mechanisms caused by high light stress (Page et al. 2012), the variation in the accumulation of AA suggests that there are complex relationships among the effects of light stress and genetic, developmental and metabolic signal transduction pathways (Solfanelli et al. 2006).

Braidot et al. (2014) found no significant difference between control and pulsed warm white LED light treatments for ascorbate in lamb's lettuce. It was demonstrated that lamb's lettuce plants under low-intensity light treatments during cold storage were able to promote photosynthesis but, at the same time, induced photodamage. In contrast, under intermittent low-intensity light cycles, the metabolism of the green tissues was still able to provide carbon moieties for the synthesis of bioactive molecules involved in delaying senescence. In addition, the variation of AA concentration was affected by lighting conditions, depending on the species. Almost no difference was found in wheatgrass and barley grass, whereas flashing low-frequency light (596 nm; 2.9 Hz) resulted in a significant decrease in vitamin C concentration in radish sprouts (Urbonavičiūtė et al. 2009c; Table 8.1). Such behaviour can be attributed to light-induced stress, which stimulates the activity of antioxidants, mostly due to natural defence mechanisms against photo-oxidative damage (Wu et al. 2007), and the response sensitivity to the light conditions might depend on the natural level of ascorbate in different plant species.

The lighting spectral range of LEDs is available from near UV to near infrared (Morrow 2008). Thus, lighting spectra can be selected according to specific requirements to obtain particular results. Wenke and Qichang (2012) described the participation of AA in protection systems against various stress factors, including UV-A. In general, UV-A irradiation had uneven effects on AA accumulations in various plants. An increase of antioxidant compounds, including AA, was observed under a low level of UV-A (Helsper et al. 2003). A significant increase of AA concentration was observed in sprouted lentil, radish and wheat seeds (Samuolienė et al. 2011b) and in winter wheat (Urbonavičiūtė et al. 2009a) grown under supplemental UV-A (about 4% of total PPFD). The supplemental UV-A light at a higher irradiance level (4% of total PPFD) had a harmful effect on the AA accumulation in basil and beet but induced its accumulation in pak choi. On the other hand, longer UV-A wavelengths resulted in an increase of AA concentrations in many cases (Brazaitytė et al. 2015b). Other authors did not find any impact of UV-A light on AA accumulation (Li and Kubota 2009; Wenke and Qichang 2012; Table 8.1).

Hogewoning and Harbinson (2007) suggested that blue light responses can be opposed by green light action and like in many cases, plant response depends on the irradiation level. Both blue and green light responses are cryptochrome-dependent (Zhang et al. 2011). Positive effects of supplemental green and blue light on AA accumulation were found in green baby leaf lettuce (Samuolienė et al. 2012b), sprouted lentil, radish and wheat seeds (Samuolienė et al. 2011b) as well as in some winter wheat varieties (Urbonavičiūtė et al. 2009a), but no positive effects on AA accumulation in red baby leaf lettuce (Li and Kubota 2009) nor in romaine lettuce (Samuolienė et al. 2013b) were found. Such unequal effects might be related to the dosage of blue (Hogewoning et al. 2010) and/or green (Folta and Maruhnich 2007) light, since in most cases positive effects on AA accumulation were achieved when these components were up to 50% of total PPFD (Table 8.1). On the other hand, an altered AA concentration may act as a primary ‘cross talk’ signal, coordinating the activity of antioxidant system defence mechanisms (Pastori et al. 2003), or may act as a key component involved in excess photonic energy in the xanthophyll cycle (Yabuta et al. 2007).

Red light treatment did not result in accumulation of AA in lettuce, spinach, komatsuna, tatsoi, beet or parsley (Ohashi-Kaneko et al. 2007; Li and Kubota 2009; Samuolienė et al. 2012a, b), whereas significant interactions among lettuce varieties, lighting treatment, season or their combination for AA were found (Samuolienė et al. 2012c). Positive effects of high-intensity red LED light on AA accumulation were found in amaranth, basil, kale, broccoli, mustard, borage and pea (Samuolienė et al. 2012a; Table 8.1). Ma et al. (2014) noticed that red LED light induced AA accumulation in broccoli compared to blue LED or dark treatments. It was suggested that modulation of AA reduction by a modified white LED light treatment was highly regulated at the transcriptional level. The up-regulation of the AA biosynthetic genes (*BO-VTC2* and *BO-GLDH*) and AA regeneration genes (*BO-MDAR1* and *BO-MDAR2*) contributed to a higher AA content in plants of the modified white LED light treatment during the first and second days after harvest (Ma et al. 2014).

Such results indicate that light effects on AA accumulation may be species- or even variety-dependent. In addition, AA biosynthesis also may be under the control of carbohydrate pools, as conversion of glucose to AA occurs via hexoses (Smirnoff and Wheeler 2000). Furthermore, the combination of light spectra and/or intensity with other unfavourable environmental conditions may accelerate reactive oxygen species production, and photo-oxidative damage can occur.

## 8.6 Conclusions

An overview of the research has shown that the flexibility in selecting LED light parameters is useful for finding the optimal or stressful light conditions that induce enhanced accumulation of plant metabolites and improve nutritional quality of various plants, especially vegetables. Moreover, LED-based technology is also useful for better understanding of the photophysiological responses caused by different light parameters. The most overarching lighting strategies concerning nutritional quality of various vegetables are based on red and blue LED illumination. However, a number of studies have shown that other wavelengths of LEDs, such UV-A, far-red or green, could be applied in combination with LEDs for improving nutritional quality of vegetables. Generally, while some supplemental wavelengths of LEDs caused positive effects on improvement of nutritional quality, others resulted in a decreased content of one or multiple phytochemicals, which have health-promoting properties. Furthermore, plant reactions to changes of light spectra were species- and/or cultivar-dependent. On the other hand, investigations were carried out with different photoperiod, temperature, fertility, etc., conditions, which also could lead to changes in the contents of bioactive compounds. This indicates that although it is known that some processes concerning changes of phytochemical contents in plants are dependent upon light quality and quantity, in-depth explanations of the physiological, biochemical and molecular mechanisms of these processes are still lacking. In conclusion, it is still difficult to define the common patterns in various plants, and further investigation is necessary concerning the effects of LEDs and their large-scale applications in controlled environments for producing vegetables with high nutritional quality.

## References


Abbasi ASR, Hajirezaei M, Hofius D, Sonnewald U, Voll LM (2007) Specific roles of α- and γ-tocopherol in abiotic stress responses of transgenic tobacco (*Nicotiana tabacum* L.). Plant Physiol 143:1720–1738

Agati G, Galardi C, Gravano E, Romani A, Tattini M (2002) Flavonoid distribution in tissues of *Phillyrea latifolia* as estimated by microspectrofluorometry and multispectral fluorescence microimaging. Photochem Photobiol 76:350–360

Agati P, Matteini P, Goti A, Tattini M (2007) Chloroplast-located flavonoids can scavenge singlet oxygen. New Phytol 174:77–89

Agati G, Biricolti S, Guidi L, Ferrini F, Fini A, Tattini M (2011) The biosynthesis of flavonoids is enhanced similarly by UV radiation and root zone salinity in *L. vulgare* leaves. J Plant Physiol 168:204–212

Agati G, Azzarello E, Pollastri S, Tattini M (2012) Flavonoids as antioxidants in plants: location and functional significance. Plant Sci 196:67–76

Agati G, Brunetti C, Di Ferdinando M, Ferrini F, Pollastri S, Tattini M (2013) Functional roles of flavonoids in photoprotection: new evidence, lessons from the past. Plant Physiol Biochem 72:35–45

Avercheva OV, Berkovich YA, Erokhin AN, Zhigalova TV, Pogosyan SI, Smolyanina SO (2009) Growth and photosynthesis of chinese cabbage plants grown under light-emitting diode-based light source. Russ J Plant Physiol 56:14–21

Avercheva O, Berkovich YA, Smolyanina S, Bassarskaya E, Pogosyan S, Ptushenko V, Erokhin A, Zhigalova T (2014) Biochemical, photosynthetic and productive parameters of Chinese cabbage grown under blue-red LED assembly designed for space agriculture. Adv Space Res 53:1574–1581

Balasundram N, Sudram K, Samman S (2006) Phenolic compounds in plants and agri-industrial by-products: antioxidant activity, occurrence, and potential uses. Food Chem 99:191–203

Bian ZH, Yang QC, Liu WK (2015) Effects of light quality on the accumulation of phytochemicals in vegetables produced in controlled environments: a review. J Sci Food Agric 95:869–877

Bliznikas Z, Žukauskas A, Samuolienė G, Viršilė A, Brazaitytė A, Jankauskienė J, Duchovskis P, Novičkovas A (2012) Effect of supplementary pre-harvest led lighting on the antioxidant and nutritional properties of green vegetables. Acta Hortic 939:85–91

Botella-Pavia P, Rodriguez-Concepcion M (2006) Carotenoid biotechnology in plants for nutritionally improved foods. Physiol Plant 126:369–381

Bouly J, Schleicher E, Dionisio-Sese M, Vandenbussche F, Van Der Straeten D, Bakrim N, Meier S, Batschauer A, Galland P, Bittl R, Ahmad M (2007) Cryptochrome blue light photoreceptors are activated through interconversion of flavin redox states. J Biol Chem 282:9383–9391

Braidot E, Petrussa E, Peresson C, Patui S, Bertolini A, Tubaro F, Wählby U, Coan M, Vianello A, Zancani M (2014) Low-intensity light cycles improve the quality of lamb's lettuce (*Valerianella olitoria* L. Pollich) during storage at low temperature. Postharvest Biol Technol 90:15–23

Brazaitytė A, Jankauskienė J, Novičkovas A (2013) The effects of supplementary short-term red LEDs lighting on nutritional quality of *Perilla frutescens* L. microgreens. Rural Dev 6:54–58

Brazaitytė A, Sakalauskienė S, Samuolienė G, Jankauskienė J, Viršilė A, Novičkovas A, Sirtautas R, Miliauskienė J, Vaštakaitė V, Dabašinskas L, Duchovskis P (2015a) The effects of

LED illumination spectra and intensity on carotenoid content in *Brassicaceae* microgreens. Food Chem 173:600–606

Brazaitytė A, Viršilė A, Jankauskienė J, Sakaulauskienė S, Samuolienė G, Sirtautas R, Novičkovas A, Dabašinskas L, Miliauskienė J, Vaštakaitė V, Bagdonavičienė A, Duchovskis P (2015b) Effect of supplemental UV-A irradiation in solid-state lighting on the growth and phytochemical content of microgreens. Int Agrophys 29:13–22

Brazaitytė A, Jankauskienė J, Viršilė A, Samuolienė G, Sakalauskienė S, Sirtautas R, Novičkovas A, Dabašinskas L, Vaštakaitė V, Miliauskienė J, Bagdonavičienė A, Duchovskis P (2015c) Response of *Brassicaceae* microgreens to supplemental UV-A exposure. Nordic View to Sustain Rural Development. 25th NJF Congress, Riga, Latvia, p 52

Brazaitytė A, Sakalauskienė S, Viršilė A, Jankauskienė J, Samuolienė G, Sirtautas R, Vaštakaitė V, Miliaukienė J, Duchovskis P, Novičkovas A, Dabašinskas L (2016a) The effect of short-term red lighting on *Brassicaceae* microgreens grown indoors. Acta Hortic 1123:177–183

Brazaitytė A, Viršilė A, Samuolienė A, Jankauskienė J, Sakalauskienė S, Sirtautas R, Novičkovas A, Dabašinskas L, Vaštakaitė V, Miliauskienė J, Duchovskis P (2016b) Light quality: growth and nutritional value of microgreens under indoor and greenhouse conditions. Acta Hortic 1134:277–284

Carvalho IS, Cavaco T, Brodelius M (2011a) Phenolic composition and antioxidant capacity of six artemisia species. Ind Crops Prod 33:382–388

Carvalho R, Takaki M, Azevedo R (2011b) Plant pigments: the many faces of light perception. Acta Physiol Plant 33:241–248

Chalker-Scott L (1999) Environmental significance of anthocyanins in plant stress responses. Photochem Photobiol 70:1–9

Chen X, Guo W, Xue X, Wang L, Qiao X (2014) Growth and quality responses of 'Green Oak Leaf' lettuce as affected by monochromic or mixed radiation provided by fluorescent lamp (FL) and light-emitting diode (LED). Sci Hortic 172:168–175

Chen X, Xue X, Guo W, Wang L, Qiao X (2016) Growth and nutritional properties of lettuce affected by mixed irradiation of white and supplemental light provided by light-emitting diode. Sci Hortic 200:111–118

Cheng Z, Moore J, Yu L (2006) High-throughput relative DPPH radical scavenging capacity assay. J Agri Food Chem 54:7429–7436

Costa L, Millan Montano Y, Carrión C, Rolny N, Guiamet JJ (2013) Application of low-intensity light pulses to delay postharvest senescence of *Ocimum basilicum* leaves. Postharvest Biol Technol 86:181–191

Cuttriss AJ, Cazzonelli CI, Wurtzel ET, Pogson BJ (2011) Carotenoids. Adv Bot Res 58:1–36

D'Souza C, Yuk H-G, Khoo GH, Zhou W (2015) Application of light-emitting diodes in food production, postharvest preservation, and microbiological food safety. Compr Rev Food Sci Food Saf 14:719–740

Darko E, Heydarizadeh P, Schoefs B, Sabzalian MR (2014) Photosynthesis under artificial light: the shift in primary and secondary metabolism. Philos Trans R Soc B Biol Sci. doi:10.1098/rstb.2013.0243

DellaPenna D, Maeda H (2007) Tocopherol functions in photosynthetic organisms. Curr Opin Plant Biol 10:260–265

Demmig-Adams B, Gilmore AM, Adams WW (1996) *In vivo* function of carotenoids in higher plants. FASEB J 10:403–412

Długosz-Grochowska O, Kołton A, Wojciechowska R (2016) Modifying folate and polyphenol concentrations in Lamb's lettuce by the use of LED supplemental lighting during cultivation in greenhouses. J Funct Foods 26:228–237

Dorais M, Gosselin A (2002) Physiological response of greenhouse vegetable crops to supplemental lighting. Acta Hortic 280:59–67

Dougher TA, Bugbee B (2001) Evidence for yellow light suppression of lettuce growth. Photochem Photobiol 73:208–212

Du J, Cullen JJ, Buettner GR (2012) Ascorbic acid: chemistry, biology and the treatment of cancer. Biochem Biophys Rev Can 1826:443–457

Fan X, Zang J, Xu Z, Guo S, Jiao X, Liu X, Gao Y (2013) Effects of different light quality on growth, chlorophyll concentration and chlorophyll biosynthesis precursors of non-heading Chinese cabbage (*Brassica campestris* L.). Acta Physiol Plant 35:2721–2726

Flores-Perez U, Rodriguez-Concepcion M (2012) Carotenoids. In: Salter A, Wiseman H, Tucker GA (eds) Phytonutrients. Wiley, New York, pp 89–109

Folta KM, Maruhnich SA (2007) Green light: a signal to slow down or stop. J Exp Bot 58:3099–3111

Franklin KA, Whitelam GC (2004) Light signals, phytochromes and cross-talk with other environmental cues. J Exp Bot 55:271–276

Gautier H, Massot C, Stevens R, Sérino S, Génard M (2009) Regulation of tomato fruit ascorbate content is more highly dependent on fruit irradiance than leaf irradiance. Ann Bot 103:495–504

Gruda N (2005) Impact of environmental factors on product quality of greenhouse vegetables for fresh consumption. Cr Rev Plant Sci 24:227–247

Grusak MA (2002) Phytochemicals in plants: genomics-asisted plant improvement for nutritional and health benefits. Curr Opin Biotech 13:508–511

Grusak MA, DellaPenna D (1999) Improving the nutrient composition of plants to enhance human nutritiona and health. Ann Rev Plant Physiol Plant Mol Biol 50:133–161

Havaux M, Kloppstech K (2001) The protective functions of carotenoid and flavonoid pigments against excess visible radiation at chilling temperature investigated in *Arabidopsis* npq and tt mutants. Planta 213:953–966

Heim KE, Tagliaferro AR, Bobilya DJ (2002) Flavonoid antioxidants: chemistry, metabolism and structure-activity relationships. J Nutr Biochem 13:572–584

Helsper JPFG, Ric de Vos CH, Maas FM, Jonker HH, Van Den Broeck HC, Jordi W, Sander Pot C, Paul Keizer LC, Schapendonk Ad HCM (2003) Response of selected antioxidants and pigments in tissues of *Rosa hybrida* and *Fuchsia hybrida* to supplemental UV-A exposure. Physiol Plant 117:171–178

Heuvelink E, Bakker MJ, Hogendonk L, Janse J, Kaarsemaker R, Maaswinkel R (2006) Horticultural lighting in the Netherlands: new developments. Acta Hortic 711:25–33

Hogewoning SW, Harbinson J (2007) Insights on the development, kinetics, and variation of photoinhibition using chlorophyll fluorescence imaging of a chilled, variegated leaf. J Exp Bot 58:453–463

Hogewoning SW, Trouwborst G, Maljaars H, Poorter H, van Ieperen W, Harbinson J (2010) Blue light dose-responses of leaf photosynthesis, morphology, and chemical composition of *Cucumis sativus* grown under different combinations of red and blue light. J Exp Bot 61:3107–3117

Iwai M, Ohta M, Tsuchiya H (2010) Enhanced accumulation of caffeic acid, rosmarinic acid and luteolin-glucoside in red perilla cultivated under red diode laser and blue LED illumination followed by UV-A irradiation. J Funct Foods 2:66–70

Johkan M, Shoji K, Goto F, Hashida S, Yoshihara T (2010) Blue light-emitting diode light irradiation of seedlings improves seedling quality and growth after transplanting in red leaf lettuce. HortScience 45:1809–1814

Khanam UKS, Oba S, Yanase E, Murakami Y (2012) Phenolic acids, flavonoids and total antioxidant capacity of selected leafy vegetables. J Funct Foods 4:979–987

Kim HH, Goins GD, Wheeler RM, Sager JC (2004) Green-light supplementation for enhanced lettuce growth under red- and blue-light emitting diodes. HortScience 39:1617–1622

Koga R, Meng T, Nakamura E, Miura C, Irino N, Devkota HP, Yahara S, Kondo R (2013) The effect of photo-irradiation on the growth and ingredient composition of young green barley (*Hordeum vulgare*). Agric Sci 4:185–194

Kokalj D, Hribar J, Cigić B, Zlatić E, Demšar L, Sinkovič L, Šircelj H, Bizjak G, Vidrih R (2016) Influence of yellow light-emitting diodes at 590 nm on storage of apple, tomato and bell pepper fruit. Food Technol Biotechnol 54:228–235

Kopsell DA, Kopsell DE (2006) Accumulation and bioavailability of dietary carotenoids in vegetable crops. Trends Plant Sci 11:499–507

Kopsell DA, Sams CE (2013) Increase in shoot tissue pigments, glucosinolates and mineral elements in sprouting broccoli after exposure to short-duration blue light from light emitting diodes. J Am Soc Hortic Sci 138:31–37

Kopsell DA, Pantanizopoulos NI, Sams CE, Kopsell DE (2012) Shoot tissue pigment levels increase in 'Florida Broadleaf' mustard (*Brassica juncea* L.) microgreens following high light treatment. Sci Hortic 140:96–99

Kopsell DA, Sams CE, Barickman TC, Morrow RC (2014) Sprouting broccoli accumulate higher concentrations of nutritionally important metabolites under narrow-band light-emitting diode lighting. JASHS 139:469–477

Kopsell DA, Sams CE, Morrow RC (2016) Interaction of light quality and fertility on biomass, shoot pigmentation and xanthophyll cycle flux in Chinese kale. J Sci Food Agric. doi:10.1002/jsfa.7814

Kwack Y, Kim KK, Hwang H, Chun C (2015) Growth and quality of sprouts of six vegetables cultivated under different light intensity and quality. Hortic Environ Biotechnol 56(4):437–443

Lee K, Lee SM, Park SR, Jung J, Moon JK, Cheong JJ, Kim M (2007a) Overexpression of *Arabidopsis* homogentisate phytyltransferase or tocopherol cyclase elevates vitamin E content by increasing gamma-tocopherol level in lettuce (*Lactuca Sativa* L.). Mol Cell 24:301–306

Lee SH, Tewari RK, Hahn EJ, Paek KY (2007b) Photon flux density and light quality induce changes in growth, stomatal development, photosynthesis and transpiration of *Withania somnifera* (L.) Dunal. Plantlets. Plant Cell Tiss Org Cult 90:141–151

Lee SW, Seo JM, Lee MK (2014) Influence of different LED lamps on the production of phenolic compounds in common and Tartary buckwheat sprouts. Ind Crops Prod 54:320–326

Lee MJ, Son KH, Oh MM (2016) Increase in biomass and bioactive compound in lettuce under various ratios of red to far-red LED light supplemented with blue LED light. Hortic Environ Biotechnol 57:139–147

Lefsrud MG, Kopsell DA, Kopsell DE, Curran-Celentano J (2006) Irradiance levels affect growth parameters and carotenoid pigments in kale and spinach grown in a controlled environment. Physiol Plant 127:624–631

Lefsrud M, Kopsell D, Wenzel A, Sheehan J (2007) Changes in kale (*Brassica oleracea* L. var. *acephala*) carotenoid and chlorophyll pigment concentrations during leaf ontogeny. Sci Hortic 112:136–141

Lefsrud MG, Kopsell DA, Sams CE (2008) Irradiance from distinct wavelength light-emitting diodes affect secondary metabolites in kale. HortScience 43:2243–2244

Li Q, Kubota C (2009) Effects of supplemental light quality in growth and phytochemicals of baby leaf lettuce. Env Exp Bot 67:59–64

Li T, Yang Q (2015) Advantages of diffuse light for horticultural production and perspectives for further research. Front Plant Sci. doi:10.3389/fpls.2015.00704

Li H, Tang C, Xu Z, Liu X, Han X (2012) Effects of different light sources on the growth of non-heading chinese cabbage (*Brassica campestris* L.). JAS 4:262–273

Lillo C, Lea US, Ruoff P (2008) Nutrient depletion as a key factor for manipulating gene expression and product formation in different branches of the flavonoid pathway. Plant, Cell Environ 31:587–601

Lin C (2002) Blue light receptors and signal transduction. Plant Cell S207–S225

Lin KH, Huang MY, Huang WD, Hsu MH, Yang ZW, Yang CM (2013) The effects of red, blue, and white light-emitting diodes on the growth, development, and edible quality of hydroponically grown lettuce (*Lactuca sativa* L. var. *capitata*). Sci Hortic 150:86–91

Lindoo SJ, Caldwell MM (1978) Ultraviolet-B radiation induced inhibition of leaf expansion and promoting of anthocyanin production. Plant Physiol 61:278–282

Liu D, Shi J, Ibarra AC, Kakuda Y, Xue SJ (2008) The scavenging capacity and synergistic effects of lycopene, vitamin E, Vitamin C, and β-carotene mixtures on the DPPH free radical. LWT—Food Sci Technol 41:1344–1349

Liu H, Liu B, Zhao C, Pepper M, Lin C (2011) The action mechanisms of plant cryptochromes. Trends Plant Sci 16:684–691

Long SP, Zhu X-G, Naidu SL, Ort DR (2006) Can improvement in photosynthesis increase crop yields? Plant, Cell Environ 29:315–330

Ma G, Zhang L, Setiawan CK, Yamawaki K, Asai T, Nishikawa F, Maezawa S, Sato H, Kanemitsu N, Kato M (2014) Effect of red and blue LED light irradiation on ascorbate content and expression of genes related to ascorbate metabolism in postharvest broccoli. Postharvest Biol Technol 94:97–103

Maiani G, Caston MJ, Catasta G, Toti E, Cambrodón IG, Bysted A, Granado- Lorencio F, Olmedilla-Alonso B, Knuthsen P, Valoti M, Böhm V, Mayer-Miebach E, Behsnilian D, Schlemmer U (2009) Carotenoids: actual knowledge on food sources, intakes, stability and bioavailability and their protective role in humans. Mol Nutr Food Res 53:1–25

Matioc-Precup MM, Cachiţă-Cosma D (2013) The content in assimilating pigments of the cotyledons of the red cabbage plantlets illuminated with LEDs. Studia Universitatis "Vasile Goldiş". Seria Ştiinţele Vieţii 23:45–48

Mitchell CA, Both AJ, Bourget CM, Burr JF, Kubota C, Lopez RG, Morrow RC, Runkle ES (2012) LEDs: the future of greenhouse lighting. Chronica Hortic 52:6–10

Mizuno T, Amaki W, Watanable H (2011) Effects of monochromatic light irradiation by LED on the growth and anthocyanin contents in leaves of cabbage seedling. Acta Hortic 907:179–184

Mohr H, Drumm-Herrel H, Oelmüller R (1984) Coaction of phytochrome and blue/UV light photoreceptors. In: Senger H (ed) Blue light effects in biological systems. Springer, Berlin, pp 6–19

Morrow RC (2008) LED lighting in horticulture. HortScience 43:1947–1950

Mou B (2009) Nutrient content of lettuces and its improvement. Curr Nutr Food Sci 5:242–248

Ogawa T, Inoue Y, Kitajima M, Shibata K (1973) Action spectra for biosynthesis of chlorophylls a and b and β-carotene. Photochem Photobiol 18:229–235

Ohashi-Kaneko K, Takase M, Kon N, Fujiwara K, Kurata K (2007) Effect of light quality on growth and vegetable quality in leaf lettuce, spinach and komatsuna. Eniviron Contr Biol 45:189–198

Ouzounis T, Parjikolaei BR, Fretté X, Rosenqvist E, Ottosen CO (2015) Pre-dawn and high intensity application of supplemental blue light decreases the quantum yield of PSII and enhances the amount of phenolic acids, flavonoids, and pigments in *Lactuca sativa*. Front Plant Sci. doi:10.3389/fpls.2015.00019

Page M, Sultana N, Paszkiewicz K, Florance H, Smirnoff N (2012) The influence of ascorbate on anthocyanin accumulation during high light acclimation in *Arabidopsis thaliana*: further evidence for redox control of anthocyanin synthesis. Plant, Cell Environ 35:388–404

Pastori GM, Kiddle G, Antoniw J, Bernard S, Veljovic-Janovic S, Verrier PJ, Noctor G, Foyer CH (2003) Leaf vitamin C contents modulate plant defence transkripts and regulate genes that control development through hormone signaling. Plant Cell 15:939–951

Qian H, Liu T, Deng M, Miao H, Cai C, Shen W, Wang Q (2016) Effects of light quality on main health-promoting compounds and antioxidant capacity of Chinese kale sprouts. Food Chem 196:1232–1238

Rice-Evans CA, Miller NJ, Paganga G (1996) Structure-antioxidant activity relationships of flavonoids and phenolic acids. Free Radical Biol Med 20:933–956

Rice-Evans CA, Miller NJ, Paganga G (1997) Antioxidant properties of phenolic compounds. Trends Plant Sci 2:152–159

Romani A, Pinelli P, Galardi C (2002) Polyphenols in greenhouse and open-air-grow lettuce. Food Chem 79:337–342

Rosales MA, Cervilla LM, Sánchez-Rodríguez E, Cervilla LM, Sánchez-Rodríquez E, Rubio-Wilhelmi Mdel M, Blasco B, Ríos JJ, Soriano T, Castilla N, Romero L, Ruiz JM (2011) The effect of environmental conditions on nutritional quality of cherry tomato fruits: evaluation of two experimental Mediterranean greenhouses. J Sci Food Agric 91:152–162

Samuolienė G, Brazaitytė A, Sirtautas R, Novičkovas A, Duchovskis P (2011a) Supplementary red-LED lighting affects phytochemicals and nitrate of baby leaf lettuce. J Food Agric Environ 9:271–274

Samuolienė G, Urbonavičiūtė A, Brazaitytė A, Šabajevienė G, Sakalauskaitė J, Duchovskis P (2011b) The impact of LED illumination on antioxidant properties of sprouted seeds. Centr Eur J Biol 6(1):68–74

Samuolienė G, Brazaitytė A, Sirtautas R, Sakalauskienė S, Jankauskienė J, Duchovskis P, Novičkovas A (2012a) The impact of supplementary short-term red led lighting on the antioxidant properties of microgreens. Acta Hortic 956:649–656

Samuolienė G, Sirtautas R, Brazaitytė A, Duchovskis P (2012b) LED lighting and seasonality effects antioxidant properties of baby leaf lettuce. Food Chem 134:1494–1499

Samuolienė G, Sirtautas R, Brazaitytė A, Viršilė A, Duchovskis P (2012c) Supplementary red-LED lighting and the changes in phytochemical content of two baby leaf lettuce varieties during three seasons. J Food Agric Environ 10:701–706

Samuolienė G, Brazaitytė A, Jankauskienė J, Viršilė A, Sirtautas R, Novičkovas A, Sakalauskienė S, Sakalauskaitė J, Duchovskis P (2013a) LED irradiance level affects growth and nutritional quality of *Brassica* microgreens. Cent Eur J Biol 8:1241–1249

Samuolienė G, Brazaitytė A, Sirtautas R, Viršilė A, Sakalauskaitė J, Sakalauskienė S, Duchovskis P (2013b) LED illumination affects bioactive compounds in romaine baby leaf lettuce. J Sci Food Agric 93:3286–3291

Samuolienė G, Brazaitytė A, Viršilė A, Jankauskienė J, Sakalauskienė S, Duchovskis P (2016) Red light-dose or wavelength-dependent photoresponse of antioxidants in herb microgreens. PLoS ONE. doi:10.1371/journal.pone.0163405

Schneider C (2005) Chemistry and biology of vitamin E. Mol Nutr Food Res 49:7–30

Seo JM, Arasu MV, Kim YB, Park SU, Kim SJ (2015) Phenylalanine and LED lights enhance phenolic compound production in Tartary buckwheat sprouts. Food Chem 177:204–213

Shalaby EA, Shanab SMM (2013) Antioxidant compounds, assays of determination and mode of action. Afr J Phar Pharmacol 7:528–539

Sharma OP, Bhat TK (2009) DPPH antioxidant assay revisited. Food Chem 113:1202–1205

Shen YZ, Guo SS, Ai WD, Tang YK (2014) Effects of illuminants and illumination time on lettuce growth, yield and nutritional quality in a controlled environment. Life Sci Space Res 2:38–42

Smirnoff N, Wheeler GL (2000) Ascorbic acid in plants: biosynthesis and function. Crit Rev Plant Sci 19:267–290

Smirnoff N, Page M, Ishikawa T (2013) Ascorbate and photosynthesis: how does *Arabidopsis* adjust leaf ascorbate concentration to light intensity? BioTechnol 94:206–214

Solfanelli C, Poggi A, Loreti E, Alpi A, Perata P (2006) Sucrose-specific induction of the anthocyanin biosynthetic pathway in *Arabidopsis*. Plant Physiol 144:637–646

Son KH, Oh MM (2013) Leaf shape, growth, and antioxidant phenolic compounds of two lettuce cultivars grown under various combinations of blue and red light-emitting diodes. HortScience 48:988–995

Stahl W, Sies H (2005) Bioactivity and protective effects of natural carotenoids. Biochim Biophys Acta 1740:101–107

Stange C, Flores C (2012) Regulation of carotenoid biosynthesis by photoreceptors. In: Najafpour M (ed) Advances in photosynthesis fundamental aspects, In Tech Europe, University Campus, Croatia, pp 77–76

Stutte GW, Edney S, Skerritt T (2009) Photoregulation of bioprotectant content of red leaf lettuce with light-emitting diodes. HortScience 44:79–82

Tamulaitis G, Duchovskis P, Bliznikas Z, Breive K, Ulinksaite R, Brazaitytė A, Novičkovas A, Žukauskas A (2005) Highpower light-emitting diode based facility for plant cultivation. J Phys D Appl Phys 38:3182–3187

Tattini M, Gravano E, Pinelli P, Mullinacci N, Romani A (2000) Flavonoids accumulate in leaves and glandular trichomes of *Phillyrea latifolia* exposed to excess solar radiation. New Phytol 148:69–77

Tilbrook K, Arongaus AB, Binkert M, Heijde M, Yin R, Ulm R (2013) The UVR8 UV-B photoreceptor: perception, signaling and response. The *Arabidopsis* Book. Am Soc Plant Biol 11:e0164

Tuan PA, Thwe AA, Kim YB, Kim JK, Kim SJ, Lee S, Chung SO, Park SU (2013) Effects of white, blue, and red light-emitting diodes on carotenoid biosynthetic gene expression levels and carotenoid accumulation in sprouts of Tartary buckwheat (*Fagopyrum tataricum* Gaertn.). J Agric Food Chem 61:12356–12361

Urbonavičiūtė A, Pinho P, Samuolienė G, Duchovskis P, Vitta P, Stonkus A, Tamulaitis G, Žukauskas A, Halonen L (2007) Effect of short-wavelength light on lettuce growth and nutritional quality. Sodininkystė ir daržininkystė 26:57–165

Urbonavičiūtė A, Samuolienė G, Brazaitytė A, Duchovskis P, Karklelienė R, Šliogerytė K, Žukauskas A (2009a) The effect of light quality on nutritional aspects of leafy radish. Sodininkystė ir daržininkystė 28:147–155

Urbonavičiūtė A, Samuolienė G, Brazaitytė A, Duchovskis P, Ruzgas V, Žukauskas A (2009b) The effect of variety and lighting quality on wheatgrass antioxidant properties. Zemdirbyste 96:119–128

Urbonavičiūtė A, Samuolienė G, Sakalauskienė S, Brazaitytė A, Jankauskienė J, Duchovskis P, Ruzgas V, Stonkus A, Vitta P, Žukauskas A, Tamulaitis G (2009c) Effect of flashing amber light on the nutritional quality of green sprouts. Agron Res 7:761–767

Vaštakaitė V, Viršilė A, Brazaitytė A, Samuolienė G, Jankauskienė J, Sirtautas R, Duchovskis P (2015a) The effect of supplemental lighting on antioxidant properties of *Ocimum basilicum* L. microgreens in greenhouse. Rural Dev doi:10.15544/RD.2015.031

Vaštakaitė V, Viršilė A, Brazaitytė A, Samuolienė G, Jankauskienė J, Sirtautas R, Novičkovas A, Dabašinskas L, Sakalauskienė S, Miliauskienė J, Duchovskis P (2015b) The effect of blue light dosage on growth and antioxidant properties of microgreens. Sodininkystė ir daržininkystė 34:25–35

Vauzour D, Vafeiadou K, Spencer JPE (2012) Polyphenols. In: Salter A, Wiseman H, Tucker G (eds) Phytonutrients. Wiley, UK, pp 110–145

Verkerke W, Labrie C, Dueck T (2015) The effect of light intensity and duration on vitamin C concentration in tomato fruits. Acta Hortic 1106:49–54

Viršilė A, Sirtautas R (2013) Light irradiance level for optimal growth and nutrient contents in borage microgreens. Rural Dev 6:272–275

Voll LM, Abbasi A-R (2007) Are there specific in vivo roles for α- and γ-tocopherol in plants? Plant Signal Behav 2:486–488

Wang Y, Folta KM (2013) Contributions of green light to plant growth and development. Am J Bot 100:70–78

Wenke L, Qichang Y (2012) Effects of day-night supplemental UV-A on growth, photosynthetic pigments ant antioxidant system of pea seedlings in glasshouse. Afr J Biotechn 11:14786–14791

Wojciechowska R, Długosz-Grochowska O, Kołton A, Żupnik M (2015) Effects of LED supplemental lighting on yield and some quality parameters of lamb's lettuce grown in two winter cycles. Sci Hortic 187:80–76

Wu YS, Tang KX (2004) MAP kinase cascades responding to environmental stress in plants. Acta Bolon Sinica 46:127–136

Wu MC, Hou CY, Jiang CM, Wang YT, Wang CY, Chen HH, Chang HM (2007) A novel approach of LED light radiation improves the antioxidant activity of pea seedlings. Food Chem 1001:1753–1758

Yabuta Y, Mieda T, Rapolu M, Nakamura A, Motoki T, Maruta T, Yoshimura K, Ishikawa T, Shigeoka S (2007) Light regulation of ascorbate biosynthesis is dependent on the photosynthetic electron transport chain but independent of sugars in *Arabidopsis*. J Exp Bot 58:2661–2671

Yadav D, Rastogi A, Szymańska R, Kruk J, Sedlářová M, Pospíšil P (2013) Singlet oxygen scavenging by tocopherol and plastohromanol under photooxidative stress in *Arabidopsis*. Bio Technol 94:7

Yeh N, Chung JP (2009) High-brightness LEDs efficient lighting sources and their potential in indoor plant cultivations. Renew Sustain Energy Rev 13:2175–2180

Zhang T, Maruhnich SA, Folta KM (2011) Green light induces shade avoidance symptoms. Plant Physiol Prev. doi:10.1104/pp.111.180661

Zhang C, Zhang W, Ren G, Li D, Cahoon RE, Chen M, Zhou Y, Yu B, Cahoon EB (2015a) Chlorophyll synthase under epigenetic surveillance is critical for vitamin E synthesis, and altered expression affects tocopherol levels in *Arabidopsis*. Plant Physiol 168:1503–1511

Zhang L, Ma G, Yamawaki K, Ikoma Y, Matsumoto H, Yoshioka T, Ohta S, Kato M (2015b) Effect of blue LED light intensity on carotenoid accumulation in citrus juice sacs. J Plant Physiol 188:58–63

Žukauskas A, Bliznikas Z, Breivė K, Novičkovas A, Samuolienė G, Urbonavičiūtė A, Brazaitytė A, Jankauskienė J, Duchovskis P (2011) Effect of supplementary pre-harvest LED lighting on the antioxidant properties of lettuce cultivars. Acta Hortic 907:87–90

# Chapter 9
# Light-Emitting Diodes in Postharvest Quality Preservation and Microbiological Food Safety

**Craig D'Souza, Hyun-Gyun Yuk, Gek Hoon Khoo and Weibiao Zhou**

## 9.1 Introduction

Light-emitting diodes (LEDs) may be correctly described as being 'ubiquitous' nowadays, yet one may not be fully conscious of the extent of its presence and function in the food industry. Previous chapters of this book have already discussed their relevance in floriculture, horticulture, in vitro plant morphogenesis, in preventing insect infestation, and in food production applications. LEDs have also been recognized as containing characteristics that render it suitable for various niche applications such as in space agriculture, high-technology farming, aquaculture, and other forms of food production (Yeh et al. 2015). The subsequent stages to food production, to describe it succinctly, encompass the storage, distribution, and consumption of nutritious and safe food. It is counterproductive to neglect the quality of food during these postharvest stages as it would ultimately lead to unwanted food losses or deterioration of value along the supply chain. As much as one-third of the world's produced food is wasted, with a significant proportion being lost during the postharvest stages (FAO 2011). In developing countries, the main reasons include the lack of technological infrastructure and

C. D'Souza · W. Zhou (✉)
Food Science and Technology Programme, C/o Department of Chemistry,
National University of Singapore, 3 Science Drive 3, Singapore 117543, Singapore
e-mail: chmzwb@nus.edu.sg

H.-G. Yuk
Department of Food Science and Technology, Korea National University of Transportation,
61 Daehak-ro, Jeungpyeong-gun, Chungbuk 27909, Republic of Korea

G.H. Khoo
Post-Harvest Technology Department, Technology and Industry Development Group,
Agri-Food and Veterinary Authority of Singapore, 2 Perahu Road, Singapore 718915,
Singapore



S. Dutta Gupta (ed.), *Light Emitting Diodes for Agriculture*,
DOI 10.1007/978-981-10-5807-3_9

facilities to further process food or to allow for an efficient cold-chain system. In industrialized nations, excess food produced is eventually not consumed and is instead disposed of. Unsafe food caused by poor handling or hygiene standards may also result in food wastage; hence, there still exists a pressing need to develop technologies that extend the shelf life of foods while keeping them safe for consumption (FAO 2011). This chapter aims to address these problems, in reference to the opportunities that LEDs offer.

It is an intuitive notion that light is necessary for healthy plant growth, hence light is intimately associated with the idea of food production. Yet it is not readily apparent how light is involved in other aspects of the food supply chain. In recent years, the importance of light in retaining the postharvest status in certain foods, particularly in leafy vegetables, has been increasingly receiving the attention of researchers. It has been known for long that light is able to mitigate senescence in growing plants, and that different quality of light was able to result in varying nutritional quality of foods (Noodén and Schneider 2004). Since there is still residual biological activity during the postharvest stage, light can still have a similar biological effect and therefore reduce the degradation of the food quality through senescence, or through nutrient loss (Zhan et al. 2012b). Furthermore, light is an integral component of photodynamic inactivation (PDI), a phenomenon which causes microbial inactivation through a combination of light, a photosensitizing agent, and oxygen (Luksiene and Brovko 2013). A major advantage of this technique is that it is considered nonthermal due to the small increase in temperature to the system being treated, compared to traditional thermal methods. As such, the technique is a possible means of treating heat-sensitive food products such as minimally processed fruits and vegetables, or even various food surfaces. The technique is also promising as an alternative novel method to deal with the proliferation of antibiotic resistance in pathogens (Hamblin and Hasan 2004). Since light is central in the above applications, it is necessary to select a suitable light technology.

The most critical requirements for such a lighting technology include the ability to adjust the spectral composition of emitted light with ease and flexibility, as well as the exclusion of heating effects through radiation. This is because plant tissues contain various components that respond to different parts of the light spectrum and thereafter activate biological responses that result in desirable effects. Similarly, unique photoactive molecules which pose a threat to pathogenic bacteria also operate most effectively under certain wavelengths of light. Since thermal treatments can result in unwanted quality changes in foods, the availability of a lighting technology that reduces thermal heating to a minimum is also desirable. For these reasons, LEDs are well suited in the application of light for postharvest preservation and microbiological inactivation (D'Souza et al. 2015). The current availability of a great number of studies that have investigated the effectiveness of utilizing LEDs in the areas of postharvest preservation and food safety gives a clearer picture to their industrial, commercial, or potentially even personal application in homes, such as in the household refrigerator. This chapter highlights the relevant studies which have shown how LEDs perform this function of keeping food safe for as long as possible, in the period after it has left the 'farm.'

It must be noted that the relevant studies in the literature have been slightly limited in their scope. For example, postharvest studies are often conducted on fruits and vegetables; very few studies have investigated their effect on meat yet. In terms of food safety, this chapter focuses primarily on microbiological food safety in line with the current trend of such LED-related studies. The food types studied in this application are more varied, from fruits and vegetables, to beverages and even chicken. Even so, there is a rich amount of knowledge that can be gleaned from these studies, and the documentation of these studies will hopefully motivate the transferal of such knowledge to other related applications in the near future. Henceforth, the proceeding chapters will delve into the unique application of LEDs in postharvest preservation and microbiological food safety.

## 9.2 Brief Recapitulation of LED Technology and the Measurement of Light

Previous chapters have discussed in depth the properties and features of LED devices (see Chaps. 1 and 2 for more details). This section recapitulates these points in order to relate them to their application in postharvest and food safety techniques. Briefly, LEDs are semiconductor diodes which produce light through the process of electroluminescence. Depending on the material of the semiconductor, light of distinct color is produced (Dutta Gupta and Jatothu 2013). For example, LEDs fabricated using gallium arsenide emit red light, whereas with gallium nitride and silicon carbide, blue light is emitted (Yeh et al. 2015). Due to the narrow bandwidth of wavelengths, light emitted from LEDs is said to be almost monochromatic. LEDs can also produce monochromatic light which is in the ultraviolet (UV) or infrared (IR) range. Furthermore, broad-spectrum white light can also be produced from LEDs, either by mixing light from individual red, blue, and green LEDs (DenBaars et al. 2013), or by combining a UV LED and a tricolor phosphor coating, or a blue LED with a yellow phosphor coating (Park et al. 2014). In other words, LEDs confer great flexibility over the spectral composition of light, or what is referred to as 'light quality.'

The above properties are important for several reasons. Firstly, by producing high quantities of light of the wavelengths that are desired, less energy is consumed in producing light of wavelengths that are unwanted. This is especially important in photobiological interactions in plants, which involve interactions between light and plant pigments and photoreceptors. Chlorophylls, the photosynthetic pigment which is familiar even to the layman, possess absorption peaks typically in the blue and red regions; the reason they appear green is because green light is mostly reflected away (Zhu et al. 2008). Based on this, early studies exploring the potential of using LEDs for horticulture and plant growth achieved satisfactory results using red and blue LEDs only (Massa et al. 2008). Apart from chlorophylls, a variety of other photoreceptors or pigments are responsible for sensing or absorbing energy from different regions of the light spectrum, including limited regions of the UV and IR range (Pinho et al. 2012). For example, light in the blue region is absorbed mainly by photoreceptors such as cryptochromes, phototropins, as well as pigments such as lycopene, β-carotene, and xanthophylls, whereas green light is absorbed by pigments such as certain flavonoids and betalanins. Phytochrome is well known for sensing the ratio of red to far-red radiation present in light, thereafter triggering a variety of other photomorphological processes. Although cryptochrome is known to absorb in the UV range (at around 320–400 nm), more research is required to fully understand the mechanism behind UV perception at lower wavelengths of UV radiation (Carvalho et al. 2011). With this knowledge, monochromatic LEDs can be used to study phenomena relevant to these photoreceptors and pigments, or various LEDs can be combined to produce a light of a desired spectral composition for other purposes. Similarly, when LEDs are used in inactivating pathogenic or spoilage microorganisms through direct exposure to high-intensity light, or with a photoactive molecule which is excited at particular wavelengths, the monochromatic nature of LEDs is an advantage. In contrast, broad-spectrum lighting technologies have lower photon efficiency compared to LEDs, which produce relatively lower quantities of light of the desired wavelength at the same power consumption (Nelson and Bugbee 2014).

Secondly, monochromatic light is useful in limiting the propagation of radiant heat. The production of radiant heat from broad-spectrum light is a problem, and lighting sources such as high-intensity discharge lamps produce substantial amounts of IR radiation. This may therefore cause surface heating on plants or exposed surfaces, causing unwanted effects. Since only a narrow bandwidth of wavelengths is emitted from LEDs, IR radiation is typically absent, hence less surface heating and other associated detrimental effects are caused (Morrow 2008; Mitchell et al. 2012). However, substantial heating occurs in the p-n junctions of LEDs, which is the site of electroluminescence. Higher temperatures tend to reduce luminous efficacy, hence resulting in less light being produced. This can be prevented by using devices such as heat sinks and cooling fans. For this reason, LEDs are suitable for use in cold, temperature-controlled environments such as refrigerators and hence would be appropriate for using in cold-chain storage or in transport vehicles due to their added resistance to damage from vibration and mechanical forces (US Department of Energy 2012).

There are various other advantages that LEDs possess which are superior to other forms of lighting such as high-intensity discharge lighting, fluorescent lights, and others (D'Souza et al. 2015). Other notable features include LEDs having a unique ability of reaching full output almost immediately after being switched on, with little restrike delay, and hence can be used for high-frequency pulsing and dimming to further save energy (Yeh and Chung 2009; Branas et al. 2013; US Department of Energy 2013). LEDs also have a longer life expectancy ranging from 50,000 to 100,000 h, compared to fluorescent or high-pressure sodium lamps, which range from about 10,000 to 17,000 h (Dutta Gupta and Jatothu 2013).

Finally, while the concept of light quality has already been covered, light quantity requires some discussion. The photon flux is the most commonly encountered unit of measurement of light quantity (typically in the form of μmol $m^{-2}$ $s^{-1}$). It describes the number of moles of photons received per unit area per second, regardless of the wavelength or energy carried by the photons. Therefore, it is only useful in quantifying light when conceived of in the 'particle form of light,' which is more applicable to photochemical or photobiological reactions in plants (Pinho et al. 2012). Another metric commonly used in studies related to food safety is termed 'irradiance,' which is the power of light energy received per unit area (W $m^{-2}$). Since photons of different wavelengths possess different amounts of energy, irradiance varies with the spectral composition of light. As an illustration, although a treatment of 100 μmol $m^{-2}$ $s^{-1}$ of blue light is equivalent to 100 μmol $m^{-2}$ $s^{-1}$ of red light in terms of photon flux, the blue light treatment will have a higher irradiance (in terms of W $m^{-2}$) than red light since blue photons possess more energy than red photons. A related but outdated unit is the 'Einstein,' denoted by 'E' (e.g., μE $m^{-2}$ $s^{-1}$), but its usage is discouraged due to its ambiguity: it can be interpreted either as photon flux, or as irradiance (Thimijan and Heins 1983). However, it has still been used in several recent studies (Braidot et al. 2014; Dhakal and Baek 2014a, b). Irradiance is commonly used in food safety studies as the peak wavelength of monochromatic light is usually fixed according to the photoactive molecule being used (i.e., the photosensitizer), or within the blue to near-UV region. Hence, spectral composition is not relevant. Measuring in terms of energy is also useful as microbial inactivation usually depends on the dosage (J $cm^{-2}$), which is the product of time and irradiance.

In short, LEDs are useful in postharvest and food safety applications because they are energy-efficient, reduce unwanted heating of foods, are suitable for cold storage and transport, have long-lasting life times, and are mechanically robust and compact in size and shape. Most importantly, the quality of light emitted is easily customizable, especially due to its monochromatic nature. In the next section, the effects of various light qualities and quantities will be shown to have many beneficial effects on the postharvest quality of foods, especially of fruits and vegetables.

## 9.3 LEDs in Postharvest Quality Preservation of Fruits and Vegetables

The factors that affect postharvest quality are very broad. In general, postharvest techniques aim to prevent the visual, textural, and nutritional deterioration of a food that occurs rapidly after harvesting. Furthermore, it aims to keep the levels of harmful or spoilage-related microorganisms to a minimum, as well as to control the rate of ripening so as to optimize the commercial value of an edible fruit. In short, it aims to ensure that harvested produce is in an optimal state for consumption after being transported and distributed. Critical conditions for preserving postharvest quality include using the optimal combination of temperature and relative humidity, as well as concentrations of oxygen, carbon dioxide, and ethylene (Kader and Rolle 2004).

The effect of light on the postharvest quality of leafy vegetables and fruits, in particular, has been receiving more attention in recent years. It has been generally accepted that the postharvest quality of certain leafy vegetables that are exposed to small quantities of light is better than when stored in the dark (Braidot et al. 2014). Earlier postharvest studies on vegetables focused on the use of fluorescent lights mainly, showing that light can even increase the postharvest concentration of nutrients such as ascorbic acid, phenolic compounds, sugars, carotenoids, and other bioactive compounds in vegetables like spinach (Toledo et al. 2003; Lester et al. 2010; Glowacz et al. 2015), broccoli (Zhan et al. 2012a), and romaine lettuce (Zhan et al. 2012b). However, several studies utilizing LEDs as a light source have emerged recently. In general, LEDs have been used to delay senescence in perishable fruits and vegetables, in modifying nutritional content, in manipulating the rate of ripening of fruits, and in preventing fungal infections on foods to reduce food spoilage. Table 9.1 provides a summary of the different postharvest-related functions in which LEDs have successfully been shown to impart a beneficial effect.

### *9.3.1 Delay of Senescence in Vegetables Through LEDs*

Senescence is a genetically controlled process that maximizes the survival of individual plants. Senescence allows for the conservation of available macromolecules and nutrients within the plant, by relocating them from aging plant tissue to new or developing tissue. Although it is beneficial to living and growing plants, it leads to unwanted loss of quality in harvested fruits and vegetables, which may have been detached from the rest of the plant. This in turn interrupts the transport of materials between tissues. Senescence in the postharvest stage is generally gauged in terms of characteristics which represent the marketable quality of the food; hence, this could broadly include general characteristics such as color and degree of wilting, or more specific indicators such as chlorophyll content. Based on these factors, there is notable evidence that light treatment can delay senescence in detached leaves, stems, and flowers (Pogson and Morris 2004), but light must be delivered appropriately, according to the optimal intensity, spectral composition, duration or photoperiod considerations, to the target fruit or vegetable (Noodén and Schneider 2004).

Too much light could lead to excessive photooxidative stress, which results in lower postharvest quality (Glowacz et al. 2015). Hence, selecting the correct light intensity is important. In order to determine the correct amount of light for a successful treatment, the light compensation point, which is the amount of light that results in equal rates of photosynthesis and respiration in a plant tissue, could be considered as a benchmark. Light administered in quantities below the light compensation point results in a net loss of sugars, which accelerates senescence

**Table 9.1** Effects of LED lighting in postharvest preservation and effectiveness of treatments

| Application | Food | LED (wavelength) | Intensity | Treatment time | Effectiveness | References |
|---|---|---|---|---|---|---|
| Delaying of senescence in vegetables | Broccoli (*Brassica oleracea* L. var. italica) | Red (660 nm) | 50 μmol $m^{-2}$ $s^{-1}$ | Continuous | Reduced yellowing and less ethylene production observed compared to blue and white LEDs | Ma et al. (2014) |
| | Broccoli (*Brassica oleracea* L. var. italica cv Legacy) | White and blue LED | 20 μmol $m^{-2}$ $s^{-1}$ | Continuous | Generally higher chlorophyll, carotenoid, fructose, glucose, and sucrose content compared to dark control, but antioxidant capacity (DPPH and ABTS +) was not significantly different | Hasperué et al. (2016) |
| | Lamb's lettuce (*Valerianella olitoria* L. Pollich) | Warm white | 1.4 μE$m^{-2}$ $s^{-1}$ | Eight cycles of 1 h and sixteen cycles of 0.5 h | Chlorophyll degradation was delayed; higher pheophytin levels and lower pro-oxidant capacity observed compared to dark control | Braidot et al. (2014) |
| | Lettuce (*Lactuca sativa*), butterhead and iceberg | Red (660 nm) and Blue (455 nm) | 5 μmol $m^{-2}$ $s^{-1}$ | Continuous | Overall visual quality was rated unacceptable after 15 d for butterhead lettuce irradiated with red and blue LEDs and 19 d for iceberg lettuce irradiated with blue LED | Woltering and Seifu (2015) |
| Accelerating secondary ripening processes | Strawberries (*Fragaria ananassa* Duch cv. Fengguang) | Blue (470 nm) | 40 μmol $m^{-2}$ $s^{-1}$ | Continuous | Increase in ethylene production, respiration, color development, total antioxidant activity, and antioxidant enzyme activity compared to control | Xu et al. (2014a, b) |
| | Peach (*Prunus persica* cv. Jinli) | Blue (470 nm) | 40 μmol $m^{-2}$ $s^{-1}$ | Continuous | Increase in ethylene production, total soluble solids content, color development, and decrease in firmness, titratable acidity, compared to control | Gong et al. (2015) |
| | Satsuma mandarins (*C. unshiu* Marc. 'Aoshimaunshu') | Red (660 nm) | 12 μmol$m^{-2}$ $s^{-1}$ | Continuous | Acceleration of color development in the rind of irradiated fruit compared to those stored in the dark | Yamaga et al. (2016) |

(continued)

**Table 9.1** (continued)

| Application | Food | LED (wavelength) | Intensity | Treatment time | Effectiveness | References |
|---|---|---|---|---|---|---|
| Delaying of ripening | Mature green tomatoes (*Solanum lycopersicum* L. cv. Dotaerang) | Blue (440–450 nm) | 85.7 μEm$^{-2}$ s$^{-1}$ | Continuous | A slower rate of color change from green to red and loss of firmness observed compared to red light | Dhakal and Baek (2014b) |
| Enhancing or delaying loss of postharvest nutritional content | Broccoli (*Brassica oleracea* L. var. italica) | Red (660 nm) | 50 μmolm$^{-2}$ s$^{-1}$ | Continuous | Higher ascorbic acid content observed compared to blue and white LED | Ma et al.(2014) |
| | Cabbage 'Dongdori' | White, Blue (436 nm), Green (524 nm) Red (665 nm) | Unspecified. Electrical power stated as 1.380, 1.455, 1.515, and 1.065 W for white, green, blue, and red LEDs, respectively | Continuous | All LED treatments improved the total chlorophyll, vitamin C, and total phenolics compared to dark control. Green LED was most effective for increasing chlorophyll content while blue LED increased vitamin C. All LED treatments increased phenolic content. Moisture content for all treatments decreased by less than 5% only, and pH increased at the same rate for all treatments | Lee et al. (2014) |
| | Lamb's lettuce (*V. olitoria* L. Pollich) | Warm white | 1.4 μ Em$^{-2}$ s$^{-1}$ | Sixteen cycles of 0.5 h | A slower decrease of carotenoids content observed compared to dark control | Braidot et al. (2014) |
| | Lettuce (*Lactuca sativa*), butterhead and iceberg | Green (660 nm) and Blue (455 nm) | 5 μmol m$^{-2}$ s$^{-1}$ | Continuous | Glucose, fructose, and sucrose content increased most significantly in butterhead lettuce with green LED treatment after 7 d, and in iceberg lettuce with blue LED treatment after 14 d | Woltering and Seifu (2015) |
| | Watercress (*Nasturtium officinale* R. Br.) and garden pea sprouts (*Pisum sativum* L.) | UV-A (375 nm) | 33 μmolm$^{-2}$ s$^{-1}$ | 160 min daily for three days | Higher in quercetin-glycoside content observed compared to dark control | Kanazawa et al. (2012) |
| | Chinese bayberries (*Myrica rubra*) | Blue (470 nm) | 40 μmol m$^{-2}$ s$^{-1}$ | Continuous | Greater total anthocyanin content measured compared to dark control | Shi et al. (2014) |
| | Grape berries (*Vitis labruscana* Bailey cv 'Campbell Early' and 'Kyoho') | Blue (440 nm) | 40 μmol m$^{-2}$ s$^{-1}$ | Continuous | Generally increased concentrations of stilbene compounds in the skin, compared to fluorescent light, purple or red LEDs | Ahn et al. (2015) |

(continued)

**Table 9.1** (continued)

| Application | Food | LED (wavelength) | Intensity | Treatment time | Effectiveness | References |
|---|---|---|---|---|---|---|
| | Mature green tomatoes (*Solanum lycopersicum* L. cv. Dotaerang) | Blue (440–450 nm) | 85.7 $\mu Em^{-2}\ s^{-1}$ | | Higher content of glutamic acid and γ-butyric acid measured compared to red light | Dhakal and Baek (2014a) |
| | Peach (*Prunus persica* cv. Jinli) | Blue (470 nm) | 40 $\mu molm^{-2}\ s^{-1}$ | Continuous | Greater total carotenoid, zeaxanthin and β-carotene, β-cryptoxanthin, and lutein content compared to dark control after 20 days | Cao et al. (2017) |
| | Satsuma mandarins (*Citrus unshiu* Marc.) | Red (660 nm) | 50 $\mu molm^{-2}\ s^{-1}$ | Continuous | Increased total carotenoids measured in the flavedo compared to blue LED light and dark control | Ma et al. (2011) |
| | Satsuma mandarins, Valencia oranges (*Citrus sinensis* Osbeck) and Lisbon lemons (*Citrus limon* Burm f.) | Blue (470 nm) | 50 $\mu mol\ m^{-2}\ s^{-1}$ | Continuous | Increased total carotenoids and ascorbate measured in the juice sacs compared to red LED light and dark control | Zhang et al. (2012) |
| | Strawberries (*Fragaria ananassa* cv Sulhyang) | UV-A (385 nm), Blue (470 nm), Green (525 nm), Red (630 nm) | Unspecified. 20 mA current used for each LED | Continuous | Blue, red, and green LED improved anthocyanin content of immature strawberries compared to dark storage; Blue and green LED improved the vitamin C content. Total phenolics stimulated most by blue LED, and total soluble solids improved most by green LED | Kim et al. (2011) |
| Preventing food spoilage | 'Fallglo' Tangerines | Blue (456 nm) | 40 $\mu molm^{-2}\ s^{-1}$ | Continuous | Reduced fungal colonization of *Penicillium digitatum* on surface of fruit compared to dark and white light treatments | Alferez et al. (2012); Liao et al. (2013) |
| | Satsuma mandarins (*C. unshiu* Marc. 'Aoshimaunshu') | Blue (465 nm) | 8 and 80 $\mu molm^{-2}\ s^{-1}$ | Continuous | Reduced fungal colonization and disease incidence of *Penicillium italicum* on surface of fruit compared to dark control | Yamaga et al. (2015) |
| | Strawberries (*Fragaria ananassa*) | Deep ultraviolet (272, 289, 293 nm) | 20 $mWm^{-2}$ | Continuous | Mold growth, suspected to be *Botrytis cinerea*, was absent in LED-treated samples after 9 d, whereas those stored in dark had extensive growth after 6 d | Britz et al. (2013) |

*Source* Adapted from D'Souza et al. (2015) with updates

(Noodén and Schneider 2004). However, light quality must be considered as well. A study by Costa et al. (2013) found that subjecting basil leaves (*Ocimum basilicum* L.) to pulsed white fluorescent light treatment at a photon flux below the compensation point effectively retarded senescence. The effect from the above treatment was comparable to pulsed red light produced using a white light and red filter. However, when a far-red filter was used, quality indicators suggested that senescence was proceeding, thereby indicating the involvement of phytochrome of senescence. Therefore, in the case of basil leaves, light quality was more influential than light quantity.

Conventionally, postharvest application of light in studies related to leafy vegetables did not exceed 30 μmol $m^{-2}$ $s^{-1}$ (Noichinda et al. 2007; Lester et al. 2010), and several studies have even used various forms of pulsed lighting (Costa et al. 2013; Gergoff-Grozeff et al. 2013). LEDs are well-equipped to provide such quantities of light and are far more effective than other lighting technologies at providing pulsed light. However, only a few studies which use white LED irradiation exist. In one such study, an LED produced pulses of warm white light on lamb's lettuce, at a very low average photon flux of approximately 1.4 μE$m^{-2}$ $s^{-1}$ for 8 h in total (Braidot et al. 2014). Two different pulse treatments were used: specifically, 8 cycles of 1-h pulses or 16 cycles of 0.5-h square-wave pulses. Both treatments resulted in an increase in the chlorophyll *a*/*b* ratio above the initial ratio, and slower reduction in pheophytin levels, thus suggesting a delay in senescence. Furthermore, less potential oxidative damage was observed based on the pro-oxidant capacity of lipophilic extracts. However, the treatment of 16 cycles of 0.5-h pulses slowed down the degradation of chlorophylls *a* and *b* and helped retain carotenoid levels. Glucose content in light-treated or control samples was measured to be less than the initial glucose content, suggesting that pulsed light in low doses might be insufficient for photosynthesis to occur effectively. Hence, despite a net loss of glucose, there was still a limited amount chlorophyll and carotenoids produced.

Hasperué et al. (2016) investigated the rate of postharvest senescence in broccoli (*Brassica oleracea* var. Italica cv. Legacy) when treated with 20 μmol $m^{-2}$ $s^{-1}$ of a combination of white and blue LEDs. LED-treated samples showed the least amount of yellowing, and a corresponding retention of chlorophylls *a* and *b* compared to the dark control samples. Retention of glucose, fructose, and sucrose was also observed. Moreover, sucrose was increased by LED irradiation after 35 days when stored in 5 °C. All quality indicators for senescence were better for samples irradiated by LEDs than those stored in the dark even up to 42 days when stored at 5 °C. Therefore, in general, using low quantities of light from LEDs is a good means of preventing senescence from proceeding, hence keeping produce as fresh as possible, and in good marketable condition.

### 9.3.2 Enhancement of Nutritional Status of Vegetables and Fruits Through LEDs

The previous cases have shown that white LEDs can help to retain, or slow down the degradation of certain nutrients such as ascorbate, chlorophylls, carotenoids, and sugars. However, they can also be used to increase the nutrient content of foods. Investigations into the effects of various types of light treatments involving either monochromatic lighting regimes using LEDs, or the use of LEDs supplementing traditional light sources, have been shown to produce crops with superior nutritional quality (Bian et al. 2015). For example, Lee et al. (2014) investigated the effect of white, blue (436 nm), green (524 nm), and red (665 nm) LED treatments on the nutrient content of cabbages. It was found that after 18 days, chlorophyll content was highest for samples treated with green and white LEDs, followed by red and blue LEDs. In contrast, vitamin C and total phenolic content were increased by blue and white LED treatments. Although the results demonstrated that LED treatments generally improved the nutritional quality of vegetables stored in a refrigerator, the quantity of light received by the cabbages was not specified.

Outside of the visible range, UV and IR LEDs avail more interesting potential applications in terms of nutritional enhancement. For instance, watercress and garden pea sprouts were exposed to 33 μmol $m^{-2}$ $s^{-1}$ of UV-A radiation from an LED (375 nm) for a duration of 160 min daily over 3 days and then stored in darkness (Kanazawa et al. 2012). The quercetin-glycoside content of the vegetables was found to be significantly greater than those stored in the dark after 6 days from the beginning of the treatment. Hence, the study suggested that such UV LEDs could stimulate flavonoid and phenylpropanoid production in vegetables.

Near infrared (NIR) LEDs were used to investigate the effects of NIR radiation on transpiration rates and reactive oxygen species (ROS) accumulation in 'Notip' and 'Cisco' lettuce (*Lactuca sativa* L. Crispa Group) after harvesting (Kozuki et al. 2015). The 850-nm LED produced optimal results, leading to the lowest relative transpiration rates among all the irradiated as well as the non-irradiated control samples, for irradiation durations as low as 1 min. This was attributed to stomatal closing caused by increased ROS production in response to NIR irradiation, resulting in firmer and more visually appealing samples. Although the study measured a 20% increase in ROS production in guard cells, no further study was conducted to ascertain whether there was a corresponding increase in nutrients such as antioxidants, hence such an investigation might be worth pursuing in future studies.

Other than the leaves of vegetables, other edible plant parts respond differently to different LED treatments. Red LED treatment of broccoli (*Brassica oleracea* L. var. *italica*) at 50 μmol $m^{-2}$ $s^{-1}$ for 4 days caused a slower rate of ethylene production, slower degradation of ascorbate content, and less yellowness of the treated samples compared to the blue LED treatment and dark control (Ma et al. 2014). In contrast, the study by Hasperué et al. (2016) reported that the antioxidant levels, total phenolic content, and ascorbic acid levels in treated samples were mostly equal to or less than the samples stored in the dark. However, yellowing was similarly suppressed and carotenoid content increased significantly under this treatment. The lower photon flux of 20 μmol $m^{-2}$ $s^{-1}$ was not strong enough to induce the production of antioxidants.

To account for the biological response of such foods to light, several studies have investigated the relationship between exposure to LED light and biomolecular responses in terms of gene expression. In this regards, fruits have been studied in great detail. Blue LEDs were found to effectively increase total carotenoids in the peels and pulp of two cultivars of peaches (*Prunus persica* 'Hujing' and 'Jinli'), and the necessary gene expression contributing to the increases was investigated by Cao et al. (2017). Blue (440 nm) and red (660 nm) LED treatments at 80 μmol $m^{-2}$ $s^{-1}$ increased the content of stilbenes in grape berries (*Vitis labruscana* Bailey) by appropriately regulating gene expression of key enzymes in the phenylpropanoid and stilbene biosynthesis pathways (Ahn et al. 2015). Shi et al. (2016) also showed evidence that blue LED (470 nm) irradiation at 40 μmol $m^{-2}$ $s^{-1}$ increased glucose and fructose while maintaining sucrose levels in Chinese bayberries (*Myrica rubra* Sieb. and Zucc. cv. Biqi), by upregulating genes involved in sugar metabolism such as sucrose phosphate synthase, acid invertase, glucose sensor, and cryptochrome genes.

Furthermore, citrus fruits have been extensively studied in this manner. Ma et al. (2011) outlined the effectiveness of red LED compared to blue LED irradiation on the regulation of gene expression that gave rise to an increase in β-cryptoxanthin in the flavedo of Satsuma mandarins. This effect was even greater in the flavedo of fruits treated to a combination of red LED and exogenous ethylene exposure (Ma et al. 2015). In contrast, Zhang et al. (2012) showed that blue LED treatment was more effective at increasing total carotenoids in the juice sacs of Satsuma mandarins, Valencia oranges (*Citrus sinensis* Osbeck), and Lisbon lemons (*Citrus limon* Burm. f.), and they studied the regulation of similar genes. A later study showed that blue LED treatment caused greater upregulation of gene expression for ascorbic acid biosynthetic and regeneration genes, and two types of reduced glutathione-producing genes, than did red LEDs, for the same citrus cultivars (Zhang et al. 2015). From the above studies, different LEDs would induce different biochemical responses (and hence nutritional changes) in different species of fruits. Within similar species, different LEDs may have different effects depending on the location on the fruit.

### *9.3.3 Accelerating or Delaying the Ripening of Fruits Using LEDs*

To reduce postharvest losses of fruits that are being transported over long distances or stored for long durations, manipulating the rate of ripening is a strategy that can be used. For example, the application of blue light prior to storage in the dark extended the ripening time of tomatoes (Dhakal and Baek 2014a, b). Mature green tomatoes had a slower rate of color change and were firmer when irradiated with blue light (440–450 nm) for a period of 7 days, compared to those stored in darkness or irradiated with red light (650–660 nm). Correspondingly, lycopene accumulation was reduced in response to blue light irradiation. Therefore, blue LED treatment was shown to be a convenient way of delaying the ripening of tomatoes, thereby extending their postharvest commercial value.

In contrast, blue LED light (470 nm) accelerated respiration, ethylene production, and the development of red color in strawberries (Xu et al. 2014a, b). Yet, green (525 nm) and red (630 nm) LED irradiations were also able to accelerate the increase in anthocyanins in immature strawberries to a smaller extent than blue LEDs (470 nm), suggesting that secondary ripening processes can be hastened by other LEDs if blue LEDs are unavailable (Kim et al. 2011). In a similar vein, the effect of monochromatic LED light of various wavelengths should be studied on various other climacteric fruits as this would be of immense commercial value.

### *9.3.4 Preventing Fungal Spoilage Through LEDs*

Decay by fungi such as gray mold (*Botrytis cineria*) causes a significant amount of food loss (Kader and Rolle 2004). Recently it has been shown that blue LED light can help to attenuate fungal infections in citrus fruits. Soft rot area, mycelial growth, and sporulation of *Penicillium digitatum*, *Penicillium italicum*, and *Phomopsis citri* on the surface of tangerines were reduced when treated with blue light at 40 μmol $m^{-2}$ $s^{-1}$ over 5–7 d, compared to white light LED at a similar photon flux, and dark control (Alferez et al. 2012; Liao et al. 2013). This treatment was shown through real-time qRT-PCR analysis to increase the expression of phospholipase $A_2$ ($PLA_2$), an enzyme involved in the production of lysophosphatidylcholine which increases resistance to fungal infection and growth. In contrast, red light treatment led to the down-regulation of phospholipase D (PLD), which also provides antifungal defense (Alferez et al. 2012). Other than the above phospholipases, octanal, which possesses antifungal properties as well, increased in concentration in the flavedo of 'Fallglo' tangerines and sweet oranges upon blue LED irradiation. Polygalacturonase activity in *P. digitatum*, which is critical for fungal pathogenicity, was also lowered upon blue LED irradiation (Liao et al. 2013). The effectiveness of using blue LEDs in citrus fruits was replicated in a study on Satsuma mandarins (*C. unshiu* Marc. 'Aoshimaunshu'), which showed that both 8 and 80 μmol $m^{-2}$ $s^{-1}$ of blue LED light (465 nm) were able to significantly decrease the rate of growth in the soft rot, mycelial, and sporulation zones over 6 days (Yamaga et al. 2015).

Following this, the question of whether continuous irradiation over several days is the most effective form of treatment arises. Alferez et al. (2012) found that 12-h blue LED treatments per day (followed by 12 h of darkness) were more effective at reducing mycelial growth of *P. digitatum* compared to continuous irradiation. However, these fruits were pre-treated for 3 d with blue LED light prior to inoculation, which may not be reflective of natural conditions in which the time of contamination or infection may not be known. Indeed, when fruits were inoculated immediately after harvesting, there were no significant differences between continuous treatments and 12-h treatments daily for 5 d, in terms of soft rot area of *P. digitatum* (Liao et al. 2013). However, both treatments reduced mycelial and sporulation areas to negligible after 5 d. Even so, since their effects were similar, it is worth considering using 12-h irradiation regimens for energy savings.

Further studies were performed using *P. digitatum* and *P. italicum* strains resistant to fungicides thiabendazole and imazalil, to ascertain the optimal lighting regime for inhibiting their growth in vitro. When 700 μmol $m^{-2}$ $s^{-1}$ of blue LED light was applied immediately after inoculation, colony growth was completely suppressed. When it was applied after 4 days, growth persisted but was severely limited. However, a lower photon flux of 120 μmol $m^{-2}$ $s^{-1}$ of blue LED light exerted a greater fungicidal effect when applied 4 d after inoculation. Although 700 μmol $m^{-2}$ $s^{-1}$ is a significantly high intensity of light, it was still possible to maintain the temperature of the experimental system at 20 °C throughout the duration of treatment (Lafuente and Alférez 2015). These studies exemplify how LED exposure is a viable alternative to common fungicides, as the risk of fungicide resistance increases.

Another strategy that can be employed in response to increasing fungicide resistance is the use of synergistic combinations of treatments. Yu and Lee (2013) tested the effectiveness of combining LED irradiation with the use of antagonistic bacteria, *Bacillus amyloliquefaciens* JBC36, which was applied as a biofilm to the surface of fruit. As opposed to the above studies, in vitro experiments found that the irradiation of 240 μmol $m^{-2}$ $s^{-1}$ of red LED light (645 nm) was more effective than other wavelengths in increasing the motility and biofilm formation of the bacteria. Furthermore, the LED treatment stimulated the production of iturin and fengycin, which are antifungal lipopeptides, thereby further contributing to the antifungal activity of the bacteria. Ramkumar et al. (2013) confirmed that red light exposure increased expression of *fen*A gene in *Bacillus amyloliquefaciens* JBC36, which is responsible for the synthesis of fengycin. The use of such a synergistic strategy might solve the problem of re-emergence of infections when LED treatment is discontinued, due to fungal growth below the surface of the fruit (Alferez et al. 2012). Further studies should be conducted to verify this.

UV LEDs can also be used to prevent fungal infection. A system consisting of UV LEDs of wavelengths of 272, 289, or 293 nm was used to irradiate strawberries purchased from a supermarket over 9 d at 20 $mWm^{-2}$. The treatment prevented any mold growth for the period of 9 d, whereas significant growth of mold (suspected to be *Botrytis cinerea*) was found on strawberries stored in 6 d of darkness. The UV treatment also resulted in the retention of anthocyanins and total soluble sugar levels, compared to those stored in the dark and which were found to have decreases in the above nutrients (Britz et al. 2013). LEDs with wavelength of 405 nm were also reported to prevent the growth of *B. cinerea* on detached tomato leaves, which are not usually consumed (Imada et al. 2014). This occurred through the interaction between light at that wavelength and the endogenous porphyrins in the mold, resulting in the production of toxic ROS. Using an LED within the visible range is preferable to UV LEDs of low wavelength as such UV radiation can harm the eyes and skin (Shama 2014).

### 9.3.5 *Evaluation of LEDs in Postharvest Preservation*

This section (i.e., Sect. 9.3) has focused on a few aspects of postharvest quality, namely the prevention of senescence, fungal infection and ripening, or the acceleration of ripening where applicable, as well as the enhancement of nutritional quality. With regard to nutritional quality, there is still currently a lack of studies showing the effect of various LEDs on leafy vegetables, which is surprising as there have been many studies conducted on leafy vegetables during the pre-harvest growth stage. Furthermore, there have been many postharvest studies conducted on leafy vegetables using broad-spectrum lighting. Granted that it is challenging to find the optimal lighting regime that is not excessive (hence risking oxidative damage), using low quantities of monochromatic light is still a possible path to take in initial studies. For example, Woltering and Seifu (2015) found that small quantities (5 μmol $m^{-2}$ $s^{-1}$) of red, blue, and green LED lights resulted in increased levels of glucose, fructose, and sucrose in butterhead lettuce, and marked reduction in sugars depletion in iceberg lettuce, compared to the samples stored in darkness. Moreover, since the quantity of light used was significantly below the light compensation point, it was concluded that increase in sugar levels was due to the process of gluconeogenesis, instead of photosynthesis. This seems to contradict the earlier statement by Noodén and Schneider (2004), which could be due to the fact that monochromatic light, not white light, was used in this experiment. It also means that the process of gluconeogensis could be exploited in novel ways to improve the nutritional quality of leafy vegetables using monochromatic light and hence should be investigated further.

An advantage of using low-powered LEDs is that it can potentially lead to high energy savings. Braidot et al. (2014) showed that lamb's lettuces stored at 6 °C with pulsed lighting were not significantly different in terms of postharvest quality than samples stored in the dark at 4 °C. A higher storage temperature could be conducive for long-term energy savings. Furthermore, although Lee et al. (2014) did not specify the photon flux of the various treatments, it was stated that the input electrical power ranged from 1.0 to 1.5 W for each LED system in the refrigerators. These confirm the practicality of using LEDs in cold storage facilities.

A recurring issue with the use of light in general on vegetables is the reduction in mass due to moisture loss. This is normally due to transpiration, which is aggravated by light exposure. It is well known that blue light increases stomatal conductance and transpiration in leaves (Massa et al. 2008; Muneer et al. 2014), which results in moisture loss during the postharvest storage. Lee et al. (2014) reported a lower moisture content in cabbages that were exposed to blue, green, and white LEDs over 12 days, compared to samples treated with red LEDs or kept in the dark. Low moisture content can result in wilted and less visually appealing leaves, and therefore, a lower consumer acceptance, but perhaps this could potentially be reversed by IR radiation as previously shown by Kozuki et al. (2015). Therefore, in order to preserve moisture content in leafy vegetables exposed to light, future studies could incorporate IR LEDs to retard water loss due to transpiration, while incorporating other LEDs to bring about improvements in nutritional content.

Therefore, future studies conducted using LEDs on fruits and vegetables need to account for other quality changes that might compromise consumer acceptability, such as texture (which can be measured by a texture analyzer), color, or even flavor-active compounds.

## 9.4 LEDs in Food Safety

While the previous section has covered various postharvest quality attributes that may increase the shelf life of perishable foods by slowing down degradative processes within the food, or accelerating other biological processes that increase the commercial or nutritional value of the foods, another critical aspect of the postharvest quality is the microbiological safety of produce. Food safety is of prime priority in the food industry. Food contaminated with pathogenic bacteria could result in foodborne diseases and therefore must be appropriately processed. Thermal techniques, while being the most efficacious methods of eliminating pathogens, will cause the destruction of foods such as fresh produce, juices, and ready-to-eat salads. Compounded with consumers' demand for minimally processed food free from chemical sanitizers and other additives, and an increasing risk of antimicrobial resistance in food pathogens, new forms of effective food safety technologies for food processing facilities need to be found (Capita and Alonso-Calleja 2011).

Visible light has bactericidal effects when combined with a photosensitizer and oxygen, through a phenomenon known as photodynamic inactivation (PDI). Moreover, UV radiation itself has bactericidal effects. When combined with suitable nanoparticles, UV radiation can cause bacterial death through photocatalytic oxidation. While the use of the above techniques has been researched quite widely for applications in the fields of medicine, dentistry, and water purification, recently more attention has been given to applications in food-related decontamination processes, with LEDs being widely studied as a suitable source of light. Other than the energy savings that LEDs offer, the lack of radiant heat is an attractive feature since heat can potentially accelerate the deterioration of food quality. The subsequent sections first present the foundational in vitro studies demonstrating the efficacy of using LED treatments in PDI, photocatalytic inactivation and direct UV exposure, followed by studies that have been conducted on model food systems such as beverages, or actual food matrices such as fruits and vegetables.

### 9.4.1 PDI Using Exogenous Photosensitizers

PDI is one of the most common modes of decontamination studied in food-related applications of LEDs (Luksiene and Brovko 2013). Essentially, PDI requires a photoactive molecule (also known as a photosensitizer), light, and oxygen. Excitation of the photosensitizer occurs during the interaction with a light photon. Subsequently, ROS are generated when the photosensitizer returns to ground state. This occurs through two pathways. Firstly, the Type I mechanism involves the transfer of energy to surrounding substrates, which then results in ROS generation of species such as superoxide anion ($O_2^{\bullet -}$), hydrogen peroxide ($H_2O_2$), and hydroxyl radicals ($^{\bullet}OH$). In contrast, the Type II mechanism involves the transfer of energy from the photosensitizer to the stable molecular oxygen in its triplet state ($^3O_2$), causing its excitation to the singlet state ($^1O_2$). These ROS cause extensive damage to the cellular components comprising lipids, fatty acids, peptides, and other substrates such as in the cell membrane (Kiesslich et al. 2013). Since the production of the above ROS results in the indiscriminate destruction of cellular components, it is expected that resistance to PDI is more difficult to evolve. However, the treatment provided must be sufficient to completely inactivate the target pathogen; otherwise, sublethal treatments may induce stress tolerance (St. Denis et al. 2011). Even so, experiments have shown that resistance to PDI does not develop in targeted microorganisms even after 10 cycles of PDI treatment (Tavares et al. 2010; Bartolomeu et al. 2016).

The photosensitizer is the most crucial component in PDI, and the properties of various photosensitizers have been reviewed in considerable depth by other authors such as Luksiene and Brovko (2013) and Kiesslich et al. (2013). The following is a summary of the salient points. The characteristics of a functionally effective exogenous photosensitizer include possessing a high light absorption coefficient within the wavelength range of excitation, a triplet state which reaches a high quantum yield ($\Phi_T > 0.4$), high energy ($E_T \geq 95$ kJmol$^{-1}$, and sufficiently long lifetime ($\tau_T > 1$ µs). These characteristics allow for maximum energy transfer from the photosensitizer to reactants. The lipophilicity and the ionization constant ($pK_a$) must be considered alongside the nature of the food matrix as these affect the uptake of the molecule into the target pathogen. Finally, they should not in themselves be toxic. Most photosensitizers that have been identified and validated are confined to clinical applications and may not be suitable for application in food. However, photosensitizers that are suitable and effective in food applications, found in natural sources, or have been studied in substantial depth previously include hypericin, curcumin, alpha-terthienyl, and chlorophyllin (Luksiene and Brovko 2013). Based on the photosensitizer being used, it is crucial to select a suitable light source. Various forms of lighting can be used in PDI, including broad-spectrum and pulsed lighting that provide sufficient quantity of light in the range of absorbance of the photosensitizer. However, it is more economical to use light whose peak wavelength coincides with the absorption maximum of the selected photosensitizer, and therefore monochromatic light sources such as LEDs are the most appropriate lighting source.

Conventionally, a photosensitizer is added from an external source into the media carrying the microorganism of interest, or the food matrix in question. Hence the photosensitizer is found in the exogenous environment to the pathogen, where the lethal ROS are generated. In terms of susceptibility, in vitro studies have shown that Gram-positive bacteria are more susceptible to PDI as the photosensitizer is more easily trapped in the peptidoglycan layer of the cell wall, whereas the double cell membrane structure in Gram-negative bacteria acts as a more effective barrier to photosensitizers (Demidova and Hamblin 2004). Increasing the photosensitizer concentration, or using cationic photosensitizers or photosensitizers conjugated to positively charged polymers, has been shown to improve their uptake (Luksiene and Brovko 2013). Another potential strategy that could increase susceptibility in Gram-negative species is to conjugate photosensitizers to antimicrobial peptides which bind specifically to target cells. Eosin Y was conjugated to an antimicrobial peptide, $(KLAKLAK)_2$, and was shown to target both Gram-positive and Gram-negative bacteria, as opposed to red blood cells or other mammalian cells (Johnson et al. 2012).

With sufficient knowledge of the mechanisms of inactivation that occur during PDI, qualitative and quantitative comparisons can be made between different forms of PDI treatment by using appropriate mathematical models, which can bring more clarity to the inactivation kinetics of a PDI treatment. Aponiene et al. (2015) found that the Logistic model was suitable for describing inactivation curves of *B. cereus* incubated with hypericin and exposed to a green LED of 585 nm ($R^2 > 0.97$). Furthermore, Dementavicius et al. (2016) compared three models, namely the Weibull, Logistic, and Geeraerd models, to find which one best described the inactivation of *B. cereus* and *L. monocytogenes* similarly incubated with hypericin and exposed to a green LED. The study concluded that of the three, the Logistic model gave the best fit in terms of the determination coefficient ($R^2$) and root-mean-square error (RMSE). In the Logistic model, model parameters include 'number of cells resistant to treatment,' 'shoulder parameter,' 'population reduction suddenness,' and 'maximum reduction rate'. A thorough explanation of these parameters is discussed in Dementavicius et al. (2016). The study concluded that *L. monocytogenes* was more easily inactivated by hypericin-based PDI than *B. cereus* based on the comparison of the above parameters. This shows the merits of using mathematical modeling if done appropriately and rigorously. The availability of quantitative data can give objective insights into the efficacy of a treatment, or the susceptibility of a bacterial species to the treatment. It would be useful and beneficial to the food industry to conduct such studies on actual food matrices, such as the one performed on endogenous photosensitizers by Ghate et al. (2016).

### 9.4.2 PDI Through Endogenous Photosensitizers

Instead of applying photosensitizers to food systems from an external source, the excitation of endogenous photosensitizers located intrinsically within bacterial pathogens represents another antimicrobial strategy. Endogenous photosensitizers usually exist in the form of intracellular components like 'porphyrins, cytochromes, flavins, and NADH' (Lubart et al. 2011). Research in this field has been very productive in the recent years (Table 9.2), hence posing as a valid alternative to the use of exogenous photosensitizing agents.

Without the need for photosensitizing additives, the most critical conditions for ensuring successful inactivation lie in the LED wavelength and intensity. In terms of wavelength, it has long been established that blue light or near-UV radiation, typically within the band of 400–405 nm (Soret band), is the most effective at inactivating bacteria and fungi as it coincides with the absorption maximum of photoactive porphyrins within the organisms (Maclean et al. 2008, 2014; Endarko et al. 2012; Imada et al. 2014). For a variety of Gram-positive and Gram-negative foodborne pathogens irradiated with 486 J $cm^{-2}$ from an LED of 405 nm peak wavelength, it was determined that the resulting reduction in populations was partly due to cellular membrane damage, but not DNA fragmentation (Kim et al. 2015, 2016a).

While the above mentioned studies demonstrated the effectiveness of LEDs with a peak wavelength of 405 nm, some studies that compared LEDs emitting red, blue, and green lights also confirmed that the maximum inactivation of *Salmonella typhimurium*, *Escherichia coli* O157:H7, *Listeria monocytogenes*, and *Staphylococcus aureus* was caused by blue LEDs of a slightly higher peak wavelength of 461 nm (Ghate et al. 2013). Another study confirmed the superior inactivation ability of blue LEDs over green and red LEDs on *Porphyromonas gingivalis*, *S. aureus* and *E. coli* DH5α (Kim et al. 2013). In both studies, green LEDs were also moderately effective at inactivating bacteria as light within the green region could still be absorbed by photosensitizers (Maclean et al. 2009), whereas no inactivation was observed by red LEDs. Moreover, treatment with blue LEDs resulted in the highest rate of sublethal injury to bacteria, indicating that blue light can significantly injure surviving populations of bacteria (Ghate et al. 2013). However, LEDs with peak wavelength at 405 nm were shown to be significantly more effective than LEDs at 460 nm, resulting in greater inactivation of *S. aureus*, *Lactobacillus plantarum*, and *Vibrio parahaemolyticus* after 7-h treatment at 4, 10, and 25 °C, despite the former generating a smaller maximum dosage at 7 h compared to the latter (Kumar et al. 2016). The effectiveness of the LED was attributed to the possibility that a significant proportion of the output spectrum fell within the UV range, hence compounding the killing effect on bacteria.

The effect of temperature of the system appears to have different effects on different bacteria. According to Ghate et al. (2013), at 20 °C, blue LED (461 nm) treatments halted bacterial growth of *S. typhimurium*, *E. coli* O157:H7, *L. monocytogenes*, and *S. aureus*; but when temperatures were lowered to 15 and

**Table 9.2** Effect of PDI on foodborne bacteria using endogenous photosensitizers and LED illumination in vitro

| Target bacteria | LED peak wavelength (nm) | Intensity | Treatment time | Effect | References |
|---|---|---|---|---|---|
| *Bacillus cereus* | 405 | 18 $mWcm^{-2}$ | 7.5 h | Bacterial populations in phosphate-buffered saline (PBS) held at 4 °C were reduced by 1.9 log CFU $mL^{-1}$, respectively, after 7.5 h | Kim et al. (2015) |
| | 400 | 20 $mWcm^{-2}$ | 20 min | Bacterial populations were incubated in 7.5 mM of 5-aminolevulinic acid (ALA), a non-photosensitizing metabolic precursor to endogenous photosensitizers. Irradiation for 20 min in Luria-Bertoni (LB) medium caused reduction of up to 6.3 log cycles. Treatment was carried out at 37 °C | Luksiene et al. (2009) |
| *Listeria monocytogenes* | 461 | 596.7 $Jcm^{-2}$ | 7.5 h | Bacterial populations in TSB held at 15 and 10 °C were reduced by 4.3 and 5.2 log CFU $mL^{-1}$, respectively, after 7.5 h, compared to 0.9 and 1.5 log CFU $mL^{-1}$ at 521 nm. No significant reductions at 641 nm | Ghate et al. (2013) |
| | 405 | 185 $Jcm^{-2}$ | 5 h | Bacterial populations in TSB were reduced by 5 log CFU $mL^{-1}$ to below detection limits after treatment with irradiance of 8.6 $mWcm^{-2}$ | Endarko et al. (2012) |
| | 405 | 84 $Jcm^{-2}$ | NR | When treated with light at 70 $mWcm^{-2}$ at 22 °C, bacterial populations in TSB were reduced by 5 log CFU $mL^{-1}$ to below detection levels | McKenzie et al. (2014) |
| | 405 | 18 $mWcm^{-2}$ | 7.5 h | Bacterial populations in PBS held at 4°C were reduced by 2.1 log CFU $mL^{-1}$, respectively, after 7.5 h. | Kim et al. (2015) |
| | 400 | 20 $mWcm^{-2}$ | 20 min | Bacterial populations were incubated in 7.5 mM of ALA. Irradiation for 20 min in LB medium caused reduction of up to 4 log cycles. Treatment was carried out at 37 °C | Buchovec et al (2010) |
| *Listeria innocua* | 395 | 36 $Jcm^{-2}$ | 1115 s | Bacterial populations in maximum recovery diluent were reduced by 2.74 log CFU $mL^{-1}$ | Birmpa et al. (2014) |
| *Staphylococcus aureus* | 461 | 596.7 $Jcm^{-2}$ | 7.5 h | Bacterial populations in TSB held at 15 and 10 °C were reduced by approximately 5.2 and 4.7 log CFU $mL^{-1}$, respectively, after 7.5 h, compared to 1.7 and 1.5 log CFU $mL^{-1}$ at 521 nm. No significant reductions at 641 nm | Ghate et al. (2013) |
| | 405 | 18 $mWcm^{-2}$ | 7.5 h | Bacterial populations in PBS held at 4 °C were reduced by 0.9 log CFU $mL^{-1}$ after 7.5 h | Kim et al. (2015) |
| | 405 | 24 $mWcm^{-2}$ | 7 h | Bacterial populations in TSB held at 4, 10 and 25 °C were reduced by approximately 1.0 log CFU $mL^{-1}$, compared to negligible reductions at 461 and 521 nm | Kumar et al. (2016) |

(continued)

**Table 9.2** (continued)

| Target bacteria | LED peak wavelength (nm) | Intensity | Treatment time | Effect | References |
|---|---|---|---|---|---|
| Methicillin-resistant *S. aureus* | 470 | 220 $Jcm^{-2}$ | N.R. | Bacterial populations on trypticase soy agar were reduced from 6.0 log CFU $mL^{-1}$ to below detectable levels after irradiation | Bumah et al. (2015) |
| *Campylobacter* spp. | 405 | 18 $Jcm^{-2}$ | 30 min | Bacterial populations were reduced from 5.25 log CFU $mL^{-1}$ to below detection limit | Murdoch et al. (2010) |
| | 395 | 0.06–18.00 $Jcm^{-2}$ | 5 min | At a distance of 3 cm from LED, 10 isolates of *C. jejuni* and *C. coli* were inactivated from around 6–7 log CFU $mL^{-1}$ to below detection limit after 5 min. As distance increased, treatment time required for inactivation increased. Certain strains took longer to inactivate | Haughton et al. (2012) |
| *Escherichia coli* O157: H7 | 461 | 596.7 $Jcm^{-2}$ | 7.5 h | Bacterial populations in TSB held at 15 and 10 °C were reduced by approximately 5 log CFU $mL^{-1}$ after 7.5 h, compared to 1.0 and 1.8 log CFU $mL^{-1}$, respectively, at 521 nm. No significant reductions at 641 nm | Ghate et al. (2013) |
| | 405 | 378 $Jcm^{-2}$ | NR | When treated with light at 70 $mWcm^{-2}$ at 22 °C, bacterial populations TSB were reduced by 5 log CFU $mL^{-1}$ to below detection levels | McKenzie et al. (2014) |
| | | 18 $mWcm^{-2}$ | 7.5 h | Bacterial populations in PBS held at 4 °C were reduced by 1.0 log CFU $mL^{-1}$, respectively, after 7.5 h | Kim et al. (2015) |
| | 395 | 36 $Jcm^{-2}$ | 1115 s | Bacterial populations in maximum recovery diluent were reduced by 1.37 log CFU $mL^{-1}$. | Birmpa et al. (2014) |
| *Salmonella typhimurium* | 461 | 596.7 $Jcm^{-2}$ | 7.5 h | Bacterial populations in trypticase soy broth (TSB) held at 15 and 10 °C were reduced by 5.0 and 4.6 log CFU $mL^{-1}$, respectively, after 7.5 h, compared to approximately 1.7 log CFU $mL^{-1}$ at 521 nm. No significant reductions at 641 nm | Ghate et al. (2013) |
| | 405 | 18 $mWcm^{-2}$ | 7.5 h | Bacterial populations in PBS held at 4 °C were reduced by 2.0 log CFU $mL^{-1}$, respectively, after 7.5 h | Kim et al. (2015) |
| | 400 | 20 $mWcm^{-2}$ | 20 min | Bacterial populations were incubated in 7.5 mM of 5-aminolevulinic acid (ALA). Irradiation for 20 min in LB medium caused reduction of up to 6 log CFU $mL^{-1}$. Treatment was carried out at 37 °C | Buchovec et al. (2009) |

(continued)

**Table 9.2** (continued)

| Target bacteria | LED peak wavelength (nm) | Intensity | Treatment time | Effect | References |
|---|---|---|---|---|---|
| *Salmonella typhimurium* and *Salmonella heidelberg* | 470 | 165 $Jcm^{-2}$ | N.R. | Bacterial populations on Salmonella-Shigella agar were reduced from 6.0 log CFU $mL^{-1}$ to below detectable levels after irradiation | Bumah et al. (2015) |
| *Shigella sonnei* | 405 | 18 $mWcm^{-2}$ | 7.5 h | Bacterial populations in PBS held at 4 °C were reduced by 0.8 log CFU $mL^{-1}$, respectively, after 7.5 h | Kim et al. (2015) |
| *Vibrio parahaemolyticus* | 405 | 24 $mWcm^{-2}$ | 7 h | Bacterial populations in TSB held at 4 and 10 °C were reduced by approximately 6.0 log CFU $mL^{-1}$ after 5 and 7 h, respectively, compared to negligible reductions at 521 nm | Kumar et al. (2016) |

*Source* Adapted from D'Souza et al. (2015) with updates

10 °C, inactivation was more pronounced such that blue LED treatments resulted in populations below detectable limits after 6–7.5 h. On the other hand, Kumar et al. (2016) reported greater inactivation of *L. plantarum* when illuminated by LEDs (405 nm) for 7 h at 25 °C with a dosage of approximately 600 $Jcm^{-2}$ of blue LED treatment, which was similar to the treatment used by Ghate et al. (2013). In contrast, *V. parahaemolyticus* was more effectively inactivated at 4 and 10 °C under the same conditions. However, this apparent discrepancy in efficacy could be due to the use of phosphate-buffered saline (PBS) as a bacterial medium. Unlike other growth media such as trypticase soy broth, it lacks the nutrients required by injured bacteria to recover from an injured state. Additionally, the authors cautioned that the apparent high inactivation *V. parahaemolyticus* might have been due to cells being converted into the viable but non-culturable (VBNC) state, which is worth future investigation. These results suggest that different bacterial species will respond differently to varying temperatures, as determined by the adaptability of the bacteria's membrane fluidity to temperature and/or the dependence of the bacteria's self-repair system on temperature. While adapting to lower temperatures, the cell membranes of such bacteria may become composed of a greater proportion of unsaturated fatty acids which are more susceptible to ROS damage (Ghate et al. 2013; Kumar et al. 2016). More extensive studies will be required on different bacterial strains; but in general, inactivation was substantial at typical refrigerator temperatures of between 4 and 10 °C, and, hence, PDI inactivation of foodborne pathogens can readily be carried out in a refrigerator fitted with suitable LEDs, for example.

Bacterial susceptibility to PDI through endogenous photosensitizers varies significantly among, and within, bacterial species. For example, *Campylobacter jejuni* required a much lower dosage of blue light at 405 nm than *Salmonella enteritidis* and *E. coli*. This might be because *C. jejuni*, which is a microaerophilic species, is naturally more susceptible to damage via ROS. However, authors cautioned that this apparent susceptibility might be due to the ability of *Campylobacter* spp. to become VBNC (Murdoch et al. 2010), hence leading to an overestimation in its susceptibility. Further studies are therefore required to confirm this. In contrast, a study using a LED of 405 nm showed that *Listeria* spp. were most easily inactivated, followed by *E.coli*, *Shigellasonnei*, and *S. enteritidis* (Endarko et al. 2012). While several authors suggested that Gram-positive bacteria were more susceptible than Gram-negative species (Maclean et al. 2009; Birmpa et al. 2014), others observed that susceptibility was not determined by Gram nature (Ghate et al. 2013). Moreover, it was shown that there were differences in susceptibility between various strains of *Campylobacter* spp. isolates when exposed to the same treatment, and this example of intraspecies variation in susceptibility was thought to be due to different concentrations of endogenous porphyrin within species (Maclean et al. 2009; Haughton et al. 2012). On this note, Kumar et al. (2015) showed a correlation between the higher susceptibility of Gram-positive species of bacteria and the quantity of intracellular coproporphyrins. However, within Gram-positive species, there was no direct and strong correlation between coproporphyrin content and susceptibilities, possibly due to other components in bacterial cells that are capable of quenching ROS, such as pyocyanin in *P. aeruginosa*. The interactions between ROS produced by photosensitizing intracellular components and other such radical-scavenging components in cells suggest that future investigations should be orientated toward characterizing such components and studying their effect on the overall success of a PDI treatment.

One way of increasing susceptibility by increasing photosensitizing intracellular components is through the external addition of 5-aminolevulinic acid (ALA), a non-photoactive metabolic precursor in heme biosynthesis which can give rise to various endogenous photosensitizing porphyrins. The addition of ALA is suitable for food applications because ALA is colorless and tasteless, while being effective against a range of foodborne pathogens, yeasts and fungi, viruses, and even certain protozoa (Harris and Pierpoint 2012; Luksiene and Brovko 2013). It has been shown to inactivate not only vegetative *S. typhimurium* cells (Buchovec et al. 2009), but also *Bacillus cereus* spores (Luksiene et al. 2009) and *L. monocytogenes* biofilms on packaging surfaces (Buchovec et al. 2010) when treated with LED light at 400 nm for as little as 15 min.

As discussed earlier, mathematical models present us with a useful means of evaluating the efficacy of a photosensitizing treatment in terms of its inactivation kinetics. Several studies on the inactivation kinetics of PDI through endogenous photosensitizers exist. Ghate et al. (2013), who studied the effect of wavelength, temperature, and dosage of LED treatment on the inactivation and decimal reduction values of selected pathogens, reported that D-values for treatments using LEDs at 461 nm at 10 °C ranged from 1.19 h for *L. monocytogenes* to approximately 1.4–1.5 h for *E. coli* O157:H7, *S. typhimurium*, and *S. aureus*. Kumar et al. (2015) modeled the inactivation curves of *B. cereus*, *E. coli* O157:H7, *S. aureus*, *S. typhimurium*, *L. monocytogenes*, and *P. aeruginosa* treated with 405 nm and 520 nm LEDs at 4, 10, and 25 °C. A more recent work by Kumar et al. (2016) modeled the inactivation curves of *L. plantarum*, *S. aureus*, and *V. parahaemolyticus*, while other studies described the susceptibility of *L. monocytogenes*, *B. cereus*, *S. aureus*, *S. typhimurium*, and *E. coli* O157:H7 to 405-nm LED treatments using the Weibull model (Kim et al. 2015, 2016a). Since the above were in vitro studies, more inactivation studies should be conducted on food systems, as well as packaging and contact surfaces.

Despite the success of using LEDs directly to perform PDI, this method may not be as effective as PDI using exogenous photosensitizers. For instance, a dosage of 185 $Jcm^{-2}$ was required to inactivate *L. monocytogenes* in vitro using a blue LED at 405 nm (Endarko et al. 2012), whereas 36 $Jcm^{-2}$ was sufficient for a 7 log inactivation when treating a thermo resistant *L. monocytogenes* 56 Ly strain in vitro using sodium chlorophyllin (Na–Chl) as a photosensitizer (Luksiene et al. 2010). Nevertheless, PDI through endogenous photosensitizers is probably more desirable as the treatment does not require any photosensitizing additive to function properly. Furthermore, there is still very little data available on the consequence of adding photosensitizers to the acceptability of foods from the consumer's point of view.

### 9.4.3 UV LEDs

UV radiation can be classified based on its wavelength range: a wavelength range of 200–280 nm is assigned UV-C; 280–315 nm is assigned UV-B, while 315–400 nm is assigned UV-A. UV radiation in general has a damaging effect on DNA replication and transcription. Direct exposure to UV-C or UV-B results in the inactivation of a variety of microorganisms, such as bacteria, viruses, fungi, protozoa, and several other pathogenic and parasitic organisms (Lui et al. 2014). Normally, mercury tube lamps are used to produce UV-C radiation for bactericidal purposes. In contrast, UV LEDs offer more preferable features compared to mercury tube lamps. UV LEDs have the ability to produce quick pulses with no warm-up time. Chips of various wavelengths can be constructed, as opposed to specifically fixed wavelengths of the mercury tube lamps, which usually have a peak of 254 nm. Most importantly, it contains no toxic mercury (Lui et al. 2014). Obviously, they also provide the common physical benefits of LEDs such as durability and space efficiency. Producing UV LEDs that match the efficiency of mercury tube lamps is technically challenging, but the technology is developing rapidly and is predicted to surpass mercury tube technology in the near future.

Even so, there are several studies investigating the effectiveness of direct exposure of UV radiation using UV LEDs. A UV-A LED system constructed by Hamamoto et al. (2007) could inactivate foodborne pathogens including *V. parahaemolyticus*, *S. aureus*, *S. enteritidis*, and an entero-pathogenic *E. coli* (EPEC) strain in vitro. The LEDs produced UV-A radiation at 70 m $Wcm^{-2}$ at 25 °C and inactivated up to 5–6 log cycles of the bacteria within 150 min. The most susceptible bacteria was *V. parahaemolyticus*, which went through 6 log reductions to below detection levels within 20 min, whereas EPEC and *S. aureus* were inactivated below detection limits within 60 min. The least susceptible was *S. enteritidis*, which was inactivated by 5 log cycles after 150 min. Higher levels of 8-hydroxy-2′-deoxyguanosine indicated that UV-A LED treatment resulted in greater oxidative damage to DNA than did UV-C radiation from a low-pressure mercury lamp. However, lower levels of cyclobutane pyrimidine dimer indicated that UV-A LED treatment resulted in less direct DNA damage than from UV-C radiation.

UV-C radiation is most preferred in sterilizing food systems, and its bactericidal effect is well known (Shama 2014). A study showed that a 266-nm UV LED was more efficient at inactivating 3 strains each of *E. coli* O157:H7, *S. typhimurium*, and *L. monocytogenes*, in vitro, than a conventional mercury lamp with a peak wavelength of 254 nm, resulting in as high as 6 log reductions at dosages of 0.7 m J $cm^{-2}$ (Kim et al. 2016b). According to the authors, UV lamps are point sources whereas UV LEDs have a planar configuration and hence emit light in a linear fashion toward the target area. Therefore, when both sources are activated from the same height and with the same irradiance, the target area receives a smaller intensity from UV lamps than compared to UV LEDs. Also, it must be noted that the UV lamp was covered with 52 layers of polypropylene film to reduce the intensity to match the lower intensities of the UV LEDs. This means that since UV lamps have much greater intensities, treatment times will be much lower than UV LEDs for the same magnitude of microbial inactivation, owing to the limited irradiance of UV LEDs at this current point in time. Even so, UV LEDs in the experiment were still able to cause 6 log reductions to occur, meaning that they would be practical for most sterilization situations.

UV LEDs are also capable of producing pulses of UV radiation. Pulsed UV-A LED with a maximum irradiance of 0.28 $mWcm^{-2}$ and a frequency of 100 Hz reduced biofilm populations of *E. coli* by 99% after a 60-min treatment (Li et al. 2010). Moreover, pulsing has the added advantage of lower energy consumption. Wengraitis et al. (2013) exposed *E. coli* to several pulsed-light treatments from a UV-C LED, with varying duty cycles and repetition rate frequencies. Pulsed-light treatments ranging from 0.5 to 50 Hz at a 10% duty cycle were the most energy-efficient, at a power consumption of 204 mW. On the basis of log reduction per unit energy consumed, the treatments were approximately twice as efficient compared to continuous irradiation, as well as 20 times more efficient compared to pulsed Xenon light.

### 9.4.4 Photocatalytic Oxidation Using LEDs

While UV-C LED irradiation is a good method of decontamination, UV-A is not as potent as UV-C, but combining UV-A radiation with photoactive nanoparticles results in photocatalytic oxidation, which increases the potency of UV-A radiation (Chawengkijwanich and Hayata 2008; Othman et al. 2014). Photocatalytic oxidation occurs when radiation close to the UV range (usuallyUV-A radiation at 365 nm) is irradiated onto a photoactive inorganic nanoparticle materials such as titanium dioxide ($TiO_2$), zinc oxide (ZnO), and other types of materials such as silver-titanium oxide hybrids (de Azeredo 2013). Irradiation with UV-A promotes an electron in the material's valence band to the conduction band, leading to ROS generation and subsequent inactivation of surrounding microbes such as *E. coli*, *S. aureus*, *P. aeruginosa*, *Enterococcus faecium*, *Salmonella Choleraesuis* subsp., *V. parahaemolyticus*, *L. monocytogenes*, and various other spoilage bacteria which have been experimentally investigated using non-LED sources (Kim et al. 2003; Kühn et al. 2003; Li et al. 2009; Sung et al. 2013). The main cause of death is considered to be lipid peroxidation of polyunsaturated fatty acids in cell membranes caused by ROS attack, as well as other subsequent causes such as peptidoglycan damage, enzyme and coenzyme inactivation, and nucleic acid destruction (Dalrymple et al. 2010).

There are currently several studies using UV-A LEDs as a source of irradiation for photocatalytic oxidation, and most of such studies focus on water purification (Izadifard et al. 2013). In one study, UV-A LED irradiation on $TiO_2$ film inactivated a UV-resistant strain of *E. coli* by 4 log cycles (Xiong and Hu 2013), and in another, UV-A LED irradiation on $TiO_2$-coated surfaces reduced the concentration of micropollutants in potable water (Autin et al. 2013). There is existing evidence showing the effectiveness of using UV-A radiation together with food packaging incorporating suitable photoactive nanoparticles. Several experiments investigating the effect of irradiating lettuce enclosed in $TiO_2$-coated packaging using UV-A lamps or fluorescent sources have shown that *E. coli* populations can be successfully reduced (Chawengkijwanich and Hayata 2008; Othman et al. 2014). To test the potential of using this strategy on surfaces, $TiO_2$ paste was used to inactivate *L. monocytogenes* biofilms on stainless steel and glass materials using UV-A lamps (Chorianopoulos et al. 2011).

A study by Aponiene and Luksiene (2015) attempted an innovative combination of PDI and photocatalytic oxidation using a violet LED (405 nm), chlorophyllin, and ZnO nanoparticles to inactivate *E. coli* O157:H7 in vitro. In addition, the sequence of adding the photoactive ingredients into the bacterial suspension during the dark incubation, prior to photoirradiation, was investigated. Interestingly, it was found that adding both chlorophyllin and ZnO together into the bacterial suspension prior to irradiation was not as effective as adding ZnO first, followed by chlorophyllin. The simultaneous addition of ZnO and chlorophyllin resulted in an approximate reduction of 2.7 log CFU $mL^{-1}$. In contrast, the addition of chlorophyllin for 15 min followed by ZnO (followed by further dark incubation of 15 min) resulted in a reduction of around 3 log CFU $mL^{-1}$, whereas performing the addition in the reverse order (i.e., ZnO followed by chlorophyllin) resulted in the greatest reduction of around 4.5 log CFU $mL^{-1}$. The reason for this was attributed to the initial electrostatic interactions of ZnO nanoparticles with the negatively charged cell membranes of the bacteria, after which negatively charged chlorophyllin bound to ZnO, hence increasing the overall interactions between the bacterial cell membrane and photoactive ingredients. The overall benefit of this method is that since PDI is less effective on Gram-negative species, a combination of PDI and photocatalytic oxidation could synergistically increase the success of inactivating such species.

### 9.4.5 *Effect of PDI Treatments Using LEDs on Food Products*

Recently, more studies using LEDs have been performed on real food matrices to understand their efficacy in inactivating bacteria inoculated on the surfaces of various types of foods. Table 9.3 shows a summary of PDI treatments using exogenous photosensitizers on foods.

In fruits and vegetables, reductions of around 2 log cycles of bacteria in an approximate time frame of up to an hour were generally reported. For example, treating apricots, plums and cauliflowers inoculated with *B. cereus* with hypericin as a photosensitizer, and a green LED (585 nm) light with an irradiance of 3.84 $mWcm^{-2}$ led to a significant decrease of the bacterial population after only 30 min of irradiation (Aponiene et al. 2015). Similarly, treating strawberries that

**Table 9.3** Efficacy of PDI on food systems using LEDs and exogenous photosensitizers

| Photosensitizer | Pathogen | Wavelength of LED (nm) | Intensity and duration | Food | Effect | References |
|---|---|---|---|---|---|---|
| Curcumin-polyvinylpyrrolidone (PVP-C) and NovaSol®-Curcumin formulation (NovaSol®-C) | *S. aureus* | 435 | 9.4 $mWcm^{-2}$ for 24 h | Cucumber (*Cucumvis sativus*) | Reduction of 2.6 log CFU achieved relative to control when concentration of 50 or 100 μM of PVP-C was used | Tortik et al. (2014) |
| | | | | Peppers (*Capsicum* spp.) | Reduction of 2.5 log CFU achieved relative to control when concentration of 50 μM of PVP-C was used | |
| | | | | Chicken meat | Reduction of 1.7 log CFU achieved relative to control when concentration of 50 or 100 μM of NovaSol®-C was used | |
| Hypericin | *B. cereus* | 585 | 3.84 $mWcm^{-2}$ for 30 min | Apricots (*Prunus armeniaca*); Plums (*Prunus domestica*); Cauliflower (*Brassica oleracea*) | Reduction of 1.1, 0.7, and 1.3 log CFU $g^{-1}$ on surface of apricots, plums, and cauliflower, respectively, compared to initial inoculated concentration. No significant change in antioxidant content detected in extracts | Aponiene et al. (2015) |
| Na–Chlorophyllin (Na–Chl) | *L. monocytogenes* | 400 | 12 $mWcm^{-2}$ for 20 min | Strawberries (*F. ananassa* Dutch) | Reduction of 1.8 log CFU of *L. monocytogenes* achieved. Mesophils were reduced by 1.7 log, while yeast and molds were reduced by 0.86 log cycles. Surface | Luksiene and Paskeviciute (2011b) |

(continued)

**Table 9.3** (continued)

| Photosensitizer | Pathogen | Wavelength of LED (nm) | Intensity and duration | Food | Effect | References |
|---|---|---|---|---|---|---|
| | | | | | temperature remained under 27 °C. There was significant increase in antioxidant activity, but no change in total soluble phenolics or anthocyanins | |
| Chlorophyllin-chitosan complex | *S. typhimurium* and yeast/molds | 405 | 11 $mWcm^{-2}$ for 60 min | Strawberries (*F. ananassa* Dutch) | Reduction of 2.2 log CFU $g^{-1}$ of *S. typhimurium* and 1.4 log CFU $g^{-1}$ achieved. Visual appearance of strawberries superior to untreated samples, which were moldy. | Buchovec et al. (2016) |

*Source* Adapted from D'Souza et al. (2015) with updates

had been inoculated with *L. monocytogenes* with Na–Chl as photosensitizer combined with irradiation with a blue LED (400 nm) light at 12 $mWcm^{-2}$ resulted in a decrease of the bacterial population after a 20-min treatment (Luksiene and Paskeviciute 2011b).

Meat products have also been investigated. The population of *S. aureus* inoculated onto chicken meat was reduced by 1.7 log cycles after illumination of blue LED (435 nm) with a curcumin-based photosensitizer (Tortik et al. 2014). Without the use of added photosensitizers, *Campylobacter* spp.was inactivated effectively by near-UV LED (395 nm) in vitro and on chicken meat, with as little as 0.12 $Jcm^{-2}$ required (Haughton et al. 2012), as shown in Table 9.4. However, the use of blue LED light at 405 nm was significantly less effective in reducing the population of *Campylobacter* spp. inoculated onto chicken skin using chicken exudate (Gunther et al. 2016), requiring a higher dosage of up to approximately 180 $Jcm^{-2}$. The less effectiveness was attributed to the higher optical density of chicken exudate, hence requiring a higher level of irradiation in order to penetrate more effectively.

Another study was performed on orange juice inoculated with a *Salmonella* cocktail, which was illuminated with blue LEDs of 460 nm (Ghate et al. 2016). The greatest reduction in bacterial populations was observed when a treatment comprising of an irradiance of 92.0 $mWcm^{-2}$ for 13.58 h at 20 °C led to a 4.8 log reduction of the bacteria. However, the treatment with the lowest *D*-value (in terms of $Jcm^{-2}$) was the one comprising of an irradiance of 92.0 $mWcm^{-2}$ at 12 °C.

UV LEDs have also been tested in various food matrices. One such experiment studied the ability of UV-A LEDs (365 nm) to inactivate *E. coli* DH5α in beverages by using drinks with artificial colorants in varying concentrations, and commercially available orange juice. Lower concentrations of colorants in the solutions resulted in greater inactivation of the bacteria. As for the orange juice samples subjected to similar treatment, a lower rate of inactivation was reported compared to the control containing phosphate buffer solution. For one brand of juice, a log reduction of approximately 0.5 log cycles was reported, whereas a log reduction of 2.5 log cycles was reported in the second brand of juice. The large variation observed in inactivation between solutions containing different colorants and concentrations was suggested to be due to the fact that not all colorants possess an absorbance band which overlapped at 365 nm. Furthermore, colorants which possess antioxidant properties might have been able to quench ROS produced during the process, hence lowering the efficacy. Also, pigments and particles such as fiber might scatter, reflect, or absorb light, causing less UV radiation to penetrate into the drink (Lian et al. 2010). Although it would be preferable for more studies to be conducted to verify these claims, the optical properties of food matrices, as influenced by the presence of various food compounds or ingredients, as well as the possibility of quenching of ROS by food components, are worth paying attention to in future studies.

Apart from drinks, the effect of UV-A LED (365 nm) treatment on *E. coli* DH5α inoculated onto lettuce and cabbage leaves was studied. Irradiation of 90 min with an irradiance of 125 $mWcm^{-2}$ resulted in a decrease of 3.5 log cycles, with

**Table 9.4** Efficacy of PDI on food systems using LEDs and endogenous photosensitizers

| Pathogen | Wavelength of LED | Intensity and duration | Food/surface | Effect | References |
|---|---|---|---|---|---|
| *Campylobacter* spp. | 395 nm | 7 $mWcm^{-2}$ for 5 min | Skinless chicken fillet | Reduction of 1.43 log CFU $g^{-1}$ of pathogen on surface of chicken compared to initial microbial load. Minimal increase in L* value measured for color | Haughton et al. (2012) |
| *Campylobacter* spp. | 405 nm | 306 $mWcm^{-2}$ for 10 min | Chicken skin | A 184–186 $Jcm^{-2}$ treatment resulted in reduction of 1.7 log CFU $g^{-1}$ of *C. jejuni* and 2.1 log CFU $g^{-1}$ of *C. coli* on surface of chicken skin compared to initial microbial load. High powered LEDs resulted in temperatures close to 50 °C which may have caused thermal inactivation | Gunther et al. (2016) |
| *Salmonella* spp. | 460 nm | 92 $mWcm^{-2}$ for 13.58 h | Orange juice | Reduction of 3.6 and 4.8 log CFU $mL^{-1}$ when incubated at 12 and 20 °C, respectively. Treatments resulted in significant color changes as detected via colorimeter | Ghate et al. (2016) |
| *E. coli* O157:H7, *S. typhimurium* and *L. monocytogenes* | 266, 270, 275, 279 nm | 3 mJ $cm^{-2}$, duration not reported | Cheese slice | Reduction of 4.04–4.88 log $CFUg^{-1}$ for *E. coli* O157:H7, 3.91–4.72 log $CFUg^{-1}$ for *S. typhimurium*, and 3.24–4.88 log $CFUg^{-1}$ for *L. monocytogenes*. Lower wavelengths resulted in higher reductions | Kim et al. (2016b) |

*Source* Adapted from D'Souza et al. (2015) with updates

negligible loss of vitamin C, no formation of nitrites or nitrates, and less than 5% loss in moisture content (Aihara et al. 2014). Moreover, the effect of UV-C LEDs on the inactivation of foodborne pathogens inoculated onto cheese slices was studied, showing that 3 mJ $cm^{-2}$ of the 266 nm UV LEDs resulted in approximately 4.5 log reduction in *E. coli* O157:H7 and *S. typhimurium*, as well as 3.3 log reduction in *L. monocytogenes* (Kim et al. 2016b).

As detailed in the previous sections, LED treatments during the postharvest stages of fruits and vegetables may activate certain biological processes that could either lead to further nutritional degradation, or increase in nutrient value. Apricots, plums, and cauliflower treated with hypericin and green LED light were found to have no significant differences with the control samples in terms of antioxidant activity and color (Aponiene et al. 2015). The short illumination time of 30 min was negligible compared to the much longer hours typically used in postharvest applications, hence the duration could have been insufficient to cause the degradation or stimulation of antioxidant compounds. In contrast, an increase in total antioxidant capacity was reported in strawberries treated with Na–Chl and LEDs, although anthocyanin and total soluble phenolics content did not increase (Luksiene and Paskeviciute 2011b). In this case, it is not certain whether the increase in antioxidant capacity was due to LED illumination, as the concentration of photosensitizer used in the study by Luksiene and Paskeviciute (2011b) was almost 100 times greater compared to that used by Aponiene et al. (2015). Furthermore, Na–Chl, which was the photosensitizer used by Luksiene and Paskeviciute (2011b), possesses high antioxidant capacity (Luksiene and Paskeviciute 2011a). Hence, the increase in antioxidant activity was more likely due to the addition of Na–Chl as opposed to being a biological response to LED light. However, the addition of a photosensitizer which is high in nutritional value is also an attractive idea, as it provides the benefits of increased safety as well as a more nutritious product. In contrast, Zhang et al. (2015) showed that pulsed blue LED lighting set at a period of 400 μs and a duty of 50%, with intensity of 100 μmol $m^{-2}$ $s^{-1}$, resulted in a substantially larger increase in the ascorbate content of citrus fruit after 4 weeks of irradiation compared to storing in the dark. Although the objective of the research was to study nutritional changes, it showed that since pulsed lighting using LEDs was also a viable means of photoirradiation, the nutritional quality of food can be simultaneously improved while being kept safe.

Due to the minimal radiant heat emitted, LEDs cause minimal increase in temperature on the surface or the interior of foods. This prevents the degradation of nutrients and organoleptic properties of such foods, as well as preventing the thermal degradation of nutrients. In milk, Srimagal et al. (2016) reported small increases in temperature ranging from 1 to 2 °C when illuminated by various LEDs (405, 430, and 460 nm) over 60 min at initial temperatures of 5, 10, and 15 °C. Similarly, the surface temperature of various fruits and vegetables was increased from 20 °C to a maximum of 25 °C after up to 30 min of illumination, which was observed in apricots, plums, and cauliflowers (Aponiene et al. 2015) as well as strawberries (Luksiene and Paskeviciute 2011b). A dose of 4.2 $Jcm^{-2}$ from a near-UV LED (395 nm) caused the surface temperature of skinless chicken fillet to increase from approximately 25–30 °C (Haughton et al. 2012). Hence, it is confirmed that LED treatments are considered nonthermal due to the minimal increase in temperature from such treatments.

### 9.4.6 PDI in Decontamination of Food Surface Through Packaging Materials Using LEDs

Photosensitizers can also be incorporated onto the surfaces of packaging materials or food contact surfaces. Irradiation of chlorophyllin-based photosensitizers incorporated onto polyolefin packaging materials using LEDs with wavelength of 405 nm for 15 min at an irradiance of 20 mWcm$^{-2}$ inactivated *L. monocytogenes* (Luksiene et al. 2010) and *B. cereus* (Luksiene and Paskeviciute 2011a) by approximately 4 log cycles. Table 9.5 summarizes similar findings, where PDI was used to decontaminate food surfaces through packaging materials using LEDs.

Several studies also attempted to sterilize the contaminated surface of food contact materials using LEDs without adding exogenous photosensitizers (Table 9.6).

Apart from incorporating photosensitizers onto food contact surfaces or packaging materials, Luksiene and Brovko (2013) suggested exploring the incorporation of photosensitizers such as chlorophyllin onto various polymer-based films and coatings that are commonly used on foods like meat and poultry. Upon irradiation, PDI would be initiated on the surface of the food to ensure its microbial safety. A chlorophyllin-chitosan complex was used to coat strawberries and was then subjected to irradiation by LED with wavelength of 405 nm to test the efficacy of the treatment on inoculated *S. typhimurium* and yeasts and molds. The population count of *S. typhimurium* fell from around 5.4 to 3.2 log CFU g$^{-1}$, while the number of yeasts and mold fell from 4.0 to 2.6 log CFU g$^{-1}$. Yet the appearance of strawberries was less moldy after the experimental period. As strawberries tend to spoil quickly, this method could be a potential way to lengthen the commercial viability of strawberries in the market (Buchovec et al. 2016).

A previous study by López-Carballo et al. (2008), who used a quartz/halogen lamp instead of LEDs to provide light for photoirradiation of cooked frankfurters containing chlorophyllin-coated gelatin film or coating, reported a small reduction in the populations of *S. aureus* and *L. monocytogenes* by approximately 1.5 log cycles each. In spite of the low efficacy of the method, it is worth exploring the use of such coatings in conjunction with LED illumination to further inhibit the growth of low microbial loads of pathogens on meats kept in cold storage. However, more studies are needed to understand the effect of such films and coatings on the organoleptic properties and acceptability of such foods.

**Table 9.5** Efficacy of PDI in decontamination of food surface through packaging materials incorporated with exogenous photosensitizers using LEDs

| Photosensitizer | Pathogen | Wavelength of LED (nm) | Intensity and duration | Packaging film/material | Effect | References |
|---|---|---|---|---|---|---|
| 5-aminolevulinic acid | *B. cereus* spores | 400 | 20 $mWcm^{-2}$ for 15 min | Polyolefine packing trays | Reduction of spores from approximately 6 log $CFUcm^{-2}$ to 3.3 log $CFUcm^{-2}$ after LED treatment with 7.5 mM of ALA | Luksiene et al. (2009) |
| 5-aminolevulinic acid | *L. monocytogenes* | 400 | 20 $mWcm^{-2}$ for 15 min | Polyolefine packing trays | Reduction of planktonic cells by 3.7 log $CFUcm^{-2}$ after LED treatment with 10 mM of 5-aminolevulinic acid solution and incubated for 60 min. *L. monocytogenes* biofilms were reduced by 3.0 log $CFUcm^{-2}$ after LED treatment with 5-aminolevulinic acid solution | Buchovec et al. (2010) |
| Na–Chl | *L. monocytogenes* $ATC_{L3}C$ 7644 | 405 | 20 $mWcm^{-2}$for 5 min | Polyolefine packing trays | Planktonic cells attached to surface were reduced by 4.5 log $CFUcm^{-2}$ after treatment with $1.5 \times 10^{-7}$ M of Na–Chl solution and LED. Biofilms attached to surface were reduced by 4.5 log CFU $mL^{-1}$ after LED treatment with higher concentration of $7.5 \times 10^{-4}$ M Na–Chl solution | Luksiene et al.(2010) |
| Na–Chl | *L. monocytogenes* $ATC_{L3}C$ 7644 | 405 | 20 $mWcm^{-2}$ for 5 min | Polyolefine packing trays | Planktonic cells attached to surface were reduced by 4.5 log CFU $cm^{-2}$ after treatment with $7.5 \times 10^{-7}$ M of Na–Chl solution and LED. Biofilms attached to surface were reduced by 4.5 log CFU $mL^{-1}$ after LED treatment with higher concentration of $1.5 \times 10^{-4}$ M Na–Chl solution | Luksiene and Paskeviciute (2011a) |
| Na–Chl | *B. cereus* ATCC 12826 | 405 | 20 $mWcm^{-2}$ for 5 min | Polyolefine packing trays | Planktonic cells attached to surface were reduced by 4.5 log $CFU.cm^{-2}$ after treatment with $7.5 \times 10^{-7}$ M of Na–Chl solution and LED. Spores attached to surface were reduced by approximately 5 log $CFUcm^{-2}$ after LED treatment with higher concentration of $7.5 \times 10^{-5}$ M Na–Chl solution | Luksiene and Paskeviciute (2011a) |

*Source* Adapted from D'Souza et al. (2015) with updates

**Table 9.6** Efficacy of PDI in sterilizing the contaminated surface of food contact materials using LEDs without exogenous photosensitizers

| Pathogen | Wavelength of LED (nm) | Intensity and duration | Surface | Effect | References |
|---|---|---|---|---|---|
| *Campylobacter* spp. | 395 | Minimum of 0.12 $Jcm^{-2}$, time not determined | Stainless steel and polyvinylchloride cutting board | Population reduced from an initial inoculated microbial load of 4 log CFU $cm^{-2}$ to no detectable pathogen | Haughton et al. (2012) |
| *Campylobacter* spp. | 405 | 306 $mWcm^{-2}$ for 10 min | Stainless steel | A 181–183 $Jcm^{-2}$ treatment resulted in reduction of 4.9 log CFU $g^{-1}$ of *C. jejuni* and 5.9 log CFU $g^{-1}$ of *C. coli* on surface of chicken skin compared to initial microbial load | Gunther et al. (2016) |
| *Salmonella enteritidis, L. monocytogenes* | 405 | 110 $mWcm^{-2}$, variable | Acrylic and Polyvinyl chloride surfaces | On PVC, *S. enterica* was fully inactivated by 2.19 log CFU per plate, while *L. monocytogenes* was reduced by 0.90 log CFU per plate after treatment of 7.5 min (dosage of 45 $Jcm^{-2}$). On acrylic, *S. enterica* was reduced by 1.63 log CFU per plate, while *L. monocytogenes* was reduced by 0.42 log CFU per plate after treatment of 10 min (dosage of 60 $Jcm^{-2}$) | Murdoch et al. (2012) |
| *E. coli, L. monocytogenes* | 405 | 36 $Jcm^{-2}$ | Nitro cellulose membrane | *E. coli* and *L. monocytogenes* were reduced by 26% and 13%, respectively, upon exposure to light with irradiance of 60 $mWcm^{-2}$. When pre-treated with acid at pH 3, inactivation was 95% and 99%, respectively | McKenzie et al. (2014) |

*Source* Adapted from D'Souza et al. (2015) with updates

### 9.4.7 Evaluation of Role of LEDs in Microbiological Food Safety

It has been sufficiently shown that inactivation methods employing LEDs possess several useful advantages, including preventing the formation of resistant strains, the absence of toxic mercury, and the ability to design a compact source of radiation, compared to conventional and bulky low-pressure mercury lamps. Pulsing can also bring about energy savings. However, one obvious and major shortfall of radiation in the visible or UV range is the low penetration depth into food, which might limit decontamination to only the surface of vegetables, fruits and some meats, or non-opaque liquid food products (D'Souza et al. 2015). Even so, LEDs can effectively be used as a component in the hurdle technology framework for those types of food which are detrimentally affected by thermal processes.

A noteworthy observation from the study by Ghate et al. (2016) on the inactivation behavior in orange juice was that a photobiological response was not independent of treatment duration and irradiance, and similar dosages won't result in a similar photobiological response. In other words, at the same temperature and dosage of blue LED light treatment, a 92.0 mWcm$^{-2}$ treatment resulted in greater inactivation than a 254.7 mWcm$^{-2}$ treatment, although it would be expected to be similar. This could be due to the mechanism of action that applied stress through PDI, or through extrinsic factors present in the orange juice matrix. It highlights the importance of conducting proper studies on food matrices as their variety and complexity can lead to unexpected deviations from the trend. Few studies have tested the law of reciprocity within food-related studies and hence would be useful for future research.

An important challenge worth considering with regard to actual food matrices is the presence of constituents that may also be photosensitizing, hence being able to contribute to PDI. Irradiation of 80 mWcm$^{-2}$ up to 15 min using an LED with wavelength of 400 nm resulted in a 5 log CFU mL$^{-1}$ reduction of *S. aureus* growth suspended in 4 mM solution of gallic acid (Nakamura et al. 2012). Similarly, solutions of various polyphenols such as caffeic acid, gallic acid, epigallocatechin, epigallocatechin gallate, and chlorogenic acid were shown to be conducive photosensitizers to the inactivation of various species of bacteria, including *E. faecalis*, *S. aureus*, *Streptococcus mutans*, *Aggregatibacter actinomycetemcomitans*, *E. coli*, and *P. aeruginosa* (Nakamura et al. 2015). The group further demonstrated the practical implications of these findings by showing that irradiation of aqueous extracts of crushed grapes using and LED with peak wavelength of 400 nm could reduce a population of 8 log CFU mL$^{-1}$ of *S. aureus* to below detectable levels in 20 min. Photooxidation of phenolic compounds present in the system produced hydroxyl radicals which then resulted in bacterial cell death (Tsukada et al. 2016). Hydroxyl radicals were similarly produced in brandy through photooxidation of gallic acid when irradiated by white LED, suggesting the practicality of utilizing photodynamic inactivation as a means of sterilizing beverages with high content of gallic acid (Espejo and Armada 2014).

On the other hand, it is also crucial to determine whether photosensitive food components are not degraded during irradiation, resulting in loss of nutritional quality or acceptability. Manzocco (2015) described the various effects of photoirradiation on protein structures in food such as protein unfolding, aggregation, and fragmentation, which may result in both advantages or disadvantages to a food system. Riboflavin is a vitamin found in foods such as milk and possesses photosensitizing, and hence, antibacterial properties under blue LED light of 462 nm but consequently decomposes into lumiflavin and lumichrome after treatment (Liang et al. 2013). Apart from nutrient loss, the decomposition of riboflavin leads to undesirable effects in terms of appearance and flavor, due to resultant lipid oxidation in foods such as milk, beer, and cheese (Cardoso et al. 2012) and hence needs proper evaluation. However, Srimagal et al. (2016), whose work optimized the wavelength, temperature, and treatment time for blue LED treatment on milk, reported that for a successful log reduction greater than 5 log CFU $mL^{-1}$ of a surrogate *E. coli* type, quality indicators such as color, moisture, viscosity, pH, titratable acidity, fat, protein, and carbohydrate content were not significantly different from non-irradiated control samples. These data probably suggest no discernible sensory differences after such a treatment, which proves that with successful optimization of the treatment, undesirable organoleptic changes can be avoided or minimized. Moreover, UV LEDs (266, 270, 275, and 279 nm peak wavelengths) were used to irradiate cheese slices and no significant difference in color was reported between treated samples and the control (Kim et al. 2016b). Therefore, successful treatments do exist in the literature, but nevertheless, proper evaluation of the food matrix is necessary to ensure that unintended quality degradation of the food does not occur.

Ultimately, a postharvest treatment which results in food that is not acceptable to consumers is deemed counterproductive. Conducting proper sensory studies would confirm that the taste and flavor of treated foods are preserved after a treatment. So far, the only evidence that foods treated with PDI using exogenous photosensitizer is indistinguishable from the control by a sensory panel was conducted in a simple and small-scale preliminary sensory study on strawberries treated with Na–Chl and irradiated with blue light (Luksiene and Paskeviciute 2011b). Therefore, it would be worthwhile to conduct proper sensory studies using trained sensory panels.

## 9.5 Conclusion

The ultimate goal of the food supply chain is to strengthen food security for the human population, and this is achieved through a two-pronged approach of increasing food supply and reducing food losses (FAO 2011). While previous chapters have addressed the former method, this chapter has demonstrated how LED technology can be incorporated into the latter, specifically in the postharvest phase of the food supply chain. LEDs can be used to delay senescence and limit the growth of spoilage or pathogenic microorganisms on foods, thereby extending the shelf life of foods and preventing foodborne disease in human populations. In addition, LEDs can perform other postharvest functions such as controlling the nutritional content and commercial maturity of foods. Therefore, the utility of LEDs is not limited to just food production or agriculture, but truly extends from 'farm to fork.'

However, it is believed that the studies presented in this chapter only represent a small fraction of the amount of effort that is required to make LED technology increasingly potent and practical. As fruits and vegetables are so diverse in their biology, understanding how to manipulate them with the correct light quality and quantity will require much effort. However, as has been shown, there are studies emerging, which make use of more advanced techniques that track the biochemical response of such foods to various kinds of light treatments, especially at the genetic level. When the exact mechanisms have been understood to a greater degree, it will become more feasible to determine what form of light treatment is suitable for various plants.

With regard to food safety, techniques such as PDI, photocatalytic oxidation, and direct UV inactivation using LEDs are still maturing. Although LEDs have been shown to be effective in many in vitro studies, more studies are required to be performed on actual food matrices. Radiation emitted by LEDs has limited penetrating power; hence, their application can potentially only be limited to surface decontamination of foods, or in non-opaque liquid foods. Moreover, coatings and packaging films can be used to enhance the application of LEDs, although very little research has been conducted so far. Other challenges that are anticipated include the sensitivity of food components to photosensitization: on the one hand, they might enhance the antibacterial effect of the treatment, but on the other hand they may provoke quality defects. However, with diligent optimization, such problems can be minimized, as demonstrated by Srimagal et al. (2016).

Finally, there is still little known about the effect of such treatments on the overall acceptability to consumers. Although conducting trained sensory panel studies requires time and resources, it can reveal important details regarding the resultant organoleptic properties of such foods after treatment. In parallel to this, objective means of measuring quality parameters related to organoleptic or sensory properties can be used, such as texture analyzers, colorimeters, moisture analyzers, and others.

As LED technology progresses and becomes more economical, attempts should be made to develop LED systems that can be used in developing countries where there is a critical lack of technological level or infrastructure to support a safe, hygienic, and efficient food supply chain. The integration of LEDs and photovoltaics has been shown to provide safe drinking water by drawing energy from the sun to be used by LEDs (Lui et al. 2014). Such combinations of technology can hopefully be transferred to food safety-related applications as well, for the betterment of society.

In conclusion, although there are still gaps in knowledge to be filled, there is much certainty that not only LEDs are useful for growing food, but also the very same features can be used in postharvest applications, as well as for the assurance of microbiological food safety. As LED technology continues to progress as expected, LEDs have the potential to become more ubiquitous than they already seem to be, as they aid in delivering safe and nutrient-rich food from the farm to fork.

## References


Ahn SY, Kim SA, Choi S-J, Yun HK (2015) Comparison of accumulation of stilbene compounds and stilbene related gene expression in two grape berries irradiated with different light sources. Horticulture Environ Biotechnol 56:36–43

Aihara M, Lian X, Shimohata T (2014) Vegetable surface sterilization system using UVA light-emitting diodes. J Med Invest 61:285–290

Alferez F, Liao H-L, Burns JK (2012) Blue light alters infection by *Penicillium digitatum* in tangerines. Postharvest Biol Technol 63:11–15

Aponiene K, Luksiene Z (2015) Effective combination of LED-based visible light, photosensitizer and photocatalyst to combat Gram (-) bacteria. J Photochem Photobiol B: Biol 142:257–263

Aponiene K, Paskeviciute E, Reklaitis I, Luksiene Z (2015) Reduction of microbial contamination of fruits and vegetables by hypericin-based photosensitization: comparison with other emerging antimicrobial treatments. J Food Eng 144:29–35

Autin O, Romelot C, Rust L, Hartd J, Jarvisa P, MacAdama J, Parsonsa SA, Jeffersona B (2013) Evaluation of a UV-light emitting diodes unit for the removal of micropollutants in water for low energy advanced oxidation processes. Chemosphere 92:745–751

Bartolomeu M, Rocha S, Cunha Â, Neves MGPMS, Faustino MAF, Almeidal A (2016) Effect of photodynamic therapy on the virulence factors of *Staphylococcus aureus*. Front Microbiol 7:267–278

Bian ZH, Yang QC, Liu WK (2015) Effects of light quality on the accumulation of phytochemicals in vegetables produced in controlled environments: a review. J Sci Food Agric 95:869–877

Birmpa A, Vantarakis A, Paparrodopoulos S, Whyte P, Lyng J (2014) Efficacy of three light technologies for reducing microbial populations in liquid suspensions. Biomed Res Int 2014:1–9

Braidot E, Petrussa E, Peresson C, Patuia S, Bertolinia A, Tubarob F, Wählbyc U, Coanc M, Vianelloa A, Zancania M (2014) Low-intensity light cycles improve the quality of lamb's lettuce (*Valerianella olitoria* [L.] Pollich) during storage at low temperature. Postharvest Biol Technol 90:15–23

Branas C, Azcondo FJ, Alonso JM (2013) Solid-state lighting: a system review. Ind Electron Mag IEEE 7:6–14

Britz S, Gaska I, Shturm Bilenko Y, Shatalov M, Gaska R (2013) Deep ultraviolet (DUV) light-emitting diodes (LEDs) to maintain freshness and phytochemical composition during postharvest storage. CLEO, Optical Society of America, San Jose, California

Buchovec I, Vaitonis Z, Luksiene Z (2009) Novel approach to control *Salmonella enterica* by modern biophotonic technology: photosensitization. J Appl Microbiol 106:748–754

Buchovec I, Paskeviciute E, Luksiene Z (2010) Photosensitization-based inactivation of food pathogen *Listeria monocytogenes* in vitro and on the surface of packaging material. J Photochem Photobiol B: Biol 99:9–14

Buchovec I, Lukseviciute V, Marsalka A, Reklaitis I, Luksiene Z (2016) Effective photosensitization-based inactivation of Gram (-) food pathogens and molds using the chlorophyllin-chitosan complex: towards photoactive edible coatings to preserve strawberries. Photochem Photobiol Sci 15:506–516

Bumah VV, Masson-Meyers DS, Enwemeka CS (2015) Blue 470 nm light suppresses the growth of *Salmonella enterica* and methicillin-resistant *Staphylococcus aureus* (MRSA) in vitro. Lasers Surg Med 47:595–601

Cao S, Liang M, Shi L, Shaob J, Songb C, Bianb K, Chenb W, Yangb Z (2017) Accumulation of carotenoids and expression of carotenogenic genes in peach fruit. Food Chem 214:137–146
Capita R, Alonso-Calleja C (2011) Antibiotic-resistant bacteria: a challenge for the food industry. Crit Rev Food Sci Nutr 53:11–48
Cardoso DR, Libardi SH, Skibsted LH (2012) Riboflavin as a photosensitizer. Effects on human health and food quality. Food Func 3:487–502
Carvalho R, Takaki M, Azevedo R (2011) Plant pigments: the many faces of light perception. Acta Physiol Plant 33:241–248
Chawengkijwanich C, Hayata Y (2008) Development of $TiO_2$ powder-coated food packaging film and its ability to inactivate Escherichia coli in vitro and in actual tests. Int J Food Microbiol 123:288–292
Chorianopoulos NG, Tsoukleris DS, Panagou EZ, Falarasb P, Nychasa GJE (2011) Use of titanium dioxide ($TiO_2$) photocatalysts as alternative means for *Listeria monocytogenes* biofilm disinfection in food processing. Food Microbiol 28:164–170
Costa L, Millan Montano Y, Carrión C, Rolnya N, Guiameta JJ (2013) Application of low intensity light pulses to delay postharvest senescence of *Ocimum basilicum* leaves. Postharvest Biol Technol 86:181–191
Dalrymple OK, Stefanakos E, Trotz MA, Goswami DY (2010) A review of the mechanisms and modeling of photocatalytic disinfection. Appl Catal B Environ 98:27–38
de Azeredo HMC (2013) Antimicrobial nanostructures in food packaging. Trends Food Sci Technol 30:56–69
Dementavicius D, Lukseviciute V, Gómez-López VM, Luksiene Z (2016) Application of mathematical models for bacterial inactivation curves using Hypericin-based photosensitization. J Appl Microbiol 120:1492–1500
Demidova TN, Hamblin MR (2004) Photodynamic therapy targeted to pathogens. Int J Immunopathol Pharmacol 17:245–254
DenBaars SP, Feezell D, Kelchner K, Pimputkar S, Pan C, Yen C, Tanaka S, Zhao Y, Pfaff N, Farrell R, Iza M, Keller S, Mishra U, Speck JS, Nakamura S (2013) Development of gallium-nitride-based light-emitting diodes (LEDs) and laser diodes for energy-efficient lighting and displays. Acta Mater 61:945–951
Dhakal R, Baek K-H (2014a) Metabolic alternation in the accumulation of free amino acids and γ-aminobutyric acid in postharvest mature green tomatoes following irradiation with blue light. Horticulture Environ Biotechnol 55:36–41
Dhakal R, Baek K-H (2014b) Short period irradiation of single blue wavelength light extends the storage period of mature green tomatoes. Postharvest Biol Technol 90:73–77
D'Souza C, Yuk H-G, Khoo GH, Zhou W (2015) Application of light-emitting diodes in food production, postharvest preservation, and microbiological food safety. Comp Rev Food Sci Food Safety 14:719–740
Dutta Gupta S, Jatothu B (2013) Fundamentals and applications of light-emitting diodes (LEDs) in in vitro plant growth and morphogenesis. Plant Biotechnol Rep 7:211–220
Endarko E, Maclean M, Timoshkin IV, MacGregor JS, Anderson JG (2012) High-intensity 405 nm light inactivation of *Listeria monocytogenes*. Photochem Photobiol 88:1280–1286
Espejo F, Armada S (2014) Colour changes in brandy spirits induced by light-emitting diode irradiation and different temperature levels. Food Bioprocess Technol 7:2595–2609
FAO (2011) Global food losses and food waste—extent, causes and prevention. Rome
Gergoff-Grozeff GE, Chaves AR, Bartoli CG (2013) Low irradiance pulses improve postharvest quality of spinach leaves (*Spinacia oleraceae* L. cv Bison). Postharvest Biol Technol 77:35–42
Ghate VS, Ng KS, Zhou W (2013) Antibacterial effect of light emitting diodes of visible wavelengths on selected foodborne pathogens at different illumination temperatures. Int J Food Microbiol 166:399–406
Ghate V, Kumar A, Zhou W, Yuk HG (2016) Irradiance and temperature influence the bactericidal effect of 460-nanometer light-emitting diodes on *Salmonella* in orange juice. J Food Prot 79:553–560

Glowacz M, Mogren LM, Reade JPH, Cobb AH, Monaghn JH (2015) High- but not low-intensity light leads to oxidative stress and quality loss of cold-stored baby leaf spinach. J Sci Food Agric 95:1821–1829

Gong D, Cao S, Sheng T (2015) Effect of blue light on ethylene biosynthesis, signalling and fruit ripening in postharvest peaches. Sci Hortic 197:657–664

Gunther NW, Phillips JG, Sommers C (2016) The Effects of 405-nm visible light on the survival of Campylobacteron chicken skin and stainless steel. Foodborne Path Dis 13:245–250

Hamamoto A, Mori M, Takahashi A (2007) New water disinfection system using UVA light-emitting diodes. J Appl Microbiol 103:2291–2298

Hamblin MR, Hasan T (2004) Photodynamic therapy: a new antimicrobial approach to infectious disease? Photochem Photobiol Sci 3:436–450

Harris F, Pierpoint L (2012) Photodynamic therapy based on 5-aminolevulinic acid and its use as an antimicrobial Agent. Med Res Rev 32:1292–1327

Hasperué JH, Guardianelli L, Rodoni LM, Chavesa AR, Martínezc GA (2016) Continuous white–blue LED light exposition delays postharvest senescence of broccoli. LWT Food Sci Technol 65:495–502

Haughton PN, Grau EG, Lyng J (2012) Susceptibility of Campylobacter to high intensity near ultraviolet/visible 395 ± 5 nm light and its effectiveness for the decontamination of raw chicken and contact surfaces. Int J Food Microbiol 159:267–273

Imada K, Tanaka S, Ibaraki Y, Yoshimura K, Ito S (2014) Antifungal effect of 405-nm light on *Botrytis cinerea*. Lett Appl Microbiol 59:670–676

Izadifard M, Achari G, Langford C (2013) Application of photocatalysts and LED light sources in drinking water treatment. Catalysts 3:726–743

Johnson GA, Muthukrishnan N, Pellois J-P (2012) Photoinactivation of gram positive and gram negative bacteria with the antimicrobial peptide $(KLAKLAK)_2$ conjugated to the hydrophilic photosensitizer Eosin Y. Bioconjugate Chem 24:114–123

Kader AA, Rolle RS (2004) The role of post-harvest management in assuring the quality and safety of horticultural produce. FAO, Rome

Kanazawa K, Hashimoto T, Yoshida S, Sungwon P, Fukuda S (2012) Short photoirradiation induces flavonoid synthesis and increases its production in postharvest vegetables. J Agric Food Chem 60:4359–4368

Kiesslich T, Gollmer A, Maisch T, Berneburg M, Plaetzer K (2013) A comprehensive tutorial on in vitro characterization of new photosensitizers for photodynamic antitumor therapy and photodynamic inactivation of microorganisms. Biomed Res Int 2013:17

Kim B, Kim D, Cho D, Cho S (2003) Bactericidal effect of $TiO_2$ photocatalyst on selected food-borne pathogenic bacteria. Chemosphere 52:277–281

Kim B, Lee H, Kim J, Kwon KH, Cha HS, Kim JH (2011) An effect of light emitting diode (LED) irradiation treatment on the amplification of functional components of immature strawberry. Horticulture Environ Biotechnol 52:35–39

Kim S, Kim J, Lim W (2013) In vitro bactericidal effects of 625, 525, and 425 nm wavelength (red, green, and blue) light-emitting diode irradiation. Photomed Laser Surgery 31:554–562

Kim M-J, Mikš-Krajnik M, Kumar A, Ghate V, Yuk HG (2015) Antibacterial effect and mechanism of high-intensity 405 ± 5 nm light emitting diode on *Bacillus cereus*, *Listeria monocytogenes*, and *Staphylococcus aureus* under refrigerated condition. J Photochem Photobiol B Biol 153:33–39

Kim M-J, Mikš-Krajnik M, Kumar A, Yuk HG (2016a) Inactivation by 405 ± 5 nm light emitting diode on *Escherichia coli* O157:H7, *Salmonella typhimurium*, and *Shigella sonnei* under refrigerated condition might be due to the loss of membrane integrity. Food Control 59:99–107

Kim S-J, Kim D-K, Kang D-H (2016b) Using UVC light-emitting diodes at wavelengths of 266 to 279 nanometers to inactivate foodborne pathogens and pasteurize sliced cheese. Appl Environ Microbiol 82:11–17

Kozuki A, Ishida Y, Kakibuchi K, Mishimaa T, Sakuraic N, Muratab Y, Nakanob R, Ushijimab K, Kubo Y (2015) Effect of postharvest short-term radiation of near infrared light on transpiration of lettuce leaf. Postharvest Biol Technol 108:78–85

Kühn KP, Chaberny IF, Massholder K, Sticklerc M, Benzc VW, Sonntaga HG, Erdingera L (2003) Disinfection of surfaces by photocatalytic oxidation with titanium dioxide and UVA light. Chemosphere 53:71–77

Kumar A, Ghate V, Kim M-J, Zhou W, Khoo GH, Yuk HG (2015) Kinetics of bacterial inactivation by 405 nm and 520 nm light emitting diodes and the role of endogenous coproporphyrin on bacterial susceptibility. J Photochem Photobiol B Biol 149:37–44

Kumar A, Ghate V, Kim M-J, Zhou W, Khoo GH, Yuk HG (2016) Antibacterial efficacy of 405, 460 and 520 nm light emitting diodes on *Lactobacillus plantarum*, *Staphylococcus aureus* and *Vibrio parahaemolyticus*. J Appl Microbiol 120:49–56

Lafuente MT, Alférez F (2015) Effect of LED blue light on *Penicillium digitatum* and *Penicilliumitalicum* strains. Photochem Photobiol 91:1412–1421

Lee Y, Ha J, Oh J, Cho MS (2014) The effect of LED irradiation on the quality of cabbage stored at a low temperature. Food Sci Biotechnol 23:1087–1093

Lester GE, Makus DJ, Hodges DM (2010) Relationship between fresh-packaged spinach leaves exposed to continuous light or dark and bioactive contents: effects of cultivar, leaf size, and storage duration. J Agric Food Chem 58:2980–2987

Li X, Xing Y, Jiang Y, Ding Y, Li W (2009) Antimicrobial activities of ZnO powder-coated PVC film to inactivate food pathogens. Int J Food Sci Tech 44:2161–2168

Li J, Hirota K, Yumoto H, Matsuo T, Miyake Y, Ichikawa T (2010) Enhanced germicidal effects of pulsed UV-LED irradiation on biofilms. J Appl Microbiol 109:2183–2190

Lian X, Tetsutani K, Katayama M, Nakano M, Mawatari K, Harada N, Hamamoto A, Yamato M, Akutagawa M, Kinouchi Y, Nakaya Y, Takashi A (2010) A new colored beverage disinfection system using UV-A light-emitting diodes. Biocontrol Sci 15:33–37

Liang J-Y, Yuann J-MP, Cheng C-W, Jian HL, Lin C, Chen L (2013) Blue light induced free radicals from riboflavin on *E. coli* DNA damage. J Photochem Photobiol B Biol 119:60–64

Liao HL, Alferez F, Burns JK (2013) Assessment of blue light treatments on citrus postharvest diseases. Postharvest Biol Technol 81:81–88

López-Carballo G, Hernández-Muñoz P, Gavara R, Ocio MJ (2008) Photoactivated chlorophyllin-based gelatin films and coatings to prevent microbial contamination of food products. Int J Food Microbiol 126:65–70

Lubart R, Lipovski A, Nitzan Y, Friedmann H (2011) A possible mechanism for the bactericidal effect of visible light. Laser Therapy 20:17–22

Lui GY, Roser D, Corkish R (2014) Photovoltaic powered ultraviolet and visible light-emitting diodes for sustainable point-of-use disinfection of drinking waters. Sci Total Environ 493:185–196

Luksiene Z, Brovko L (2013) Antibacterial photosensitization-based treatment for food safety. Food Eng Rev 5:185–199

Luksiene Z, Paskeviciute E (2011a) Novel approach to decontaminate food-packaging from pathogens in non-thermal and not chemical way: chlorophyllin-based photosensitization. J Food Eng 106:152–158

Luksiene Z, Paskeviciute E (2011b) Novel approach to the microbial decontamination of strawberries: chlorophyllin-based photosensitization. J Appl Microbiol 110:1274–1283

Luksiene Z, Buchovec I, Paskeviciute E (2009) Inactivation of food pathogen *Bacillus cereus* by photosensitization in vitro and on the surface of packaging material. J Appl Microbiol 107:2037–2046

Luksiene Z, Buchovec I, Paskeviciute E (2010) Inactivation of several strains of *Listeria monocytogenes* attached to the surface of packaging material by Na–Chlorophyllin-based photosensitization. J Photochem Photobiol B Biol 101:326–331

Ma G, Zhang L, Kato M, Yamawaki K, Kiriiwa Y, Yahata M, Ikoma Y, Matsumoto H (2011) Effect of blue and red LED light irradiation on β-cryptoxanthin accumulation in the flavedo of citrus fruits. J Agric Food Chem 60:197–201

Ma G, Zhang L, Setiawan CK, Yamawaki K, Asai T, Nishikawa F, Maezawa S, Sato H, Kanemitsu N, Kato M (2014) Effect of red and blue LED light irradiation on ascorbate content and expression of genes related to ascorbate metabolism in postharvest broccoli. Postharvest Biol Technol 94:97–103

Ma G, Zhang L, Kato M, Yamawaki K, Kiriiwa Y, Yahata M, Ikoma Y, Matsumoto H (2015) Effect of the combination of ethylene and red LED light irradiation on carotenoid accumulation and carotenogenic gene expression in the flavedo of citrus fruit. Postharvest Biol Technol 99:99–104

Maclean M, MacGregor SJ, Anderson JG, Woolsey G (2008) High-intensity narrow-spectrum light inactivation and wavelength sensitivity of *Staphylococcus aureus*. FEMS Microbiol Lett 285:227–232

Maclean M, MacGregor SJ, Anderson JG, Woolsey G (2009) Inactivation of bacterial pathogens following exposure to light from a 405-nanometer light-emitting diode array. Appl Environ Microbiol 75:1932–1937

Maclean M, McKenzie K, Anderson JG, Gettinby G, MacGregora SJ (2014) 405 nm light technology for the inactivation of pathogens and its potential role for environmental disinfection and infection control. J Hosp Infect 88:1–11

Manzocco L (2015) Photo-induced modification of food protein structure and functionality. Food Eng Rev 7:346–356

Massa GD, Kim H-H, Wheeler RM, Mitchell CA (2008) Plant productivity in response to LED lighting. HortScience 43:1951–1956

McKenzie K, Maclean M, Timoshkin IV, MacGregor SJ, Anderson JG (2014) Enhanced inactivation of *Escherichia coli* and *Listeria monocytogenes* by exposure to 405 nm light under sub-lethal temperature, salt and acid stress conditions. Int J Food Microbiol 170:91–98

Mitchell CA, Both A-J, Bourget CM, Burr JF, Kubota C, Lopez RG, Morrow RC, Runkle ES (2012) LEDs: The future of greenhouse lighting! Chronic Horticulture 52:5–13

Morrow RC (2008) LED lighting in horticulture HortScience 43:1947–1950

Muneer S, Kim EJ, Park JS (2014) Influence of green, red and blue light emitting diodes on multiprotein complex proteins and photosynthetic activity under different light intensities in lettuce leaves (*Lactuca sativa* L.). Int J Mol Sci 15:4657–4670

Murdoch LE, Maclean M, MacGregor SJ, Anderson JG (2010) Inactivation of *Campylobacter jejuni* by exposure to high-intensity 405-nm visible light. Foodborne Patho Dis 7:1211–1216

Murdoch LE, Maclean M, Endarko E, MacGregor SJ, Anderson JG (2012) Bactericidal effects of 405 nm light exposure demonstrated by inactivation of *Escherichia*, *Salmonella*, *Shigella*, *Listeria*, and *Mycobacterium* species in liquid suspensions and on exposed surfaces. The Sci World J 2012:8

Nakamura K, Yamada Y, Ikai H, Kanno T, Sasaki K, Niwano Y (2012) Bactericidal action of photoirradiated gallic acid via reactive oxygen species formation. J Agric Food Chem 60:10048–10054

Nakamura K, Ishiyama K, Sheng H, Ikai H, Kanno T, Niwano Y (2015) Bactericidal activity and mechanism of photoirradiated polyphenols against gram-positive and gram-negative bacteria. J Agric Food Chem 63:7707–7713

Nelson JA, Bugbee B (2014) Economic analysis of greenhouse lighting: light emitting diodes vs. high intensity discharge fixtures. PLoS ONE 9:e99010

Noichinda S, Bodhipadma K, Mahamontri C, Narongruka T, Ketsab S (2007) Light during storage prevents loss of ascorbic acid, and increases glucose and fructose levels in Chinese kale (*Brassica oleracea* var. alboglabra). Postharvest Biol Technol 44:312–315

Noodén LD, Schneider MJ (2004) Light control of senescence. In: Noodén LD (ed) Plant cell death processes. Academic Press, San Diego, pp 375–383

Othman SH, Abd Salam NR, Zainal N (2014) Antimicrobial activity of $TiO_2$ nanoparticle-coated film for potential food packaging applications. Int J Photoenergy 2014:6

Park JY, Lee JH, Raju GSR, Moon BK, Jeong JH, Choi BC, Kim JH (2014) Synthesis and luminescent characteristics of yellow emitting $GdSr_2AlO_5:Ce^{3+}$ phosphor for blue light based white LED. Ceramics Int 40:5693–5698

Pinho P, Jokinen K, Halonen L (2012) Horticultural lighting—present and future challenges. Lighting Res Technol 44:427–437

Pogson BJ, Morris SC (2004) Postharvest senescence of vegetables and its regulation. In: Noodén LD (ed) Plant cell death processes. Academic Press, San Diego, pp 319–329

Ramkumar G, Yu S-M, Lee Y (2013) Influence of light qualities on antifungal lipopeptide synthesis in *Bacillus amyloliquefaciens* JBC36. Eur J Plant Pathol 137:243–248

Shama G (2014) Ultraviolet Light. In: Batt CA, Tortorello ML (eds) Encyclopedia of food microbiology, 2nd edn. Academic Press, Oxford, pp 665–671

Shi L, Cao S, Chen W, Yang Z (2014) Blue light induced anthocyanin accumulation and expression of associated genes in Chinese bayberry fruit. Sci Hortic 179:98–102

Shi L, Cao S, Shao J, Chenb W, Yang Z, Zheng Y (2016) Chinese bayberry fruit treated with blue light after harvest exhibit enhanced sugar production and expression of cryptochrome genes. Postharvest Biol Technol 111:197–204

Srimagal A, Ramesh T, Sahu JK (2016) Effect of light emitting diode treatment on inactivation of *Escherichia coli* in milk. LWT Food Sci Technol 71:378–385

St Denis TG, Dai T, Izikson L, Astrakas C (2011) All you need is light. Virulence 2:509–520

Sung S-Y, Sin LT, Tee T-T, Bee S, Rahmat AR, Rahman WAWA, Tan A-C, Vikhraman M (2013) Antimicrobial agents for food packaging applications. Trends Food Sci Technol 33:110–123

Tavares A, Carvalho CMB, Faustino MA, Neves M, Tomé J, Tomé A, Cavaleiro J, Cunha A, Gomes N, Alves E, Almeida A (2010) Antimicrobial photodynamic therapy: study of bacterial recovery viability and potential development of resistance after treatment. Mar Drugs 8:91–105

Thimijan RW, Heins RD (1983) Photometric, radiometric, and quantum light units of measure: a review of procedures for interconversion. HortScience 18:818–822

Toledo MEA, Ueda Y, Imahori Y, Ayaki M (2003) L-ascorbic acid metabolism in spinach (*Spinacia oleracea* L.) during postharvest storage in light and dark. Postharvest Biol Technol 28:47–57

Tortik N, Spaeth A, Plaetzer K (2014) Photodynamic decontamination of foodstuff from *Staphylococcus aureus* based on novel formulations of curcumin. Photochem Photobiol Sci 13:1402–1409

Tsukada M, Sheng H, Tada M, Mokudai T, Oizumi S, Kamachi T, Niwano Y (2016) Bactericidal action of photo-irradiated aqueous extracts from the residue of crushed grapes from winemaking. Biocontrol Sci 21:113–121

US Department of Energy (2012) Using LEDs to their best advantage. Building technologies program: solid-state lighting technology fact sheet. http://apps1.eere.energy.gov/buildings/publications/pdfs/ssl/led_advantage.pdf. Accessed 13 Feb 2015

US Department of Energy (2013) Energy efficiency of LEDs. http://www.hiled.eu/wpcontent/themes/hiled/pdf/led_energy_efficiency.pdf. Accessed 13 Feb 2015

Wengraitis S, McCubbin P, Wade MM, Biggs TD, Hall S, Williams LI, Zulich AW (2013) Pulsed UV-C disinfection of *Escherichia coli* with light-emitting diodes, emitted at various repetition rates and duty cycles. Photochem Photobiol 89:127–131

Woltering EJ, Seifu YW (2015) Low intensity monochromatic red, blue or green light increases the carbohydrate levels and substantially extends the shelf life of fresh-cut lettuce. Acta Hortic 1079:257–264

Xiong P, Hu J (2013) Inactivation/reactivation of antibiotic-resistant bacteria by a novel UVA/LED/$TiO_2$ system. Water Res 47:4547–4555

Xu F, Cao S, Shi L, Chen W, Su X, Yang Z (2014a) Blue light irradiation affects anthocyanin content and enzyme activities involved in postharvest strawberry fruit. J Agric Food Chem 62:4778–4783

Xu F, Shi L, Chen W, Caoc S, Sud X, Yangb Z (2014b) Effect of blue light treatment on fruit quality, antioxidant enzymes and radical-scavenging activity in strawberry fruit. Sci Hortic 175:181–186

Yamaga I, Takahashi T, Ishii K, Kato M, Kobayashi Y (2015) Suppression of blue mold symptom development in satsuma mandarin fruits treated by low-intensity blue LED irradiation. Food Sci Technol Int, Tokyo 21:347–351

Yamaga I, Shirai Y, Nakajima T, Kobayashi Y (2016) Rind color development in satsuma mandarin fruits treated by low-intensity red light-emitting diode (LED) irradiation. Food Sci Technol Res 22:59–64

Yeh N, Chung J-P (2009) High-brightness LEDs-energy efficient lighting sources and their potential in indoor plant cultivation. Renew Sust Energ Rev 13:2175–2180

Yeh N, Ding TJ, Yeh P (2015) Light-emitting diodes' light qualities and their corresponding scientific applications. Renew Sust Energ Rev 51:55–61

Yu S-M, Lee YH (2013) Effect of light quality on *Bacillus amyloliquefaciens* JBC36 and its biocontrol efficacy. Biol Control 64:203–210

Zhan L, Hu J, Li Y (2012a) Combination of light exposure and low temperature in preserving quality and extending shelf-life of fresh-cut broccoli (*Brassica oleracea* L.). Postharvest Biol and Technol 72:76–81

Zhan L, Li Y, Hu J, Pang L, Fan H (2012b) Browning inhibition and quality preservation of fresh-cut romaine lettuce exposed to high intensity light. Innovative Food Sci Emerg Technolog 14:70–76

Zhang L, Ma G, Kato M, Yamawaki K, Takagi T, Kiriiwa Y, Ikoma Y Matsumoto H, Yoshioka T, Nesumi H (2012) Regulation of carotenoid accumulation and the expression of carotenoid metabolic genes in citrus juice sacs in vitro. J Exp Bot 63:871–886

Zhang L, Ma G, Yamawaki K (2015) Regulation of ascorbic acid metabolism by blue LED light irradiation in citrus juice sacs. Plant Sci 233:134–142

Zhu X-G, Long SP, Ort DR (2008) What is the maximum efficiency with which photosynthesis can convert solar energy into biomass? Curr Opin Biotechnol 19:153–159

# Chapter 10
# Regulation of Gene Expression by LED Lighting

**S. Dutta Gupta and S. Pradhan**

## 10.1 Introduction

Light plays a major role in plant growth and development, as a predominant energy source for photosynthesis and as an essential signaling inducer for various photomorphogenic responses (Lee et al. 2007; Lau and Deng 2010). In particular, spectral quality and photosynthetic photon flux density (PPFD) of light sources used to irradiate plants are known to affect both photosynthesis and photomorphogenesis. Plants have a specialized light-sensing system, the photoreceptors which absorb the light energy and stimulate the signaling network to influence plant growth and development (Fraikin et al. 2013; Hayes et al. 2014). The spectral distribution of the sun's radiation belongs to a broad range of wavelengths (around 300–1000 nm), and only 50% of the radiation reaching the earth surface is photosynthetically active radiation (PAR). The spectra of sunlight that includes red and blue wavelengths are important for photosynthesis. The other wavelengths such as ultraviolet and far red that are absorbed by specific photoreceptors act as signaling inducers for various developmental pathways. Due to the specific nature of plant photoreceptors in order to recognize specific wavelength of light, the irradiance emitted from sun is unable to regulate specific biological processes in controlled fashion. Thus, the use of artificial light is a common practice to substitute or compensate the low availability of daylight to grow a variety of plant species under controlled environment agriculture (Dorais and Gosselin 2002; Heuvelink et al. 2006). The spectral quality as well as the intensity of the radiation adopted in the

S. Dutta Gupta (✉)
Agricultural and Food Engineering Department, Indian Institute of Technology Kharagpur, Kharagpur 721302, India
e-mail: sdg@agfe.iitkgp.ernet.in

S. Pradhan
Advanced Laboratory for Plant Genetic Engineering, Indian Institute of Technology Kharagpur, Kharagpur 721302, India



S. Dutta Gupta (ed.), *Light Emitting Diodes for Agriculture*,
DOI 10.1007/978-981-10-5807-3_10

plant production systems drives the growth and development of the crops. Conventional light sources cannot be controlled with the desired spectral output due to limited and inefficient utilization of additional filters, and also prone to the lack of intelligent control ability of lighting regimes. Therefore, light-emitting diodes (LEDs) and related solid-state lighting (SSL) have emerged as a promising artificial light source for controlled plant production system.

LEDs are the smart choice for next-generation lighting source, because of their significant advantages on energy efficiency, compactness, durability, long lifetime, zero mercury, low $CO_2$, and low heat emissions (Massa et al. 2008; Kami et al. 2010; Nelson and Bugbee 2014; see Chaps. 1 and 5 for more details on LED lighting). Considering the economical and ecological impacts, LED luminaries have brought new opportunities in plant production system. LEDs can provide precision delivery of photons in crop canopy and, thus, provide an option for energy efficient sole or supplemental artificial lighting in greenhouse or plant factories (Nelson and Bugbee 2014).

The light sensors of plants can be fine-tuned with specific radiation as per with the guided instructions in order to regulate the growth and development of plants and production of bioactive metabolites. In effect, coordinated control of light treatment allows researchers and growers to essentially produce their desirable products on plants (Chen et al. 2004). Although there have been many discussions on LED-regulated plant morphogenesis, flowering, nutritional quality, postharvest quality, etc. (described in other chapters of this book), only a few reports describe the effect of LEDs on gene expression of different metabolic pathways of plants. In the present chapter, we summarize the basic findings related to the LED-induced gene expression in plants, focusing on genes regulating metabolisms of carotenoids, flavonoids, and ascorbate. Genes involved in light signaling, auxin responsive factors and plant defense, and disease resistance have also been discussed.

## 10.2 LED-Regulated Gene Expression

Despite the numerous accomplishments on the photoregulation of plant development, very little information is available on the specific effect of light quality provided by LED source regulating gene expression. LED-regulated gene expression has been studied with respect to photoreceptors and auxin responsive factors, carotenoid biosynthesis pathway, flavonoid pathway, ascorbate metabolism, and defense-related genes (Fig. 10.1). Specifically, blue, red, and white LEDs, individually and/or in combination, regulate the expression of key regulatory genes involved in various metabolic pathways of plants. Table 10.1 summarizes the effect of various LED irradiations on the expression of genes involved in biosynthetic pathways of carotenoids, flavonoids, and ascorbate metabolism along with their involvement in modulating photoreceptor genes, auxin responsive factors, and defense-related genes.

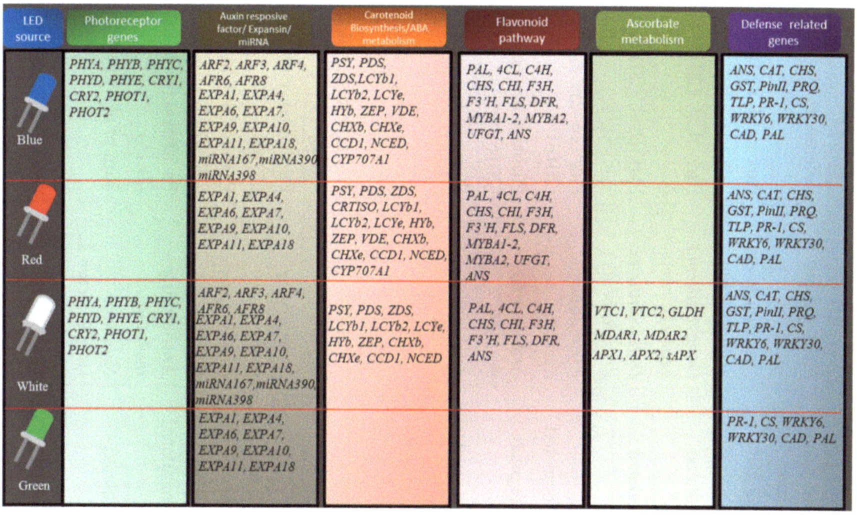

| LED source | Photoreceptor genes | Auxin resposive factor/ Expansin/ miRNA | Carotenoid Biosynthesis/ABA metabolism | Flavonoid pathway | Ascorbate metabolism | Defense related genes |
|---|---|---|---|---|---|---|
| Blue | *PHYA, PHYB, PHYC, PHYD, PHYE, CRY1, CRY2, PHOT1, PHOT2* | *ARF2, ARF3, ARF4, AFR6, AFR8 EXPA1, EXPA4, EXPA6, EXPA7, EXPA9, EXPA10, EXPA11, EXPA18, miRNA167,miRNA390 miRNA398* | *PSY, PDS, ZDS,LCYb1, LCYb2, LCYe, HYb, ZEP, VDE, CHXb, CHXe, CCD1, NCED, CYP707A1* | *PAL, 4CL, C4H, CHS, CHI, F3H, F3'H, FLS, DFR, MYBA1-2, MYBA2, UFGT, ANS* | | *ANS, CAT, CHS, GST, PinII, PRQ, TLP, PR-1, CS, WRKY6, WRKY30, CAD, PAL* |
| Red | | *EXPA1, EXPA4, EXPA6, EXPA7, EXPA9, EXPA10, EXPA11, EXPA18* | *PSY, PDS, ZDS, CRTISO, LCYb1, LCYb2, LCYe, HYb, ZEP, VDE, CHXb, CHXe, CCD1, NCED, CYP707A1* | *PAL, 4CL, C4H, CHS, CHI, F3H, F3'H, FLS, DFR, MYBA1-2, MYBA2, UFGT, ANS* | | *ANS, CAT, CHS, GST, PinII, PRQ, TLP, PR-1, CS, WRKY6, WRKY30, CAD, PAL* |
| White | *PHYA, PHYB, PHYC, PHYD, PHYE, CRY1, CRY2, PHOT1, PHOT2* | *ARF2, ARF3, ARF4, AFR6, AFR8 EXPA1, EXPA4, EXPA6, EXPA7, EXPA9, EXPA10, EXPA11, EXPA18, miRNA167,miRNA390, miRNA398* | *PSY, PDS, ZDS, LCYb1, LCYb2, LCYe, HYb, ZEP, CHXb, CHXe, CCD1, NCED* | *PAL, 4CL, C4H, CHS, CHI, F3H, F3'H, FLS, DFR, ANS* | *VTC1, VTC2, GLDH MDAR1, MDAR2 APX1, APX2, sAPX* | *ANS, CAT, CHS, GST, PinII, PRQ, TLP, PR-1, CS, WRKY6, WRKY30, CAD, PAL* |
| Green | | *EXPA1, EXPA4, EXPA6, EXPA7, EXPA9, EXPA10, EXPA11, EXPA18* | | | | *PR-1, CS, WRKY6, WRKY30, CAD, PAL* |

**Fig. 10.1** LED-induced regulation of different types of genes involved in plant metabolic pathways, light signaling, and defense

### *10.2.1 LED-Regulated Gene Expression of Photoreceptors and Auxin Responsive Factors*

Photoreceptors, the light-sensing molecules, are responsible for initiation of selected physiological responses via specific signaling networks. At least four diverse families of photoreceptors have been elucidated in plant system. These are phytochromes—that sense red/far-red light (Li et al. 2011), cryptochromes, and phototropins—which sense blue/UV-A light (Briggs and Christie 2002; Kharshiing and Sinha 2015) and UV-B absorbing UVR-8 (Heijde and Ulm 2012). Generally, these photoreceptors are consisted of more than one member in each group. There are five members in phytochrome (*PHYA* to *PHYE*), three in cryptochrome (*CRY1*, *CRY2*, *and CRY3*), two members in phototropin (*PHOT1 and PHOT2*), and one UVR8 photoreceptor (Wu 2014). Each individual photoreceptor is encoded by an individual gene and shared a high degree of similarity among the individual photoreceptors of the same family.

*PHYA*, *PHYB*, *CRY1*, and *CRY2* regulate flowering in response to light in *Arabidopsis* (Mockler et al. 2003; Exner et al. 2010). In other crop plants, photoreceptors have different roles in cellular processes such as regulation of photoperiods, flowering, tuberization, and fruit ripening. Manipulation of genes involved in different photoreceptors extremely influence the molecular pathways related to photosynthesis, photorespiration, biotic/abiotic stress, as well as secondary metabolism, such as biosynthesis of phenolics, phenylpropanoids, and flavonoids/anthocyanins (Lopez et al. 2012). Phototropins mediate hypocotyls or stems elongation toward light at the wavelengths between 315 and 500 nm.

**Table 10.1** LED-induced gene expression of different metabolic pathways, photoreceptors, and defense-related genes of plants

| Factors/systems/metabolic pathways | LED treatment (wavelength) | Genes regulated | Extent of gene expression | Plant species | References |
|---|---|---|---|---|---|
| Photoreceptors and auxin responsive factor-related gene expression | NI-FrL (730 nm) BL (450 nm) | *PHYA*, *CRY1* | Up-regulated | *Dendranthema grandiflorum* | Park et al. (2015) |
| | NI-BL (450 nm) RL (660 nm) | *FTL*, *CRY1* | Up-regulated | *Dendranthema grandiflorum* | Park et al. (2015) |
| | NI-RL (660 nm) FrL (730 nm) | *PHYB*, *AFT* | Up-regulated | *Dendranthema grandiflorum* | Park et al. (2015) |
| | BL (450 nm) | *PHYA*, *PHYD*, *CRY1* | Highly down-regulated | *Arabidopsis thaliana* | Pashkovskiy et al. (2016) |
| | BL (450 nm) | *PHYB*, *PHYE*, *PHOT1* | Down-regulated | *Arabidopsis thaliana* | Pashkovskiy et al. (2016) |
| | WL (445 + 660 nm) | *PHYA*, *PHYD*, *CRY1*, *PHOT1* | Down-regulated | *Arabidopsis thaliana* | Pashkovskiy et al. (2016) |
| | WL (445 + 660 nm) | *PHYC* | Up-regulated | *Arabidopsis thaliana* | Pashkovskiy et al. (2016) |
| | WL (445 + 660 nm) | *PHYB*, *PHYE*, *CRY2*, *PHOT2* | Unchanged | *Arabidopsis thaliana* | Pashkovskiy et al. (2016) |
| | BL (450 nm) | *ARF4*, *ARF8* | Highly up-regulated | *Arabidopsis thaliana* | Pashkovskiy et al. (2016) |
| | BL (450 nm) | *ARF3*, *ARF6* | Moderately up-regulated | *Arabidopsis thaliana* | Pashkovskiy et al. (2016) |
| | BL (450 nm) | *ARF2* | Down-regulated | *Arabidopsis thaliana* | Pashkovskiy et al. (2016) |
| Expansion | BL (456 nm) | *LeEXPA1*, *LeEXPA4*, *LeEXPA6*, *LeEXPA7*, *LeEXPA9*, *LeEPA10*, *LeEPA11 LeEPA18* | Up-regulated | *Solanum lycopersicum* L. | Kim et al. (2014) |

(continued)

**Table 10.1** (continued)

| Factors/systems/metabolic pathways | LED treatment (wavelength) | Genes regulated | Extent of gene expression | Plant species | References |
|---|---|---|---|---|---|
| Expression of carotenoid biosynthetic pathway genes | WL (380 nm) | *FtPSY*, *FtLCYB*, *FtLCYE*, *FtCHYB*, *FtCHYE*, *FtZEP* | Up-regulated | *Fagopyrum tataricum* | Tuan et al. (2013) |
| | RL (660 nm) | *FtLCYE*, *FtCHXB* | Down-regulated | *Fagopyrum tataricum* | Tuan et al. (2013) |
| | BL (470 nm) | *CitPSY*, *CitPDS*, *CitZDS*, *CitLCYb2*, *CitCHYb* | Up-regulated | *Citrus unshiu Marc.* (Satsuma mandarin) | Zhang et al. (2015) |
| | BL (470 nm) | *CitPSY*, *CitPDS*, *CitZDS*, *CitLCYb2*, *CitCHYb* | Up-regulated | *Citrus sinensis Osbeck* (Valencia orange) | Zhang et al. (2015) |
| | RL (660 nm) + ethylene | *CitPSY*, *CitPDS*, *CitZDS*, *CitCRTISO*, *CitLCYb1*, *CitLCYb2*, *CitLCYe*, *CitCHYb*, *CitZEP* | Up-regulated | *Citrus unshiu Marc.* (Satsuma mandarin) | Ma et al. (2015) |
| Expression of flavonoid pathway/ABA synthesis genes | RL (660 nm) | *VvNCED1*, *VvCYP707A1* | Up-regulated | *Vitis vinifera* | Kondo et al. (2014); Rodyoung et al. (2016) |
| | BL (450 nm) | *VlMYBA1-2*, *VlMYBA2*, *VvUFGT* | Up-regulated | *Vitis labruscana* | Kondo et al. (2014); Rodyoung et al. (2016) |
| | BL (450 nm) and WL (380 nm)<br>White LED | *FtPAL*, *FtF3'H*<br>*Ft DFR* | Highly up-regulated<br>Up-regulated | Buckwheat (*Fagopyrum tataricum*) | Thwe et al. (2014) |

(continued)

**Table 10.1** (continued)

| Factors/systems/metabolic pathways | LED treatment (wavelength) | Genes regulated | Extent of gene expression | Plant species | References |
|---|---|---|---|---|---|
| Gene expression in ascorbate metabolism | Modified WL (red-rich) (430-730 nm) | *BO-VTC2*, *BO-GLDH* | Delay down-regulation | *Brassica oleracea* L. var. *italica* | Ma et al. (2014) |
| | Modified WL (red-rich) (430–730 nm) | *BO-MDAR1*, *BO-MDAR2* | Delay down-regulation | *Brassica oleracea* L. var. *italica* | Ma et al. (2014) |
| | Modified WL (red-rich) (430–730 nm) | *BO-APX1*, *BO-APX2* | Up-regulated | *Brassica oleracea* L. var. *italica* | Ma et al. (2014) |
| Defense-related gene expression | RL (628.6 nm) Purple (394.6 nm) Blue (452.5) | *PR-1*, *WRKY30*, *WRKY6*, *CS*, *CAD*, *PAL* | Up-regulated Up-regulated | Cucumber (*Cucumis sativus*) against powdery mildew | Wang et al. (2010) |
| | BL (440 nm) and RL (660 nm) | *CAT*, *CHS*, *GST*, *PRQ*, *PinII*, *TLP* | Up-regulated | *Nicotiana benthamiana* against wildfire disease | Ahn et al. (2013) |

*NI* night interruption; *BL* blue LED; *RL* red LED; *FrL* far-red LED; *WL* white LED; *PHY* phytochome; *CRY* cryptochrome; *FTL* flower locus T; *AFT* Anti-florigenic; *ARF* auxin response factors; *EXPA* expansin *LCYb* lycopene β-cyclase; *LCYe* lycopene ε-cyclase; *PDS* phytoene desaturase; *PSY* phytoene synthase; *ZDS* ζ-carotene desaturase; *ZEP* zeaxanthin expoxidase; *CHYb* carotenoid β-hydroxylase; *CRTISO* carotenoid isomerase; *PAL* phenylalanine ammonia lyase; *F3'H* flavonoid 3'hydroxylase; *NCED* 9-cis-epoxycarotenoid dioxygenase; *CYP707A1* ABA 8′-hydroxylase; *UFGT* UDP-glucose-flavonoid 3-O-glucosyltransferase; *DFR* dihydroflavonol reductase; *Vl Vitis labruscana*; *Vv Vitis vinifera*; *Ft Fagopyrum tataricum L.*; *Cit Citrus*; *VTC1* GDP-D-mannose pyrophosphorylase; *VTC2* GDP-L-galactose phosphorylase; *GLDH* L-galactono-1,4-lactone dehydrogenase; *APX* ascorbate peroxidase; *MDAR* monodehydroascorbate reductase; *BO Brassica oleracea* L.; *NPR* Negatively regulated by phytochrome; *CAT* catalase; *CHS* chalcone synthase; *GST* gluthatione-S-transferase; *PRQ* pathogenesis-related protein; *PinII* proteinase inhibitor II; *TLP* thaumatin-like protein; *PHOT* phototropin; *MYBA* MYB transcription factor gene family; *WRKY* transcription factor; *PR* pathogenesis-related gene; *CS* callose synthase; *CAD* cinnamyl alcohol dehydrogenase

Phototropins also play important roles in chloroplast relocation (Wada et al. 2003; Kasahara et al. 2004) and stomatal movement (Kinoshita et al. 2001) resulting a very robust responses of plants to the surrounding light environment. Photoperiodic floral initiation is controlled by a systemic flowering inducer (florigen) and inhibitor (antiflorigen) such as the flowering locus T (*FTL*) and *Anti-florigenic FT/TFL1* family protein (*AFT*) genes, respectively, which are produced in the leaves (Higuchi et al. 2013). A large amount of data has established the relationship between light and hormonal regulation in plants (Folta et al. 2003; Alabadi and Blazquez 2009). Plant growth regulators in conjunction with the transcription factors express different genes in accordance with the reception of lights by the plants (Lau and Deng 2010). Thus, there exists photomorphogenic and phototropic interplay, via hormonal signaling. Auxin may modulate photomorphogenesis as well as phototropism by regulating the photoreceptors, AUX/IAA, and auxin response factors (ARF) gene expression (Molas and Kiss 2009; Singh et al. 2015). The ARF bind with the promoter regions of auxin-sensitive genes, and trigger their activation and repression. Among the ARF family, ARF5-8 and ARF19 play an important role in activating the auxin signal genes, whereas ARF1 and ARF2 suppress the activity of these genes (Okushima et al. 2005; Guilfoyle and Hagen 2007). Involvement of such regulatory proteins in the transduction of light signals is an important aspect of investigation.

The effect of blue LED (450 nm) was investigated on the photomorphogenic responses of *Arabidopsis thaliana* Col-0 along with the associated changes in transcript levels of several genes, including photoreceptors and *ARF* (Pashkovskiy et al. 2016). The blue (450 nm) and white LEDs have different effects on photoreceptors and *ARF* genes, The mRNA levels of several phytochromes (*PHYA*, *PHYD*) and cryptochromes (*CRY1*) were significantly reduced under blue led ($120 \pm 30$ μmol $m^{-2}$ $s^{-1}$) treatment, while mRNA levels of *PHYC*, *CRY2*, and *PHOT2* were unchanged, compared to white compact fluorescent lamp (WCFL). A similar decrease in mRNA levels was also observed with WL of same intensity. An increase in *PHYC* content was found in plants treated with WL, compared to the plants grown under BL. The red light component of WL may regulate the synthesis of *PHYC*. Therefore, the unchanged content of *PHYC* under BL may be due to the narrow-band nature of light source, without any red spectrum. Irradiation of WL has no effect on the expression of *PHYB*, *PHYE*, *CRY2*, *PHOT1*, and *PHOT2* genes. Both BL and WL down-regulated the *PHOT1* transcript, compared to WCFL. The down-regulation of *PHOT1* has been explained by the decreased necessity of phototropins during leaf senescence. It has been proposed that the reduction of photoreceptor gene expression manifests the reduced sensitivity of light-sensing molecule and its regulatory properties toward blue light (Pashkovskiy et al. 2016). The study also suggests the involvement of *ARF* genes in blue-light-mediated photomorphogenesis.

*ARFs* have a critical role in plant growth and developmental processes. There are 22 *ARF* genes discovered to date in *Arabidopsis*. Their function relies on leaf senescence and floral organ abscission, leaf polarity specification, floral meristem determinacy, etc. (Li et al. 2016). The expressions of *ARF* genes were significantly affected by blue LED in *A. thaliana*. Blue LED enhanced the transcripts level of the *ARF4* and *ARF8* genes significantly, while *ARF3* and *ARF6* genes were moderately expressed, but the gene expression of *ARF2* was reduced significantly. The *ARF* gene expression did not significantly influenced by WLED, unlike the blue LED lighting. Moreover, blue LED significantly induced the miRNA-mediated *ARF* gene silencing, which was involved in the activation of auxin-dependent genes. Presumably, the effects of blue light on *A. thaliana* are mediated by auxin signaling pathway involving miRNA-dependent regulation of *ARF* gene expression (Pashkovskiy et al. 2016).

Most angiospermic plants are dependent on light signals to trigger flowering (Somers et al. 2004; Kim et al. 2005). Manipulation of the photoperiodic conditions artificially can induce early flowering and can reduce the cost of commercial horticulture by reducing the flowering time and improving overall the quality of the crop (Warner and Erwin 2003). It has been well demonstrated that the night interruption (NP) following the use of artificial light source at night can regulate the flowering of both long-day and short-day plants (Leopold 1951; Lin 2000). In particular, the use of LEDs at night (NI) can regulate the expression of genes associated with flower development and photoreceptors (Yamada et al. 2008). The impact of shifts in the spectral quality of light on photoperiodic gene expression was investigated in *Dendranthema grandiflorum* by interrupting the circadian rhythms of plants at night using LEDs (Park et al. 2015). Photoperiodic light treatments were provided by changing the light quality of the NI with all possible combinations of blue (B; 450 nm), red (R; 660 nm), far-red (Fr; 730 nm), and white (W; 400–750 nm, with 28% B, 37% R, and 15% Fr light) light. Among the NI treatments, plants exposed to Fr light grew larger than plants in other treatments and especially flowering was promoted in the NI-BFr and NI-FrB treatments. The flowering inducer genes *PHYA* and *CRY1* showed higher expression at NI-FrB treatment. However, a high level of *PHYA* gene expression without any flowering was observed in the NI-RW treatment. Similarly, the Ni-BR treatment failed to induce flowering, in spite of high levels of *CRY1* and *FTL* gene expression. In NI-RFr treatment, there was no flowering in chrysanthemum. This happened because of higher expression of flowering inhibitor genes, like *PHYB* and *AFT*. It has been assumed that the NI with blue light promoted flowering following the stimulation of *CRY1* and *FTL* expression and exposure to R light suppressed flowering by stimulating the expression of *AFT* gene (Park et al. 2015). The overall expression patterns of photoperiodic genes suggest the potential of B light to promote flowering in potted floricultural crops.

### *10.2.2 LED-Induced Gene Expression of Carotenoid Biosynthesis*

Carotenoids are the second largest pigment group in nature, consisting of more than 700 members each with 40 carbon molecules (Britton 1998; Nisar et al. 2015). In plants, carotenoids play a critical role in regulating flower and fruit colorations and serve various essential functions in several physiological processes, like protecting the photosystem from photooxidative damage, and stabilization of membranes (Frank and Cogdell 1996; Havaux 1998). As the precursors of vitamin A, carotenoids are also important for human health and nutrition (Al-Delaimy et al. 2005). A high level of carotenoids in human plasma reduces the risk of cardiovascular diseases, cancers, and age-related diseases (Khachik et al. 2006; Pouchieu et al. 2014). Due to such importance, carotenoid biosynthetic pathway has been well investigated in plants (Cunningham and Gantt 1998; Hannoufa and Hossain 2012). Moreover, molecular characterization of genes involved in the main steps of carotenoid biosynthesis has been well investigated (Alquézar et al. 2009; Kato et al. 2004; Kato 2012). Efforts have also been made to enhance carotenoid accumulation in plants through metabolomics. Application of light with different spectral quality has been reported to affect carotenoid biosynthesis (Zhang et al. 2012, 2015). It is not only the spectral quality, light intensity and duration have also significantly influenced the carotenoid accumulation. In leaves and stems of pea seedlings, β-carotene content was found to be much higher in the red-light-treated groups than that of blue light (Wu et al. 2007). In tomatoes, red light treatment enhanced the accumulation of lycopene (Liu et al. 2012). Red-light-mediated enhancement of carotenoid content, especially the β-cryptoxanthin, was recorded in citrus fruit, while the accumulation of carotenoid content was not affected by blue LED exposure in the flavedo of Satsuma mandarin (Ma et al. 2012). The differences in the carotenoid contents have been attributed to the differential gene expression during carotenoid biosynthesis.

Tuan et al. (2013) investigated the LED-regulated expression of carotenoid biosynthetic genes along with the accumulation of carotenoids in sprouts of tartary buckwheat (*Fagopyrum tataricum*). The sprouts were irradiated with white (380 nm), blue (470 nm), and red (660 nm) LEDs with an intensity of 50 μmol $m^{-2}s^{-1}$ for 10 days, and changes in the expression levels of carotenoid biosynthetic genes were determined at interval of two days. The expression levels of *FtPSY*, *FtLCYB*, *FtLCYe*, *FtCHXB*, *FtCHXE*, and *FtZEP* were remarkably higher in sprouts grown under white LEDs than those under blue and red LEDs particularly at 8 DAS (days after sowing). Red LED treatment reduced the mRNA levels of *FtLCYe* and *FtCHXB* during sprout development (Fig. 10.2). However, there was no significant difference in the expression levels of genes under white, blue, and red LED treatments from 2 to 6 DAS. The maximum production of total carotenoids was observed at 10 DAS under white LEDs.

The blue LED regulated differential accumulation of various carotenoids, and their related expression of genes was studied by Zhang et al. (2015) in the juice sacs of two citrus varieties, Satsuma mandarin and Valencia orange. The juice sacs of

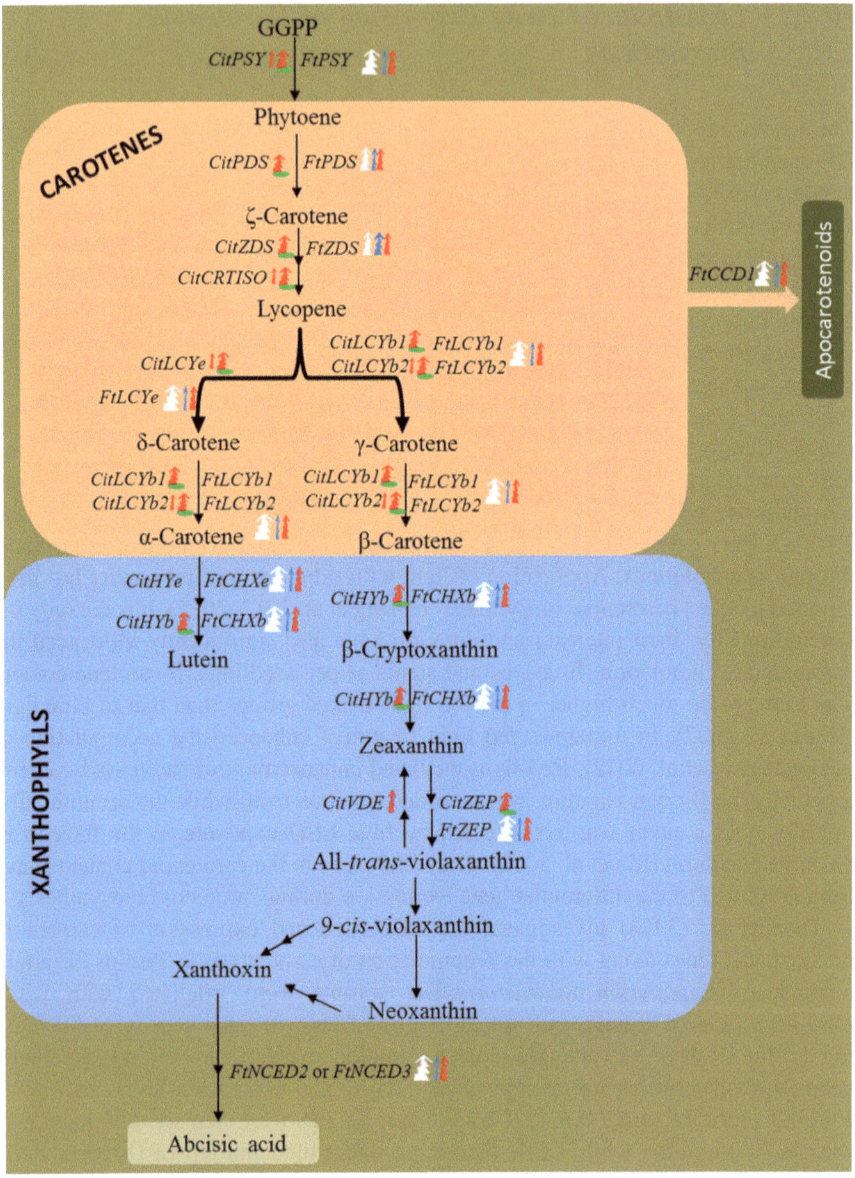


**Fig. 10.2** LED-induced gene expression in carotenoid biosynthetic pathway. Different wavebands of LEDs are marked with their respective colors. Variation in the width of the band indicates the extent of level of gene expression. *Green patch* indicates ethylene treatment along with exposure to red LED

two citrus varieties were exposed in blue LEDs for four weeks in two different intensities, viz. 50 μmol $m^{-2}$ $s^{-1}$ (50B) and 100 μmol $m^{-2}$ $s^{-1}$ (100B). In Satsuma mandarin, exposure to blue light for four weeks at 100B intensity induced the accumulation of carotenoids, especially β-cryptoxanthin. In contrast, irradiation of blue LED at 50B was found to increase the contents of all *trans*-violaxanthin and 9-*cis*-violaxanthin. The up-regulation of carotenoid biosynthetic genes (*CitPSY, CitPDS, CitZDS, CitLCYb1, CitLCYb2,* and *CitCHYb*) under 100B led to the increased content of β-cryptoxanthin in Satsuma mandarin (Fig. 10.2), whereas in Valencia orange 50B treatment was effective for inducing carotenoid accumulation with similar levels of gene expression. An increase in the expression of genes *CitPSY, CitPDS, CitZDS, CitLCYb2,* and *CitCHYb* in the fourth week of cultured Valencia orange was well correlated with the accumulation of all *trans*-violaxanthin and 9-*cis*-violaxanthin. Moreover, increased intensity of blue light from 50B to 100B changed the ratio of β,ϵ-carotenoids and β-carotenoids in both the citrus species. The findings suggest the stimulatory role of 100B treatment in shifting the pathway from β, β-branch to β,ϵ-branch. In conclusion, the study of Zhang et al. (2015) suggests the differential role of blue light intensity in the regulatory mechanism of carotenoid accumulation.

The effects of red LED light alone or in combination with ethylene on carotenoid accumulation and related gene expression have also been studied in the flavedo of citrus fruit (Ma et al. 2015). Red LED treatment up-regulated the expression of *citPSY, citCRTISO, citLCYb2, citLCYe,* and *citVDE* genes (Fig. 10.2). However, the combination of red LED light and ethylene treatment increases the expression of *citPSY, citPDS, citZDS, citCRTISO, citLCYb1, citLCYb2, citLCye, citCHYb, and citZEP*, which ultimately led to the accumulation of β-cryptoxanthin, all *trans*-violaxanthin, and lutein. With the ethylene treatment alone, the contents of lutein, all *trans*-violaxanthin, and 9-*cis*-violaxanthin were decreased. Ethylene treatment down-regulated the expression of *citLCYe*, which was associated with the decrease in lutein. The negative effect of ethylene on lutein accumulation was bypassed by red LED irradiation. These results suggest the potential role of red LED treatment along with the postharvest application of ethylene in improving the nutritional value of citrus fruit.

### *10.2.3 Regulation of Gene Expression Involved in Flavonoid Biosynthesis by LED Lighting*

Anthocyanins, the water soluble pigments, are members of the flavonoid group. A wealth of information ascribes multifarious biological roles to anthocyanins. Many of them are associated with stress responses, aposematic defensive signals against herbivory, protection of photolabile defense compounds such as thiarubrine, protection of photooxidative damage, protection against ultraviolet radiation, attraction of predators for seed dispersal, and free radical scavenging

(Long et al. 1994; Page and Towers 2002; Gould 2004; Takahama 2004). In addition, anthocyanins have a wide range of beneficial effects on human health including: antimicrobial, anti-inflammatory, antiviral, anticancer, and neuroprotective benefits (Bitsch et al. 2004; Jing et al. 2008; De Pascual-Teresa and Sanchez-Ballesta 2008). The brilliant color, high water solubility, and valuable biological properties make anthocyanins one of the most potential natural pigments, which can replace the synthetic colorants in different types of food (Camire et al. 2002).

Accumulation of anthocyanins, identification of flavonoids, and related gene expression under LED lighting have been the subject of research interest in recent years (Kondo et al. 2014; Rodyoung et al. 2016). It has been shown that light can affect the gene expression of key enzymes involved in flavonoid biosynthesis and anthocyanin contents. Flavonoids are synthesized via phenylpropanoid pathway, in which phenyl ammonia lyase (*PAL*) catalyzes the first reaction (Chang et al. 2009). The other major intermediate enzymes in flavonoid biosynthesis are chalcone synthase (*CHS*), flavanone-3-hydroxylase (F3H), dihydroflavonol reductase (*DFR*), anthocyanidine synthase (*ANS*), and anthocyanidine reductase (*ANR*). A close relationship between ABA metabolism and anthocyanin synthesis has been found in fruits of apples, grapes, and sweet cherries at veraison (Wheeler et al. 1998; Kobayashi 2009). Kondo et al. (2014) examined the effects of blue and red LED irradiation on the metabolism of abscisic acid and anthocyanin synthesis in grape berries. Endogenous ABA concentrations were higher in red-LED (660 nm)-treated grape berry skin than those in blue-LED-treated (450 nm) or untreated control skin. Simultaneously, red LED irradiation (50 μmol $m^{-2}$ $s^{-1}$) enhances the expression of 9-cis-epoxycarotenoid dioxygenase (*VvNCED1*) and ABA 8′-hydroxylase (*VvCYP707A1*), upstream enzymes in the ABA synthesis pathway. In contrast, endogenous ABA concentrations were lowest in blue-LED-treated berries, which had higher anthocyanin concentrations than the red-LED-treated berries (Kondo et al. 2014). The findings failed to establish a direct relationship between endogenous ABA concentrations and anthocyanin synthesis and suggest possible involvement of another factor that may regulate anthocyanin biosynthesis more than endogenous ABA. The expressions of *VlMYBA1-2* and *VvUFGT* were increased under both red and blue LED treatments. However, the expression of these genes did not correlate with anthocyanin concentrations. Thus, *myb*-related gene expression may not regulate the accumulation of anthocyanins in grape skin. The expression pattern of *VvUFGT* at veraison indicates its influence in anthocyanin synthesis.

The effects of light-emitting diodes on ABA synthesis and anthocyanin concentrations have also been studied in grape vines in two different seasons (Rodyoung et al. 2016). The endogenous ABA concentration varies with the growing seasons as well as LED treatments. Blue LED irradiation during the early heating culture increased the ABA concentrations, whereas increased ABA content was observed under red LED in the ordinary growing season. The genes *VvNCED1* and *VvCYP707A1* highly expressed in each treatment (red or blue LED) at veraison regardless of the growing season. In both the seasons, anthocyanin concentrations were highest under the blue LED treatment, followed by the red LED treatment, and the expressions of *VlMYBA1-2*, *VlMYBA2*, and *VvUFGT* coincided with anthocyanin concentrations (Rodyoung et al. 2016). In general, studies on grapes suggest the promoting role of blue LED irradiation in accumulating anthocyanins. The influence of blue LED in anthocyanin synthesis has also been demonstrated in strawberries, but through a *FaPHOT2* of phototropin 2 (Kadomura-Ishikawa et al. 2013).

The accumulation of phenylpropanoid compounds along with the levels of gene expression for the key enzymes under LED lighting was studied in buckwheat (*F. tataricum*) sprouts (Thwe et al. 2014). Sprouts were exposed under different LEDs with wavebands of red (660 nm), blue (450 nm), and white (380 nm). Transcript levels of key phenylpropanoid genes were studied at 2, 4, 6, 8, and 10 days after LED exposure. Expression levels for all genes peaked at two days after LED exposure. However, variability in the expression levels of genes involved in flavonoid pathway was observed among the treatments. Higher expression levels of *FtPAL* and *FtF3'H* were noted in sprouts grown under blue and white LEDs than red-LED-treated sprouts (Fig. 10.3). The expression levels of *FtC4H*, *FtCHI*, *FtFLS-2*, and *FtANS* genes were higher under blue LED than red and white LED exposure (Fig. 10.3). Compared to blue and red LED exposure, white LED enhances the expression of *FtDFR*. Overall, sprouts grown under blue and white LEDs had greater transcript numbers than red-LED-treated sprouts (Thwe et al. 2014). The study recommended the use of blue LED in order to enhance the accumulation of anthocyanins in tartary buckwheat sprouts.

### *10.2.4 LED Effects on Gene Expression Associated with Ascorbate Metabolism*

Ascorbate (AsA), a strong and active antioxidant, plays an important role in human health by preventing different diseases associated with connective tissue (e.g., scurvy). It also acts as an antioxidant, removing free radicals which could induce cancer and senescence (Iqbal et al. 2009). AsA has beneficial influences on various physiological processes of plants. It regulates cell expansion and division, controls commencement of senescence, and plays an important role in plant defense against reactive oxygen species (ROS)-mediated stress, (Smirnoff 1996; Mittler 2002). AsA can also modulate plant growth via phytohormones-mediated gene expression (Pastori et al. 2003).

Fruits and vegetables are the main sources of AsA in human diets. Senescence during postharvest storage rapidly decreases the AsA content (Nishikawa et al. 2003). A variety of postharvest management approaches such as controlled atmosphere storage/packaging, treatments with chemicals, cytokinins, and ethylene vapors have been adopted for delaying postharvest senescence (Sharma et al. 2009; Lee et al. 2012).

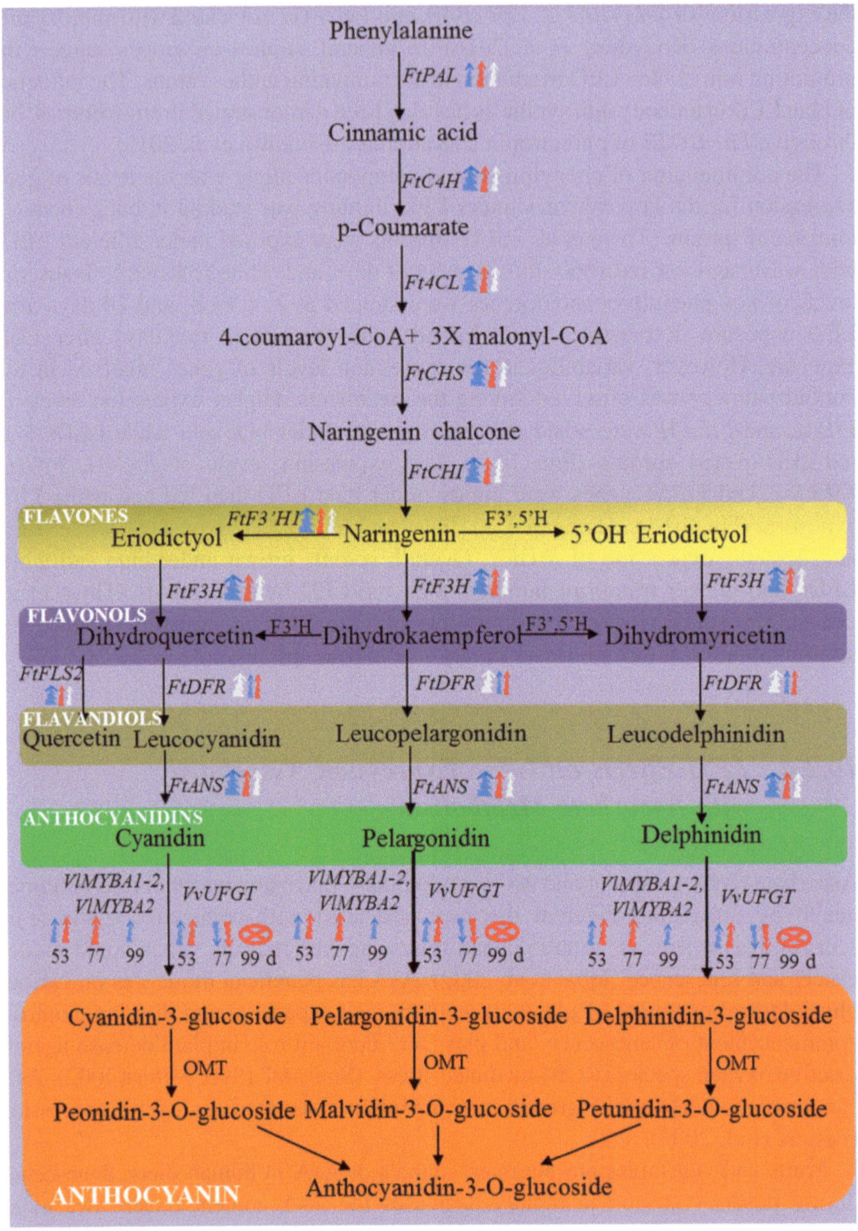


**Fig. 10.3** Influence of LED lighting on flavonoid biosynthetic pathway. Different wavebands of LEDs are marked with their respective colors. Variation in the width of the band indicates the extent of level of gene expression. Unchanged level of gene expression, compared to control at 99 d of full bloom, is represented by a *cross-circle*

In recent years, it has been demonstrated that light can prolong shelf life and preserves nutritional quality by regulating ascorbate metabolism of fresh-cut broccoli (Zhan et al. 2012; Ma et al. 2014). The effects of LED lighting on AsA metabolism and associated senescence were investigated in broccoli (*Brassica oleracea* L. var. italic) after harvest (Ma et al. 2014). Exposure to red LED (660 nm) not only delayed the senescence but it also slowed down the reduction of AsA content. In contrast, senescence process was not affected by blue LED light treatment and AsA content was found similar to that of the control. In spite of the positive effect of red LED exposure in delaying senescence and maintaining nutritional value, it was not recommended for postharvest storage due to its practical constrain. To resolve the issue, Ma et al. (2014) designed a type of modified white LED (430–730 nm) luminary, in which the ratio of red light was increased, while the proportion of blue light was decreased. Under the modified white LED light (50 μmol $m^{-2}$ $s^{-1}$), AsA content was higher than that of control on the first and second days of harvest. In broccoli, AsA metabolism is controlled by the regulation of genes which are involved in synthesis, degradation, and regeneration. The genes related to AsA metabolism have been cloned and well characterized (Nishikawa et al. 2003; Ma et al. 2012). Delayed suppression of AsA reduction on the first and second days after harvest of broccoli, under white LED, has been attributed to the up-regulation of the AsA biosynthetic genes (*BO-VTC2* and *BO-GLDH*) and AsA regeneration genes (*BO-MDAR1* and *BO-MDAR2*) (Fig. 10.4; Ma et al. 2014). The expression of AsA breakdown gene *BO-APX1* was up-regulated on the first and second day of harvest, whereas *BO-APX2* and *BO-sAPX* down-regulated only on the first day. The expression levels of these genes were not well correlated with the AsA content. Thus, their role in suppression of AsA reduction in response to the modified white LED light treatment was ruled out and the same was true with the expression of AsA regeneration gene *BO-DHAR*.

### *10.2.5 LED-Induced Defense and Transcript of Defense Genes*

The regulatory role of light in plant disease resistance other than many aspects of plant growth and development has been reported in a wide variety of plant species (Islam et al. 1998; Genoud et al. 2002; Khanam et al. 2005; Islam et al. 2008; Kudo et al. 2011). It has been suggested that light-induced signaling pathways interact with plant defense pathways, and there exists cross talk between light-mediated phytochrome signaling, development of hypersensitive response, and systematic acquired resistance (Genoud et al. 2002; Griebel and Zeier 2008). With the advancement of LED technology, there is growing interest in using this light source for the development of eco-friendly strategies to control plant disease development instead of using environmentally hazardous chemical fungicides. Wang et al. (2010) studied the effects of monochromatic LED lighting on the resistance to

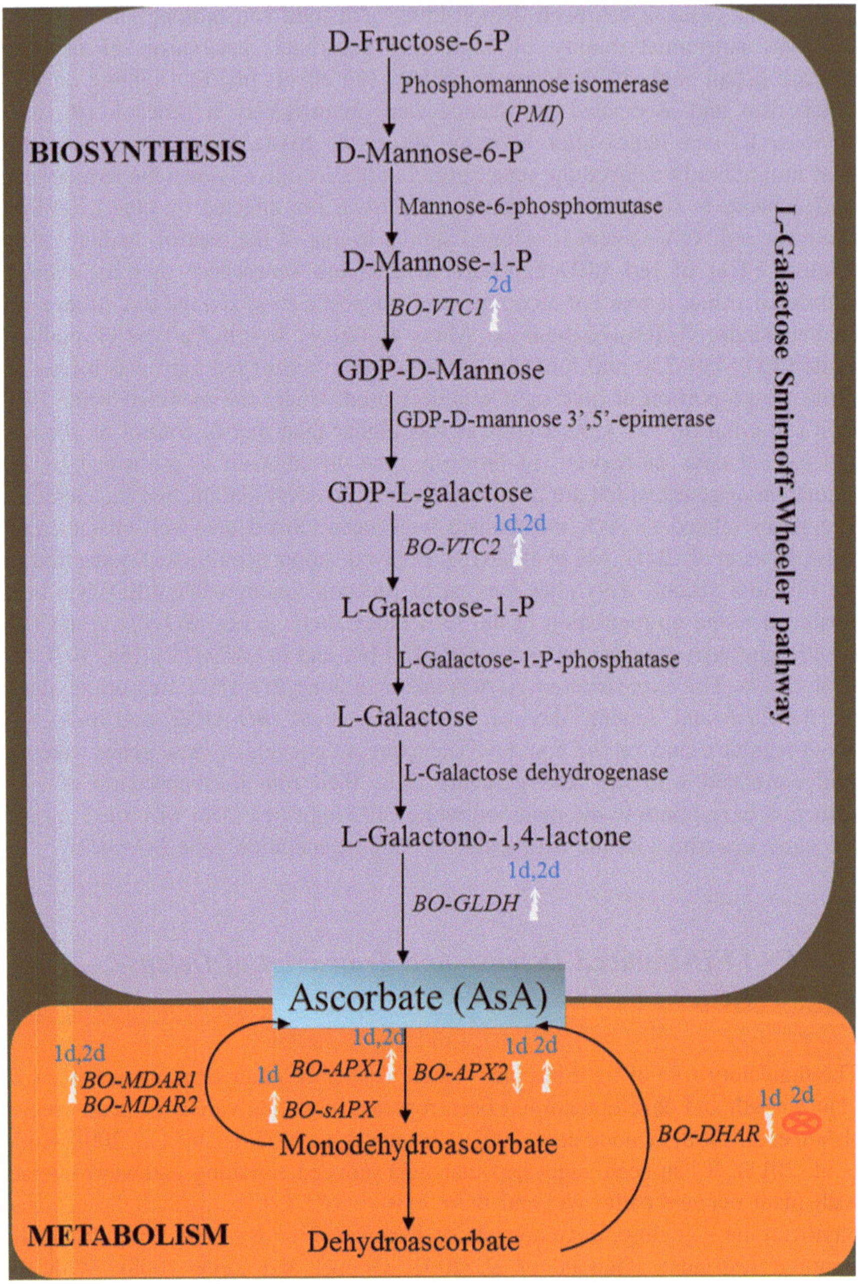


**Fig. 10.4** Effect of LED treatment on ascorbate metabolism. The *arrow* indicates up-regulation of gene expression by modified white LED (adapted from Ma et al. 2014)

powdery mildew (*Sphaerotheca fuliginea*) in cucumber (*Cucumis sativus*). The findings showed that incidence of powdery mildew significantly reduced by red LED treatment (628.6 nm), while monochromatic treatments of white; purple (394.6 nm); blue (452.5 nm); green (522.5 nm); and yellow (594.5 nm) LEDs increased the disease incidence. Exposure to red light up-regulated the expression levels of *PR-1* (a molecular marker of salicylic acid-dependent SAR pathway), *WRKY30*, and *WRKY6* (transcription factor encoding genes involved in SAR) genes, while other LED light sources down-regulated the expression of these genes (Wang et al. 2010). Plants grown under purple and blue LED showed higher incidence of disease and had higher levels of transcripts for *CS*, *CAD*, and *PAL* than those under red LED. The effects of red light to induce disease resistance have been reported in several studies (Islam et al. 2002; Rahman et al. 2003; Islam et al. 2008). The expression pattern indicates that an enhanced physical barrier or accumulation of flavonoids and total phenolics might not play a decisive role in light-induced resistance to powdery mildew in cucumber. Enhanced resistance under red LED is correlated with SA-mediated defense (Wang et al. 2010). The use of LEDs to reduce incidence of wildfire disease caused by *Pseudomonas syringae* pv. *tabaci* has been demonstrated in *Nicotiana benthamiana* (Ahn et al. 2013). Development of disease lesions on the leaves was reduced in plants irradiated with purple (380 nm), blue (440 nm), and red light (660 nm), compared to plants exposed under fluorescence light. LED irradiation induced the expression of defense-related genes such as *CHS*, *GST*, *TLP*, *PR Q*, *PINII*, and *TLP*. However, the maximum inhibition effect was obtained with blue and red light treatment. The up-regulation of PR Q, PINII, and TLP under LED before the pathogen infection suggests their role in the inhibition of disease symptom. Moreover, enhancement of expression levels of GST and TLP in plants exposed to LED irradiation inhibited the growth of *P. syringae* and suppressed the incidence of lesion development (Ahn et al. 2013).

## 10.3 Conclusions

Currently available data reveals that the custom-designed LED lighting systems are capable of regulating various valuable metabolic pathway genes, photoreceptor genes and defense-related genes. Such potential of LED lighting system opens up new avenues to enhance production of secondary metabolites of importance and to reduce disease incidence, or pathogen loads on certain crops by affecting the gene expression of cellular pathways. Most of the LED-induced gene expression studies were performed by quantitative real-time PCR, utilizing different wavebands of LEDs mainly blue, red, and white alone or in combination. LEDs with selected wavebands can stimulate early or uniform flowering in seasonal ornamentals or generate specific crops with enhanced levels of vitamins or minerals by regulating genetic factors. However, other factors are also involved in LED-regulated metabolic gene expression. Nevertheless, none of the studies peer out the LED-induced rate limiting gene expression or any feedback inhibition of metabolic pathway genes. It is feasible to trigger a cascade of positive responses by narrow-band irradiation of LEDs at certain key points in the plant life cycle. The efficient LED luminaries being developed along with the advancements in semiconductor materials and improvements in emitter design with high luminous efficacy have brought the ability to develop energy efficient lighting system. Along with breakthroughs in the molecular genetic control of plant growth and development, a coordinated effort in LED-based design of species-specific lighting system may help in maximizing the plant productivity, increasing nutritional quality of fruits and vegetables, and improvement in pharmaceutical attributes of medicinal plants, while limiting cost of production and environmental impacts of cultivation.

## References

Ahn SY, Kim SA, Baek KH, Yun HK (2013) Inhibiting wildfire and inducing defense-related gene expression by LED treatment on *Nicotiana benthamiana*. J Plant Pathol 95:477–483

Alabadi D, Blazquez MA (2009) Molecular interactions between light and hormone signaling to control plant growth. Plant Mol Biol 69:409–417

Al-Delaimy WK, Slimani N, Ferrari P, Key T, Spencer E, Johansson I, Johansson G, Mattisson I, Wirfalt E, Sieri S, Agudo A, Celentano E, Palli D, Sacerdote C, Tumino R, Dorronsoro M, Ocké MC, Bueno-De-Mesquita HB, Overvad K, Chirlaque MD, Trichopoulou A, Naska A, Tjonneland A, Olsen A, Lund E, Skeie G, Ardanaz E, Kesse E, Boutron-Ruault MC, Clavel-Chapelon F, Bingham S, Welch AA, Martinez-Garcia C, Nagel G, Linseisen J, Quirós JR, Peeters PH, van Gils CH, Boeing H, van Kappel AL, Steghens JP, Riboli E (2005) Plasma carotenoids as biomarkers of intake of fruits and vegetables: ecological-level correlations in the european prospective investigation into cancer and nutrition (EPIC). Eur J Clin Nutr 59:1397–1408

Alquézar B, Zacarías L, Rodrigo MJ (2009) Molecular and functional characterization of a novel chromoplast-specific lycopene β-cyclase from Citrus and its relation to lycopene accumulation. J Exp Bot 60:1783–1797

Bitsch R, Netzel M, Frank T, Strass G, Bitsch I (2004) Bioavailability and biokinetics of anthocyanins from red grape juice and red wine. J Biomed Biotechnol 5:293–298

Briggs WR, Christie JM (2002) Phototropins 1 and 2: versatile plant blue-light receptors. Trends Plant Sci 7:204–210

Britton G (1998) Overview on carotenoid biosynthesis. In: Britton G, Liaaen-Jensen S, Pfander H (eds) Carotenoids, 3rd edn. Birkhauser, Switzerland, pp 13–147

Camire ME, Chaovanalikit A, Dougherty MP, Briggs J (2002) Blueberry and grape anthocyanins as breakfast cereal colorants. J Food Sci 67:438–441

Chang J, Luo J, He G (2009) Regulation of polyphenols accumulation by combined overexpression/silencing key enzymes of phenylpropanoid pathway. Acta Biochim Biophys Sin (Shanghai) 41:123–130

Chen M, Chory J, Fankhauser C (2004) Light signal transduction in higher plants. Ann Rev Genet 38:87–117

Cunningham FX, Gantt E (1998) Genes and enzymes of carotenoid biosynthesis in plants. Annu Rev Plant Physiol Plant Mol Biol 49:557–583

De Pascual-Teresa S, Sanchez-Ballesta MT (2008) Anthocyanins: from plant to health. Phytochem Rev 7:281–299

Dorais M, Gosselin A (2002) Physiological response of greenhouse vegetable crops to supplemental lighting. Acta Hort 580:59–67

Exner V, Alexandre C, Rosenfeldt G, Alfarano P, Nater M, Caflisch A, Gruissem W, Batschauer A, Hennig L (2010) A gain-of-function mutation of Arabidopsis cryptochrome1 promotes flowering. Plant Physiol 154:1633–1645

Folta KM, Pontin MA, Karlin-Neumann G, Bottini R, Spalding EP (2003) Genomic and physiological studies of early cryptochrome 1 action demonstrate roles for auxin and gibberellin in the control of hypocotyl growth by blue light. Plant J 36:203–214

Frank HA, Cogdell RJ (1996) Carotenoids in photosynthesis. Photochem Photobiol 63:257–264

Fraikin GY, Strakhovskaya MG, Rubin AB (2013) Biological photoreceptors of light-dependent regulatory processes. Biochem Mosc 78:1238–1253

Genoud T, Buchala AJ, Chua NH, Metraux JP (2002) Phytochrome signalling modulates the SA perceptive pathway in *Arabidopsis*. Plant J 31:87–95

Gould KS (2004) Nature's swiss army knife: the diverse protective roles of anthocyanins in leaves. J Biomed Biotechnol 2004(5):314–320

Griebel T, Zeier J (2008) Light regulation and daytime dependency of inducible plant defenses in *Arabidopsis*: phytochrome signaling controls systemic acquired resistance rather than local defense. Plant Physiol 147:790–801

Guilfoyle TJ, Hagen G (2007) Auxin response factors. Curr Opin Plant Biol 10:453–460

Hannoufa A, Hossain Z (2012) Regulation of carotenoid accumulation in plants. Biocatal Agric Biotechnol 1:198–202

Havaux M (1998) Carotenoids as membrane stabilizers in chloroplasts. Trends Plant Sci 3:147–151

Hayes S, Velanisb CN, Jenkinsb GI, Franklina KA (2014) UV-B detected by the UVR8 photoreceptor antagonizes auxin signaling and plant shade avoidance. Proc Natl Acad Sci USA 111:11894–11899

Heijde M, Ulm R (2012) UV-B photoreceptor-mediated signalling in plants. Trends Plant Sci 17:230–237

Heuvelink E, Bakker MJ, Hogendonk L, Janse J, Kaarsemaker R, Maaswinkel R (2006) Horticultural lighting in the Netherlands: new developments. Acta Hort 711:25–34

Higuchi Y, Narumi T, Oda A, Nakano Y, Sumitomo K, Fukai S, Hisamatsu T (2013) The gated induction system of a systemic floral inhibitor, antiflorigen, determines obligate short-day flowering in chrysanthemums. Proc Natl Acad Sci USA 110:17137–17142

Iqbal Y, Ihsanullah I, Shaheen N, Hussain I (2009) Significance of vitamin C in plants. J Chem Soc Pak 31:169–170

Islam SZ, Honda Y, Arase S (1998) Light-induced resistance of broad bean against *Botrytis cinerea*. J Phyto Pathol 146:479–485

Islam SZ, Babadoost M, Honda Y, Sawa Y (2002) Characterization of antifungal glycoprotein in red-light-irradiated broadbean leaflets. Mycoscience 43:471–473

Islam SZ, Babadoost M, Bekal S, Lambert K (2008) Red light-induced systemic disease resistance against root knot nematode *Meloido gyne javanica* and *Pseudomonas syringae* pv. tomato DC 3000. J Phyto Pathol 156:708–714

Jing P, Bomser JA, Schwartz SJ, He J, Magnuson BA, Giusti MM (2008) Structure-function relationships of anthocyanins from various anthocyanin-rich extracts on the inhibition of colon cancer cell growth. J Agr Food Chem 56:9391–9398

Kadomura-Ishikawa Y, Miyawaki K, Noji S, Takahashi A (2013) Phototropin 2 is involved in blue light-induced anthocyanin accumulation in *Fragaria x ananassa* fruits. J Plant Res 126:847–857

Kami C, Lorrain S, Hornitschek P, Fankhauser C (2010) Light regulated plant growth and development. Curr Top Dev Biol 91:29–66

Kasahara M, Kagawa T, Sato Y, Kiyosue T, Wada M (2004) Phototropins mediate blue and red light-induced chloroplast movements in *Physcomitrella patens*. Plant Physiol 135:1388–1397

Kato M (2012) Mechanism of carotenoid accumulation in citrus fruit. J Jpn Soc Hortic Sci 81:219–233

Kato M, Ikoma Y, Matsumoto H, Sugiura M, Hyodo H, Yano M (2004) Accumulation of carotenoids and expression of carotenoid biosynthetic genes during maturation in citrus fruit. Plant Physiol 134:824–837

Khachik F, London E, de Moura FF, Johnson M, Steidl S, Detolla L, Shipley S, Sanchez R, Chen XQ, Flaws J, Lutty G, McLeod S, Fowler B (2006) Chronic ingestion of (3R, 3′R, 6′R)-

lutein and (3R, 3′R)-zeaxanthin in the female *Rhesus macaque*. Invest Ophthalmol Vis Sci 47:5476–5486

Khanam NN, Kihara J, Honda Y, Tsukamoto T, Arase S (2005) Studies on red light-induced resistance of broad bean to *Botrytis cinerea*: I: possible production of suppressor and elicitor by germinating spores of pathogen. J Gen Plant Pathol 71:285–288

Kharshiing E, Sinha SP (2015) Plant productivity: can photoreceptors light the way? J Plant Growth Regul 34:206–214

Kim WY, Hicks KA, Somers DE (2005) Independent roles for EARLY FLOWERING 3 and ZEITLUPE in the control of circadian timing, hypocotyl length, and flowering time. Plant Physiol 139:1557–1569

Kim EY, Park SA, Park BJ, Lee Y, Oh MM (2014) Growth and antioxidant phenolic compounds in cherry tomato seedlings grown under monochromatic light-emitting diodes. Hort Environ Biotechnol 55:506–513

Kinoshita T, Doi M, Suetsugu N, Kagawa T, Wada M, Shimazaki K (2001) Phot 1 and phot 2 mediate blue light regulation of stomatal opening. Nature 414:656–660

Kobayashi S (2009) Regulation of anthocyanin biosynthesis in grapes. J Jpn Soc Hort Sci 78:387–393

Kondo S, Tomiyama H, Rodyoung A, Okawa K, Ohara H, Sugaya S, Terahara N, Hirai N (2014) Abscisic acid metabolism and anthocyanin synthesis in grape skin are affected by light emitting diode (LED) irradiation at night. J Plant Physiol 171:823–829

Kudo R, Ishida Y, Yamamoto K (2011) Effect of green light irradiation on induction of disease resistance in plants. Acta Hort 907:251–254

Lau OS, Deng XW (2010) Plant hormone signaling lightens up: integrators of light and hormones. Curr Opin Plant Biol 13:571–577

Lee SH, Tewari RK, Hahn EJ, Paek KY (2007) Photon flux density and light quality induce changes in growth, stomatal development, photosynthesis and transpiration of *Withania somnifera* (L.) Dunal. plantlets. Plant Cell Tiss Org Cult 90:141–151

Lee J, Mattheis JP, Rudell DR (2012) Antioxidant treatment alters metabolism associated with internal browning in 'Braeburn' apples during controlled atmosphere storage. Postharvest Biol Technol 68:32–42

Leopold AC (1951) Photoperiodism in plants. Q Rev Biol 26:247–263

Li J, Lib G, Wang H, Deng XW (2011) Phytochrome signaling mechanisms. Arabidopsis Book. doi:10.1199/tab.0148

Li SB, Xie ZZ, Hu CG, Zhang JZ (2016) A review of auxin response factors (ARFs) in plants. Front Plant Sci. doi:10.3389/fpls.2016.00047

Lin C (2000) Photoreception and regulation of flowering time. Plant Physiol 123:139–150

Liu XY, Guo GR, Chang TT, Xu ZG, Takafumi T (2012) Regulation of the growth and photosynthesis of cherry tomato seedlings by different light irradiations of light emitting diodes (LED). Afr J Biotechnol 11:6169–6177

Long SP, Humphries S, Falkowski PG (1994) Photoinhibition of photosynthesis in nature. Annu Rev Plant Physiol Plant Mol Biol 45:633–662

Lopez L, Carbone F, Bianco L, Giuliano G, Facella P, Perrotta G (2012) Tomato plants over expressing cryptochrome 2 reveal altered expression of energy and stress-related gene products in response to diurnal cues. Plant, Cell Environ 35:994–1012

Ma G, Zhang LC, Kato M, Yamawaki K, Asai T, Nishikawa F, Ikoma Y, Mat-sumoto H, Yamauchi T, Kamisako T (2012) Effect of electrostatic atomization on ascorbate metabolism in postharvest broccoli. Postharvest Biol Technol 74:19–25

Ma G, Zhang L, Setiawan CK, Yamawaki K, Asai T, Nishikawa F, Maezawa S, Sato H, Kanemitsue N, Kato M (2014) Effect of red and blue LED light irradiation on ascorbate content and expression of genes related to ascorbate metabolism in post harvest broccoli. Postharvest Biol Technol 94:97–103

Ma G, Zhang L, Kato M, Yamawaki K, Kiriiwa Y, Yahata M, Ikoma Y, Matsumoto H (2015) Effect of the combination of ethylene and red LED light irradiation on carotenoid accumulation and carotenogenic gene expression in the flavedo of citrus fruit. Postharvest Biol and Technol 99:99–104

Massa GD, Kim HH, Wheeler RM, Mitchell CA (2008) Plant productivity in response to LED lighting. HortScience 43:1951–1956

Mittler R (2002) Oxidative stress, antioxidants and stress tolerance. Trends Plant Sci 7:405–410

Mockler T, Yang H, Yu W, Parikh D, Cheng Y, Dolan S, Lin C (2003) Regulation of photoperiodic flowering by *Arabidopsis* photoreceptors. Proc Natl Acad Sci USA 100:2140–2145

Molas ML, Kiss JZ (2009) Phototropism and gravitropism in plants. Adv Bot Res 49:1–34

Nelson JA, Bugbee B (2014) Economic analysis of greenhouse lighting: light emitting diodes vs. high intensity discharge fixtures. PloS One 9(e99010):1–10

Nisar N, Li L, Lu S, Khin NC, Pogson B (2015) Carotenoid metabolism in plants. Mol Plant 8:68–82

Nishikawa F, Kato M, Hyodo H, Ikoma Y, Sugiura M, Yano M (2003) Ascorbate metabolism in harvested broccoli. J Exp Bot 54:2439–2448

Okushima Y, Overvoorde PJ, Arima K, Alonso JM, Chan A, Chang C, Ecker JR, Hughes B, Lui A, Nguyen D, Onodera C (2005) Functional genomic analysis of the auxin response factor gene family members in *Arabidopsis thaliana*: unique and overlapping functions of ARF7 and ARF19. Plant Cell 17(2):444–463

Page JE, Towers GH (2002) Anthocyanins protect light sensitive thiarubrine phototoxins. Planta 215(3):478–484

Park YG, Muneer S, Jeong BR (2015) Morphogenesis, flowering, and gene expression of *Dendranthema grandiflorum* in response to shift in light quality of night interruption. Int J Mol Sci 16:16497–16513

Pashkovskiy PP, Kartashov AV, Zlobin IE, Pogosyan SI, Kuznetsov VV (2016) Blue light alters miR167 expression and microRNA-targeted auxin response factor genes in *Arabidopsis thaliana* plants. Plant Physiol Biochem 104:146–154

Pastori GM, Kiddle G, Antoniw J, Bernard S, Veljovic-Jovanovic S, Verrier PJ, Noctor G, Foyer CH (2003) Leaf vitamin C contents modulate plant defense transcripts and regulate genes that control development through hormone signaling. Plant Cell 15:939–951

Pouchieu C, Galan P, Ducros V, Latino-Martel P, Hercberg S, Touvier M (2014) Plasma carotenoids and retinol and overall and breast cancer risk: a nested case-control study. Nutr Cancer 66:980–988

Rahman MZ, Honda Y, Arase S (2003) Red-light-induced resistance in broad bean (*Vicia faba* L.) to leaf spot disease caused by *Alternaria tenuissima*. J Phytopathol 151(2):86–91

Rodyoung A, Masuda Y, Tomiyama H, Saito T, Okawa K, Ohara H, Kondo S (2016) Effects of light emitting diode irradiation at night on abscisic acid metabolism and anthocyanin synthesis in grapes in different growing seasons. Plant Growth Regul 79:39–46

Sharma RR, Singh D, Singh R (2009) Biological control of postharvest diseases of fruits and vegetables by microbial antagonists: a review. Biol Control 50:205–221

Singh D, Basu C, Meinhardt-Wollweber M, Roth B (2015) LEDs for energy efficient greenhouse lighting. Renew Sustain Energy Rev 49:139–147

Smirnoff N (1996) The function and metabolism of ascorbic acid in plants. Ann Bot 78:661–669

Somers DE, Kim WY, Geng R (2004) The F-Box protein ZEITLUPE confers dosage-dependent control on the circadian clock, photomorphogenesis, and flowering time. Plant Cell 16:769–782

Takahama U (2004) Oxidation of vacuolar and apoplastic phenolic substrates by peroxidases: physiological significance of the oxidation reactions. Phytochem Rev 3:207–219

Thwe AA, Kim YB, Li X, Seo JM, Kim SJ, Suzuki T, Chung SO, Park SU (2014) Effects of light-emitting diodes on expression of phenylpropanoid biosynthetic genes and accumulation of phenylpropanoids in *Fagopyrum tataricum* sprouts. J Agric Food Chem 62:4839–4845

Tuan PA, Thwe AA, Kim JK, Kim YB, Lee S, Park SU (2013) Molecular characterization and the light—dark regulation of carotenoid biosynthesis in sprouts of tartary buckwheat (*Fagopyrum tataricum* Gaertn.). Food Chem 141:3803–3812

Wada M, Kagawa T, Sato Y (2003) Chloroplast movement. Annu Rev Plant Biol 54:455–468

Wang H, Jiang YP, Yu HJ, Xia XJ, Shi K, Zhou YH, Yu JQ (2010) Light quality affects incidence of powdery mildew, expression of defense-related genes and associated metabolism in cucumber plants. Eur J Plant Pathol 127:125–135

Warner RM, Erwin JE (2003) Effect of photoperiod and daily light integral on flowering of five *Hibiscus* sp. Sci Hortic 97:341–351
Wheeler GL, Jones MA, Smirnoff N (1998) The biosynthetic pathway of vitamin C in higher plants. Nature 393:365–369
Wu SH (2014) Gene expression regulation in photomorphogenesis from the perspective of the central dogma. Annu Rev Plant Biol 65:311–333
Wu MC, Hou CY, Jiang CM, Wang YT, Wang CY, Chen HH, Chang HM (2007) A novel approach of LED light radiation improves the antioxidant activity of pea seedlings. Food Chem 101:1753–1758
Yamada AT, Tanigawa T, Suyama T, Matsuno T, Kunitake T (2008) Night break treatment using different light sources promotes or delays growth and flowering of *Eustoma grandiflorum* (Raf.) Shinn. J Jpn Soc Hortic Sci 77:69–74
Zhang L, Ma G, Kato M, Yamawaki K, Takagi T, Kiriiwa Y, Ikoma Y, Matsumoto H, Yoshioka T, Nesumi H (2012) Regulation of carotenoid accumulation and the expression of carotenoid metabolic genes in citrus juice sacs in vitro. J Exp Bot 63:871–886
Zhang L, Ma G, Yamawaki K, Ikoma Y, Matsumoto H, Yoshioka T, Ohta S, Kato M (2015) Effect of blue LED light intensity on carotenoid accumulation in citrus juice sacs. J Plant Physiol 188:58–63

# Chapter 11
# The Influence of Light-Emitting Diodes (LEDs) on the Growth, Antioxidant Activities, and Metabolites in Adventitious Root of *Panax ginseng* C.A. Meyer

**Bimal Kumar Ghimire, Jae Geun Lee, Ji Hye Yoo, Jae Kwang Kim and Chang Yeon Yu**

## 11.1 Introduction

*Panax ginseng* is a perennial herbaceous plant distributed in Korea, Japan, China, Europe, and North America (Mabberley 1987; Hobbs 1996; Nam 2002). The roots of these plants have been traditionally used as functional food and herbal medicine for thousands of years for treating human diseases (Chang et al. 2006). They are used to treat blood pressure (Kang and Kim 1992), improve liver functions and immune system (Kim and Jung 1987), and to treat postmenopausal symptoms (Shim and Lee 2009). Ginseng root extracts have also shown anti-stress, anti-diabetic, anti-aging, anticancer, and immunomodulatory along with neuro-protective, anti-fatigue, cardio-protective, and hepato-protective physiological and pharmacological effects (Tang and Eisenbrand 1992). Anti-inflammation and antioxidative activity of root extracts have also been reported by Shao et al. (2004),

B.K. Ghimire
Department of Applied Bioscience, Konkuk University,
Seoul 05029, Republic of Korea

C.Y. Yu (✉)
Department of Bio-Resource Sciences, Kangwon National University,
Chuncheon 24341, Republic of Korea
e-mail: cyyu@kangwon.ac.kr

J.G. Lee
Research Institute of Biotechnology, Hwajin Cosmetics, Hongcheon 25142,
Republic of Korea

J.K. Kim
Division of Life Sciences, Incheon National University, Incheon 22012,
Republic of Korea

J.H. Yoo
Bioherb Research Institute, Kangwon National University,
Chuncheon 24341, South Korea



S. Dutta Gupta (ed.), *Light Emitting Diodes for Agriculture*,
DOI 10.1007/978-981-10-5807-3_11

Xie et al. (2006), and Kim et al. (2012). Many investigators have reported the presence of various bioactive compounds in ginseng roots such as ginsenosides, flavonoids, fatty acids, mono/triterpenes, phenylpropanoids, kairomones, alkanes, alkynes, sterols, and polysaccharides (Chang et al. 2003; Ganesan et al. 2015). Among these bioactive compounds, ginsenosides are the most important due to its wide range of health benefit that includes prevention of cancer and modulation of immune system (Chiou and Zhang 2008). Ginsenosides are divided into two major groups Rb (Rb1, Rb2, Rb3, Rc, Rd, Rg3, and Rh2) and Rg (Re, Rg1, Rg2, Rf, and Rh1; Oh and Kim 2016).

Conventional methods of growing ginseng in the cultivation field are time-consuming and labor-intensive process which results into increased costs (Proctor 1996). Moreover, narrow range of soil conditions, very specific habitat, large investment, requirement of stratification to break dormancy, low germination rate, diseases, and physiological disorders causes reduction in the yield (Li 1995). In the past decade, the production of secondary metabolites by using biotechnological approaches has become indispensable and attracted as an alternative to the conventional method of cultivation (Furuya et al. 1983). The use of bioreactors for the cultivation of adventitious roots is beneficial and economical for the production of valuable bioactive compounds, with consistent quality and yield independent of the external factors (Fowler 1985; Wu et al. 2007).

Different internal and external parameters influence the in vitro growth of plant, among which light is an important factor that influences the plant growth, organogenesis/embryogenesis, and production of secondary metabolites (Shohael et al. 2006; Dutta Gupta and Jatothu 2013). In recent years, the development of LED technologies presents an enormous potential for improving plant growth and making systems more sustainable (Darko et al. 2014). Irradiation by different types of LEDs has shown to produce ginsenoside in *Panax vietnamensis* (Nhut et al. 2015), saponins in *P. ginseng* (Kim et al. 2009), flavonoid glycosides in *Petroselinum hortense* (Kreuzaler and Hahlbrock 1973), anthocyanin in *Perilla frutescens* (Zhong et al. 1991), artemisinin in *Artemisia annua* (Liu et al. 2002), and pulchelin E in *Rudbeckia hirta* (Luczkiewicz et al. 2002). Furthermore, a number of studies successfully used LEDs for in vitro culture of different plant species with a variety of purposes (Menard et al. 2006; Avercheva et al. 2009; Li et al. 2010; Brazaityte et al. 2010; Liu et al. 2011; Dutta Gupta and Jatothu 2013; Dutta Gupta and Sahoo 2015) and identified various physiological, morphological, and metabolic effects (Shin et al. 2003; Kim et al. 2008).

Various abiotic factors including photo-stress are responsible for the oxidative stress in plant. Oxidative stress causes generation of free radicals (ROSs) in the plant and causes deterioration and damage to the cellular components including DNA, lipids, and protein (Sharma et al. 2012). The accumulation of phenolics by in vitro cultured tissues potentially combat oxidative stress and play important role in the removal of toxic substances (Low and Merida 1996; Rice-Evans et al. 1996; Di Carlo et al. 1999; Gould et al. 2000; Close and McArthor 2002; Sgherri et al. 2004). Moreover, appropriate wavelengths of LEDs have been successfully used to increase antioxidative actions and to protect against abiotic stresses. LEDs as source of light attracted considerable interest due to its unique features including less energy consumption, heat generation, long life expectancy and ease of control, high efficiency, low operating temperature, PAR efficiency, high fluence rate at specific wave length (Bula et al. 1991; Yeh and Chung 2009; Yu et al. 2005). In contrast, conventional fluorescent, metal-halide, and incandescent lamps utilize more energy and produce more heat energy, causing unnecessary photo-stress and heat stress to the in vitro plants. Moreover, wide range of wavelength of light may not be necessary to promote growth and development of in vitro plants (Kim et al. 2004). Here, we examined the impact of LEDs on the production of high-value secondary metabolites, growth, and antioxidant properties in *P. ginseng* using LED lights at various wavelengths.

## 11.2 Culture Establishment and Light Treatments

Cultured adventitious roots of *P. ginseng* were kindly supplied by the Oriental Medicinal Materials and Processing Department, Kyung Hee University, Korea. Adventitious roots were cultivated in bioreactor placed in a controlled chamber free from spectral interference of LEDs. Initially, the adventitious roots of ginseng were cultured in 250 ml shake flask for five weeks and cut into 0.5–2.0 cm slices and inoculated aseptically into the bioreactor containing 10 L of SH (Schenk and Hildebrandt 1972) liquid medium supplemented with 30 g/l sucrose and 2 mg/l indole-3-butyric acid (IBA); the pH was then adjusted to 5.7–5.8 and the contents are transferred to growth chamber for five weeks equipped with red-LED (630 nm) and blue-LED (465 nm) light spectra (KM Electronic Supplier, KM1235; Seoul, Korea). Fluorescent light was used as control (Fig. 11.1a–f). All cultures were grown under the photon flux density of 24 $\mu$mol m$^{-2}$ s$^{-1}$.

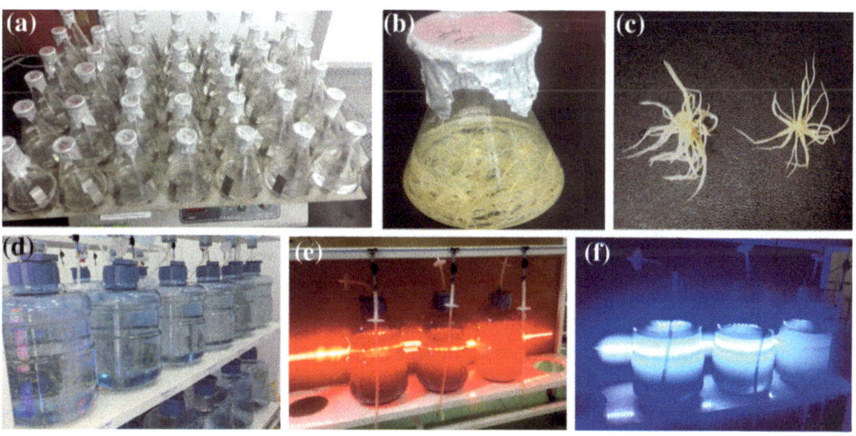


**Fig. 11.1** Adventitious root culture of *P. ginseng* under LEDs. **a** Adventitious root culture of ginseng in shake flask. **b**, **c** AR of *P. ginseng* after 8 weeks. **d** Bioreactor culture system of AR of *P. ginseng* after one week by treating fluorescent light, **e** *red*-LED, **f** *blue*-LED

## 11.3 Measurement of Electron Donation Ability and Analysis of Metabolites

Free radical scavenging activity was evaluated using ascorbic acid and butylated hydroxy toluene (BHT) as standard antioxidants. The radical scavenging activity was measured using the stable radical 1,1-diphenyl–2-picrylhydrazyl (DPPH). The radical scavenging activity of adventitious roots was expressed as the percent ratio of the absorption of DPPH in the presence and absence of the compound. Calculated $RC_{50}$ values indicate the concentration of sample required to scavenge 50% of the DPPH radical. DPPH activity was calculated as follows:

$$RC_{50} = (A_{\text{blank}} - A_{\text{sample}})/A_{\text{blank}} \times 100,$$

where $A_{\text{blank}}$ is the absorbance of the control reaction (containing all reagents except the test compound) and $A_{\text{sample}}$ is the absorbance of the test compound.

Concentration of various phenolic acids was determined by following the methods described by (Park et al. 2014) with slight modifications. AR of ginseng (0.1 g) from each LED treatment were extracted by sonication for 5 min at room temperature ($25 \pm 1°$) and incubated at 30 °C for 10 min with 1 ml of 85% methanol containing 2 g/l BHA. After centrifugation at 13,000 rpm for 10 min at 4 °C, the combined extracts and residue were analyzed to determine the phenolic acids. Fifty microliter of 3,4,5-trimethoxycinnamic acid (100 μg/ml) was added as an internal standard (IS), and the mixture was hydrolyzed with 1 ml 5 N NaOH at 30 °C under nitrogen gas for 4 h. Each hydrolyzed sample was adjusted to pH 1.5–2.0 with 6 M HCl, extracted with ethyl acetate, and evaporated in a centrifugal concentrator (CVE-2000; Eyela, Tokyo, Japan). Identification of metabolites was performed by GC-TOFMS method described previously by Park et al. (2014). The amounts of policosanols, tocopherols, and phytosterols were determined by following the methods described by Kim et al. (2012).

## 11.4 Influence of LEDs on the Growth of Adventitious Roots

Impact of LED irradiation of various types on the fresh and dry weights of AR of ginseng was evaluated and compared with control (Fig. 11.2a–i).

AR treated with red-LED irradiation resulted in higher fresh and dry weights compared to blue-LED and fluorescent light treatments (Fig. 11.3). This result is consistent with the report of Yu et al. (2005). In contrast, total dry weight of root was lower under red-LED compared to blue-LED (Brown et al. 1995; Yorio et al. 2001; Shohael et al. 2006; Shin et al. 2008). We also assessed the morphological

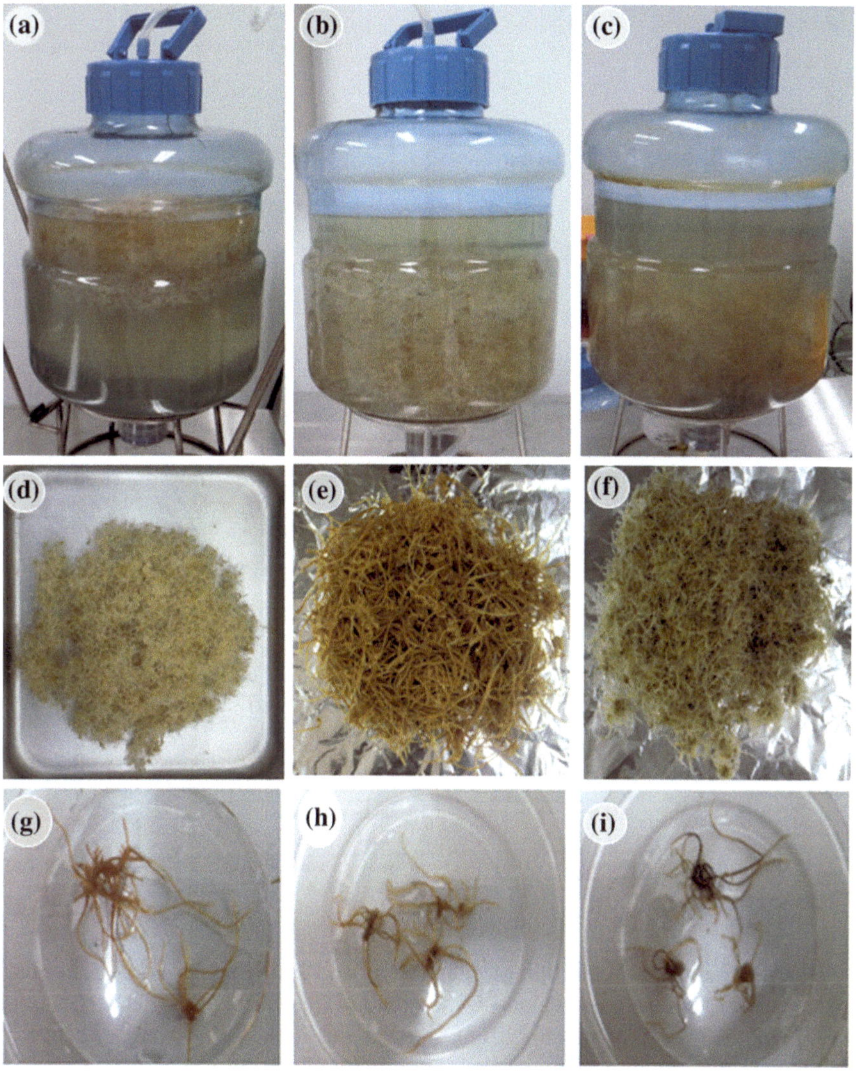


**Fig. 11.2** Bioreactor culture system of AR of *P. ginseng* after eight weeks cultured by treating **a** fluorescent light, **b** *red*-LED, **c** blue-LED; **d** harvesting of AR of *P. ginseng* after eight weeks cultured by treating fluorescent light, **e** *red*-LED, **f** *blue*-LED; morphological character of AR of *P. ginseng* under fluorescent light (**g**), *red*-LED (**h**), *blue*-LED (**I**)

variations in the AR grown under different LEDs. AR grown under blue-LED resulted in increased diameter of roots (Fig. 11.4a). However, the diameter of AR under red-LED was not significantly different from fluorescent light-treated AR. Red-LED irradiation showed an increased root length compared to blue-LED and fluorescent lights (Fig. 11.4b).

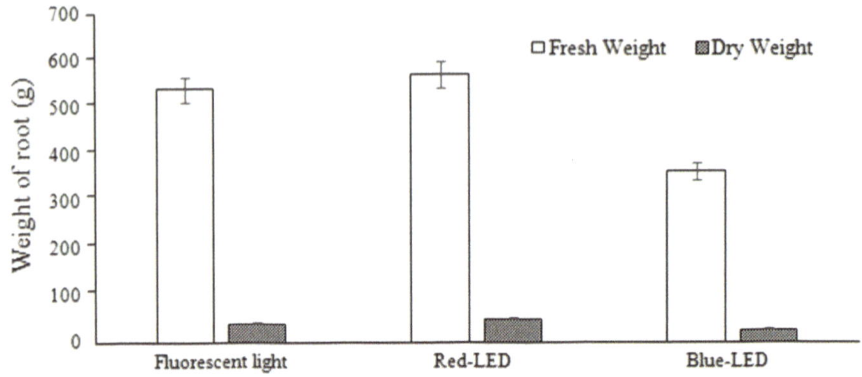


**Fig. 11.3** Fresh weight and dry weight of AR of *P. ginseng* treated with LEDs

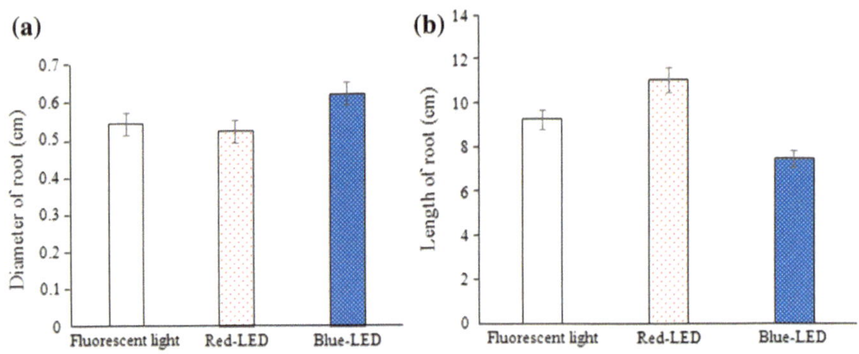


**Fig. 11.4** Morphological analysis of AR of *P. ginseng* treated with LEDs. **a** Diameter of root, **b** length of root

Blue-LED-induced inhibition of plant height and leaf expansion has been noted in *A. vidalii* (Moreira da Silva and Debergh 1997) and in pepper (Schuerger et al. 1997). In contrast, Liu et al. (2012) showed that blue-LEDs were effective in accumulating plant biomass in cherry tomato seedlings. In a similar study, Kim et al. (2009) revealed no significant differences in root length and diameter of ginseng grown in different LED light conditions. Poudel et al. (2008) reported an increase in the rooting percentage and root numbers of grape by treating red-LED (*Vitis vinifera*). Similar results were reported in Anthurium (*Anthurium andraeanum*; Budiarto 2010), cotton (*Gossypium hirutum*; Li et al. 2010), and chrysanthemum (*Chrysanthemum morifolium*; Kurilčik et al. 2008), where red LEDs were also found to stimulate root formation. Other studies revealed an optimum plant growth and development of plant by combining effect of different LEDs at the appropriate ratio (Lian et al. 2002; Nhut et al. 2001, 2002; Shohael et al. 2006; Poudel et al. 2008; Shin et al. 2008; Li et al. 2010).

## 11.5 Analysis of Phenolic Acids in Adventitious Roots

The intensity and quality of light have been considered as one of the critical factors that impacts on metabolism of plant. It has been argued that light source may act as an elicitor in the plant defense system and stimulate secondary metabolite production (Shohael et al. 2006). Moreover, light source can cause stress on the growing plant and induce secondary metabolite to prevent oxidative damage by removing or neutralizing toxic substance (Gould et al. 2000; Close and McArthor 2002). In this study, the distribution of phenolic compounds of AR was investigated under different LEDs. Among the identified phenolic compounds, ferulic acid, p-coumaric acid, sinapic acid, vanillic acid, and syringic acid were dramatically increased under blue-LED treatment than fluorescent and red-LED lights (Fig. 11.5).

Consistent with our results, Johkan et al. (2010) also noted increased concentration of phenolic compounds by treating blue-LED light in red leaf lettuce. In another study, the amount of ginsenoside (Rg1) was significantly increased in the ginseng root treated with blue-LED (Kim et al. 2009; Son et al. 2012). It has been argued that the wavelength of blue light is close to ultraviolet light, which triggers the biosynthesis of phenolic compounds with antioxidant properties (Ryan et al. 2002; Ebisawa et al. 2008). On the contrary, ginseng root cultured in fluorescent

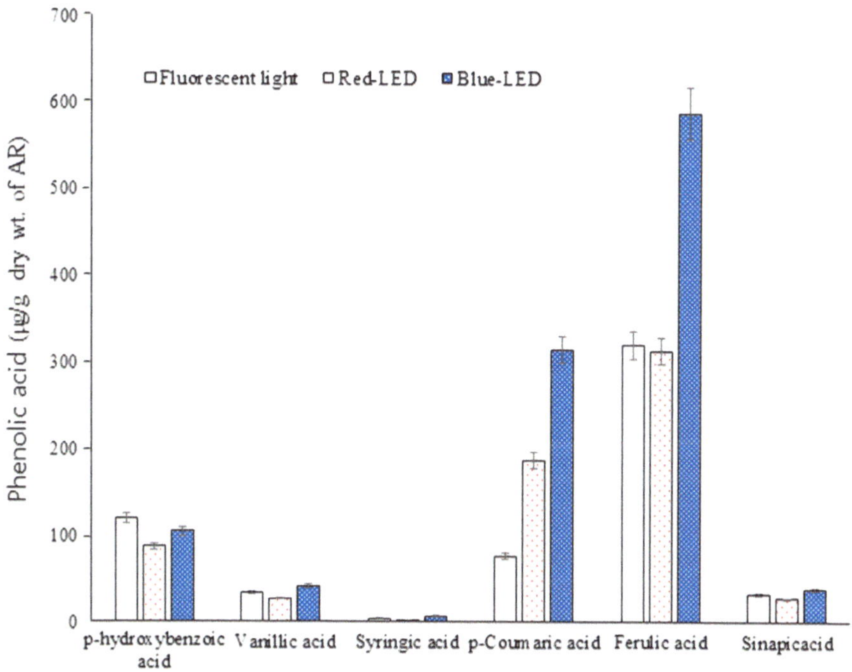


**Fig. 11.5** Total phenolic acid composition of AR of *P. ginseng* treated with LEDs

light resulted in the optimum production of ginsenoside (Yu et al. 2005). In the present study, AR treated with red-LED light notably decreased the ferulic acid, sinapic acid, syringic acid, vanillic acid, and *p*-hydroxybenzoic acid than fluorescent-irradiated AR. The lowest concentrations of 3,4-dihydroxybenzoic acid and ferulic acid were also reported in *Protea cynaroides* irradiated with red-LED (Wu and Lin 2012). Eleutheroside B and E contents increased in the ginseng root treated with blue-LED, whereas eleutheroside E1 was the highest under fluorescent light (Jeong et al. 2009). The total phenol content was found lower in wheatgrass and green barley leaves treated with red-LED light (Urbonaviciute et al. 2009a, b). They indicated that the reduction in the phenolic compounds in particular ferulic acid and 3,4-dihydroxybenzoic acid could inversely relate to the root growth and the number of roots. A similar result was also observed in *P. cynaroides* (Wu et al. 2007). It is likely that the reduction of *p*-hydroxybenzoic acid and ferulic acid in the red-LED-treated AR could be responsible for higher fresh weight and dry weight compared to blue-LED and fluorescent light-treated AR. Mucciarelli et al. (2000) argued that 3,4-dihydroxybenzoic acid could influence directly and independently on cell differentiation, indicating that there could exist a relation between phenolic compound concentration and growth of roots. Therefore, the variation in the concentration of secondary metabolites under different LEDs spectra can be attributed to the different action of light during the growth of AR. Overall, results indicated that light quality and wavelength remarkably influenced the accumulation of phenolic compounds, thereby influencing growth and antioxidant activity of ginseng root.

## 11.6 Analysis of Lipophilic Compounds in Adventitious Roots

The composition of lipophilic compounds in AR of ginseng treated with various LED light treatments was determined and compared with fluorescent light-treated AR (Fig. 11.6a). Different types of policosanols, such as eicosanol, heneicosanol, docosanol, tricosanol, tetracosanol, and octacosanol were identified. Total policosanol contents were higher in red-LED treatment of ginseng root. Among the policosanol, the amount of eicosanol and docosanol significantly increased in the presence of red-LED light than in blue-LED and fluorescent lights (Fig. 11.6b). Total tocopherol contents were higher in AR treated with blue-LED. Here, we observed an increase in the α-tocopherol and β-tocopherol which appear only in blue-LED-treated AR (Fig. 11.6c). AR cultured under red-LED produced significantly higher campesterol, cholesterol, β-sitosterol, and stigmasterol. Concentration of β-amyrin increased under the blue-LED irradiation (Fig. 11.6d).

The result indicates that the contents of policosanols, tocopherols, and sterols of AR culture could be influenced by spectral quality of different LEDs. It is possible that LED light may stimulate enzyme activities in the biosynthetic pathway of

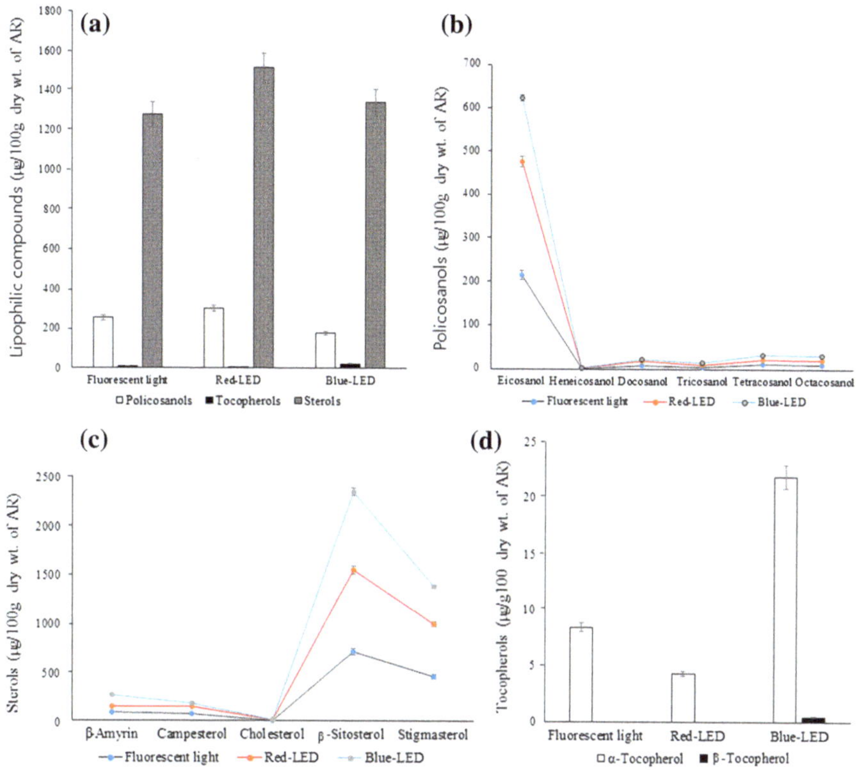


**Fig. 11.6** **a** Total lipophilic compounds in AR of *P. ginseng* treated with LEDs, **b** composition of policosanol, **c** composition of sterols, **d** composition of tocopherols in AR of *P. ginseng* treated with LEDs

phytochemicals. However, further studies on detail mechanism on the influence of LEDs or its impact on the growth and phytochemical pathway of bioactive compounds of AR ginseng would be necessary.

## 11.7 Effect of Different LEDs on Radical Scavenging Activity

We assessed the radical scavenging activity of adventitious roots of ginseng cultured under different LED lights with varying wavelengths. The results revealed that the antioxidant activity of adventitious roots greatly influenced by the types of LED light. In particular, blue-LED light showed significantly ($p < 0.05$) higher antioxidant activity than that of red-LED and fluorescent lights (Fig. 11.7). We also observed a significant increase in the phenolic compounds such as ferulic acid and

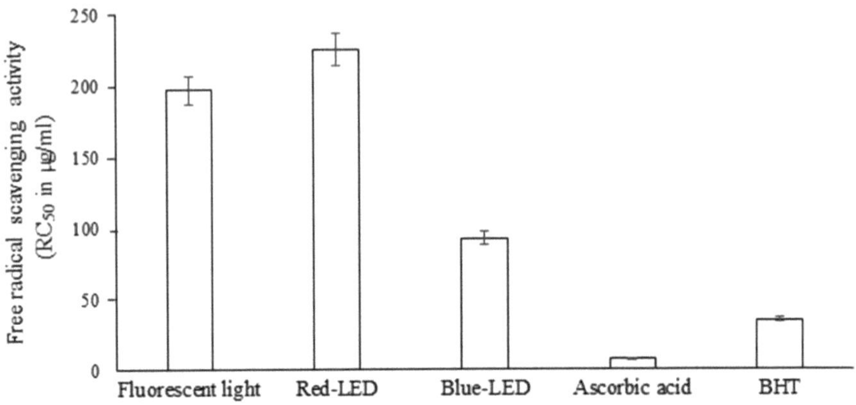


**Fig. 11.7** DPPH radical scavenging activity of AR of *P. ginseng*

*p*-coumaric acid in the AR treated with blue-LED, indicating that an increase in the phenolic acid may be responsible for enhancing the antioxidant activity of AR.

The increase in the radical scavenging activity is in agreement with Johkan et al. (2010) in which they found higher antioxidant activity in lettuce leaf treated with blue-LED. Moreover, various studies on medicinal plants have noted the association between the antioxidant activity and phenolic compounds (Dorman et al. 2004; Li et al. 2008). A number of studies revealed that the phenolic compounds such as ferulic acid and *p*-coumaric acid possess higher scavenging potential of free radicals (Kikuzaki et al. 2002; Masek et al. 2016). These phenolic acids can also up-regulate the hemeoxygenase–biliverdin reductase system used in the "indirect" free radical scavenging activity process (Calabrese et al. 2008; Barone et al. 2009; Fetoni et al. 2010). The presence of these compounds in AR indicates that the exposure of AR of ginseng to LED resulted in disruption of the metabolic pathway of phenolic compounds which can lead to an increase in the accumulation of phenolic compounds and antioxidant activity. A number of studies reported the antioxidative properties of plant sterols, policosanol, and tocopherol (Noa et al. 2001; Wang et al. 2002; Chauhan and Chauhan 2015). Thus, it is very possible that the higher level of tocopherol, policosanol, and sterol in the LED-treated AR could be responsible partly for elevating the antioxidant properties. Therefore, exposure to an appropriate wavelength of LEDs to the AR of ginseng may allow higher production of pharmacological components with higher commercial value.

## 11.8 Conclusion

Present results exhibit that the application of LED light is useful for increasing the growth, accumulation of metabolites, and antioxidant properties of AR of ginseng. LED-treated AR had higher antioxidant activity, growth of root, phenolic acid, and lipophilic compounds compared to fluorescent light. Further study is required to adequately explain the phenolic compounds' biosynthesis mechanism and its relation with root growth and antioxidant activity of ginseng grown under different LED treatments.

## References

Avercheva OV, Berkovich YA, Erokhin AN, Zhigalova TV, Pogosyan SI, Smolyanina SO (2009) Growth and photosynthesis of chinese cabbage plants grown under light-emitting diode-based light source. Russ J Plant Physiol 56:14–21

Barone E, Calabrese V, Mancuso C (2009) Ferulic acid and its therapeutic potential as a hormetin for age-related diseases. Biogerontology 10:97–108

Brazaitytė A, Duchovskis P, Urbonavičiūtė A, Samuolienė G, Jankauskienė J, Sakalauskaitė J, Šabajevienė G, Sirtautas R, Novičkovas A (2010) The effect of light-emitting diodes lighting on the growth of tomato transplants. Zemdirbyste 97:89–98

Brown CS, Schuerger AC, Sager JC (1995) Growth and photomorphogenesis of pepper plants under red light-emitting diodes with supplemental blue or far-red lighting. J Am Soc Hortic Sci 120:808–813

Budiarto K (2010) Spectral quality affects morphogenesis on *Anthurium* plantlet during in vitro culture. Agrivita 32:234–240

Bula RJ, Morrow RC, Tibbitts TW, Barta DJ, Ignatius RW, Martin TS (1991) Light-emitting diodes as a radiation source for plants. HortScience 26:203–205

Calabrese V, Calafato S, Puleo E, Cornelius C, Sapienza M, Morganti P, Mancuso C (2008) Redox regulation of cellular stress response by ferulic acid ethyl ester in human dermal fibroblasts: role of vitagenes. Clin Dermatol 26:358–363

Chang YS, Chang YH, Sung JH (2006) The effect of ginseng and caffeine products on the antioxidative activities of mouse kidney. J Ginseng Res 30:15–21

Chang YS, Seo EK, Gyllenhaal C, Block KI (2003) *Panax ginseng*: a role in cancer therapy? Integr Cancer Sci Ther 2:13–33

Chauhan K, Chauhan B (2015) Policosanol: natural wax component with potent health benefits. Int J Med Pharm Sci 5:15–24

Chiou WF, Zhang JT (2008) Comparison of the pharmacological affects of *Panax ginseng* and *Panax quinquefolium*. Acta Pharmacol Sin 29:1103–1108

Close DC, McArthor C (2002) Rethinking the role of many plant phenolic protection against photo damage not herbivores. Oikos 99:166–172

Darko E, Heydarizadeh P, Schoefs B, Sabzalian MR (2014) Photosynthesis under artificial light: the shift in primary and secondary metabolism. Philos Trans R Soc Lond B Biol Sci 369 (1640):20130243

Di Carlo G, Mascolo N, Izzo AA, Capasso F (1999) Flavonoids: old and new aspects of a class of natural therapeutic drugs. Life Sci 65:337–353

Dorman HD, Bachmayer O, Kosar M, Hiltunen R (2004) Antioxidant properties of aqueous extracts from selected Lamiaceae species grown in Turkey. J Agric Food Chem 52:762–770

Dutta Gupta S, Jatothu B (2013) Fundamentals and applications of light-emitting diodes (LEDs) in in vitro plant growth and morphogenesis. Plant Biotechnol Rep 7:211–220

Dutta Gupta S, Sahoo TK (2015) Light emitting diode (LED)-induced alteration of oxidative events during in vitro shoot organogenesis of *Curculigo orchioides* Gaertn. Acta Physiol Plant 37:233

Ebisawa M, Shoji K, Mieko KATO, Shimomura K, Yoshihara T (2008) Supplementary ultraviolet radiation B together with blue light at night increased quercetin content and flavonol synthase gene expression in leaf lettuce (*Lactuca sativa* L.). Environ Cont Biol 46:1–11

Fetoni AR, Mancuso CESARE, Eramo SLM, Ralli M, Piacentini ROBERTO, Barone EUGENIO, Paludetti GAETANO, Troiani D (2010) In vivo protective effect of ferulic acid against noise-induced hearing loss in the guinea-pig. Neurosci 169:1575–1588

Fowler MW (1985) Problems in commercial exploitation of plant tissue cultures. In: Neumann KH, Barz W, Reinhardt E (eds) Primary and secondary metabolism of plant cell cultures. Springer, Berlin, pp 362–378

Furuya T, Yoshikawa T, Orihara Y, Oda H (1983) Saponin production in cell suspension cultures of *Panax ginseng*. Planta Med 48:83–87

Ganesan P, Ko HM, Kim IS, Choi DK (2015) Recent trends of nano bioactive compounds from ginseng for its possible preventive role in chronic disease models. RSC Adv 5(119): 98634–98642

Gould KS, Markham KR, Smith RH, Goris JJ (2000) Functional role of anthocyanins in the leaves of *Quintinia serrata* A. Cunn J Exp Bot 51:1107–1115

Hobbs C (1996) Ginseng: the energy herb. Botanica Press, Loveland

Jeong JH, Kim YS, Moon HK, Hwang SJ, Choi YE (2009) Effects of LED on growth, morphogenesis and eleutheroside contents of in vitro cultured plantlets of *Eleutherococcus senticosus* Maxim. Hanguk Yakyong Changmul Hakhoe Chi 17:39–45

Johkan M, Shoji K, Goto F, Hashida SN, Yoshihara T (2010) Blue light-emitting diode light irradiation of seedlings improves seedling quality and growth after transplanting in red leaf lettuce. HortScience 45:1809–1814

Kang SY, Kim ND (1992) The antihypertensive effect of red ginseng saponin and the endothelium-derived vascular relaxation. Korean J Ginseng Sci 16:175–182

Kikuzaki H, Hisamoto M, Hirose K, Akiyama K, Taniguchi H (2002) Antioxidant properties of ferulic acid and its related compounds. J Agric Food Chem 50:2161–2168

Kim MJ, Jung NP (1987) The effect of ginseng saponin on the mouse immune system. Korean J Ginseng Sci 11:130–135

Kim HH, Goins GD, Wheeler RM, Sager JC (2004) Green-light supplementation for enhanced lettuce growth under red-and blue-light-emitting diodes. HortScience 39:1617–1622

Kim BG, Kim JH, Kim J, Lee C, Ahn J (2008) Accumulation of flavonols in response to ultraviolet-B irradiation in soybean is related to induction of flavanone 3-beta-hydroxylase and flavonol synthase. Mol Cells 25:247–252

Kim MJ, Li X, Han JS, Lee SE, Choi JE (2009) Effect of blue and red LED irradiation on growth characteristics and saponin contents in *Panax ginseng* CA Meyer. Korean J. Medicinal Crop Sci 17:187–191

Kim CK, Cho DH, Lee KS, Lee DK, Park CW, Kim WG, Lee SJ, Ha KS, Goo Taeg O, Kwon YG, Kim YM (2012) Ginseng berry extract prevents atherogenesis via anti-inflammatory action by upregulating phase II gene expression. Evidence-Based Complementary Altern Med 2012:490301. doi:10.1155/2012/490301

Kreuzaler F, Hahlbrock K (1973) Flavonoid glycosides from illuminated cell suspension cultures of *Petroselinum hortense*. Phytochemistry 12:1149–1152

Kurilčik A, Miklušytė-Čanova R, Dapkūnienė S, Žilinskaitė S, Kurilčik G, Tamulaitis G, Duchovskis P, Žukauskas A (2008) In vitro culture of Chrysanthemum plantlets using light-emitting diodes. Cent Eur J Biol 3:161–167

Li TS (1995) Asian and American ginseng—a review. Horttechnolgy 5:27–34

Li H, Xu Z, Tang C (2010) Effect of light-emitting diodes on growth and morphogenesis of upland cotton (*Gossypium hirsutum* L.) plantlets in vitro. Plant Cell Tiss Org Cult 103:155–163

Li HB, Wong CC, Cheng KW, Chen F (2008) Antioxidant properties in vitro and total phenolic contents in methanol extracts from medicinal plants. LWT 41:385–390

Lian ML, Murthy HN, Paek KY (2002) Effects of light emitting diodes (LEDs) on the in vitro induction and growth of bulblets of *Lilium* oriental hybrid 'Pesaro'. Sci Hortic 94:365–370

Liu CZ, Guo C, Wang YC, Ouyang F (2002) Effect of light irradiation on hairy root growth and artemisinin biosynthesis of *Artemisia annua* L. Process Biochem 38:581–585

Liu XY, Guo SR, Xu ZG, Jiao XL, Takafumi T (2011) Regulation of chloroplast ultrastructure, cross-section anatomy of leaves, and morphology of stomata of cherry tomato by different light irradiations of light-emitting diodes. HortScience 46:217–221

Liu XY, Guo SR, Chang TT, Xu ZG, Takafumi T (2012) Regulation of the growth and photosynthesis of cherry tomato seedlings by different light irradiations of light emitting diodes (LED). Afr J Biotechnol 11:6169–6177

Low PS, Merida JR (1996) The oxidative burst in plant defense: function and signal transduction. Physiol Plant 96:533–542

Łuczkiewicz M, Zárate R, Dembińska-Migas W, Migas P, Verpoorte R (2002) Production of pulchelin E in hairy roots, callus and suspension cultures of *Rudbeckia hirta* L. Plant Sci 163:91–100

Mabberley DJ (1987) The plant-book: A portable dictionary of the higher plants. Cambridge University Press, New York, p 706

Masek A, Chrzescijanska E, Latos M (2016) Determination of antioxidant activity of caffeic acid and p-coumaric acid by using electrochemical and spectrophotometric assays. Int J Electrochem Sci 11:10644–10658

Ménard C, Dorais M, Hovi T, Gosselin A (2006) Developmental and physiological responses of tomato and cucumber to additional blue light. Acta Hort (711):291–296

Moreira da Silva MH, Debergh PC (1997) The effect of light quality on the morphogenesis of in vitro cultures of *Azorina vidalii* (Wats.) Feer. Plant Cell Tiss Org Cult 51:187–193

Mucciarelli M, Gallino M, Maffei M, Scannerini S (2000) Effects of 3,4-dihydroxybenzoic acid on tobacco (*Nicotiana tabacum* L.) cultured in vitro. Growth regulation in callus and organ cultures. Plant Biosyst 134:185–192

Nam KY (2002) Clinical applications and efficacy of Korean ginseng. J Ginseng Res 26:111–131

Nhut DT, Takamura T, Watanabe H, Tanaka M (2001) Efficiency of a novel culture system by using light-emitting diode (LED) on in vitro and subsequent growth of micropropagated banana plantlets. Acta Hort 616:121–127

Nhut DT, Takamura T, Watanabe H, Murakami A, Murakami K, Tanaka M (2002) Sugar-free micropropagation of *Eucalyptus citriodora* using light-emitting diodes (LEDs) and film-rockwool culture system. Environ Cont Biology (Japan) 140:147–155

Nhut DT, Huy NP, Tai NT, Nam NB, Luan VQ, Hien VT, Tung HT, Vinh BT, Luan TC (2015) Light-emitting diodes and their potential in callus growth, plantlet development and saponin accumulation during somatic embryogenesis of *Panax vietnamensis* Ha et Grushv. Biotechnol Biotechnol Equip 29:299–308

Noa M, Mendoza S, Mas R, Mendoza N (2001) Effect of D-003, a mixture of high molecular weight primary acids from sugar cane wax, on CL4C-induced liver acute injury in rats. Drugs Exp Clin Res 28:177–183

Oh J, Kim JS (2016) Compound K derived from ginseng: neuroprotection and cognitive improvement. Food Funct 7:4506–4515

Park SY, Kim JK, Lee SY, Oh SD, Lee SM, Jang JS, Yang CI, Won YJ, Yeo Y (2014) Comparative analysis of phenolic acid profiles of rice grown under different regions using multivariate. Plant Omics 7:430–437

Poudel PR, Kataoka I, Mochioka R (2008) Effect of red-and blue-light-emitting diodes on growth and morphogenesis of grapes. Plant Cell Tiss Org Cult 92:147–153

Proctor JTA (1996) Ginseng: old crop, new directions. In: Janick J (ed) Progress new crops. ASHS Press, Arlington, VA, pp 565–577

Rice-Evans CA, Miller NJ, Paganga G (1996) Structure-antioxidant activity relationships of flavonoids and phenolic acids. Free Radic Biol Med 20:933–956

Ryan KG, Swinny EE, Markham KR, Winefield C (2002) Flavonoid gene expression and UV photoprotection in transgenic and mutant petunia leaves. Phytochemistry 59:23–32

Schenk RU, Hildebrandt AC (1972) Medium and techniques for induction and growth of monocotyledonous and dicotyledonous plant cell cultures. Can J Bot 50:199–204

Schuerger AC, Brown CS, Stryjewski EC (1997) Anatomical features of pepper plants (*Capsicum annuum* L.) grown under red light-emitting diodes supplemented with blue or far-red light. Ann Bot 79:273–282

Sgherri C, Stevanovic B, Navari-Izzo F (2004) Role of phenolics in the antioxidative status of the resurrection plant *Ramonda serbica* during dehydration and rehydration. Physiol Plant 122:478–485

Shao ZH, Xie JT, Hoek TLV, Mehendale S, Aung H, Li CQ, Qin Y, Schumacker PT, Becker LB, Yuan CS (2004) Antioxidant effects of American ginseng berry extract in cardiomyocytes exposed to acute oxidant stress. Biochim Biophys Acta 1670:165–171

Sharma P, Jha AB, Dubey RS, Pessarakli M (2012) Reactive oxygen species, oxidative damage, and antioxidative defense mechanism in plants under stressful conditions. J Bot 2012:217037. doi:10.1155/2012/217037

Shim M, Lee Y (2009) Ginseng as a complementary and alternative medicine for postmenopausal symptoms. J Ginseng Res 33:89–92

Shin KS, Murthy HN, Heo JW, Paek KY (2003) Induction of betalain pigmentation in hairy roots of red beet under different radiation sources. Biol Plant 47:149–152

Shin KS, Murthy HN, Heo JW, Hahn EJ, Paek KY (2008) The effect of light quality on the growth and development of in vitro cultured doritaenopsis plants. Acta Physiol Plant 30:339–343

Shohael AM, Ali MB, Yu KW, Hahn EJ, Islam R, Paek KY (2006) Effect of light on oxidative stress, secondary metabolites and induction of antioxidant enzymes in *Eleutherococcus senticosus* somatic embryos in bioreactor. Process Biochem 41:1179–1185

Son KH, Park JH, Kim D, Oh MM (2012) Leaf shape index, growth, and phytochemicals in two leaf lettuce cultivars grown under monochromatic light-emitting diodes. Kor J Hort Sci Technol 30:664–672

Tang W, Eisenbrand G (1992) *Panax ginseng* C. A. Mayer. Chinese drugs of plant origin. Springer, Berlin, pp 710–737

Urbonavičiūtė A, Samuolienė G, Brazaitytė A, Duchovskis P, Ruzgas V, Žukauskas A (2009a) The effect of variety and lighting quality on wheatgrass antioxidant properties. Zemdirbyste 96:119–128

Urbonavičiūtė A, Samuolienė G, Brazaitytė A, Ruzgas V, Šabajevienė G, Šliogerytė K, Sakalauskaitė J, Duchovskis P, Žukauskas A (2009b) The effect of light quality on the antioxidative properties of green barely leaves. Scientific Works of the Lithuanian Institute of Horticulture and Lithuanian University of Agriculture. Sodinink Darzinink 28:153–161

Wang T, Hicks KB, Moreau R (2002) Antioxidant activity of phytosterols, oryzanol, and other phytosterol conjugates. J Am Oil Chem Soc 79:1201–1206

Wu HC, Lin CC (2012) Red light-emitting diode light irradiation improves root and leaf formation in difficult-to-propagate *Protea cynaroides* L. plantlets in vitro. HortScience 47:1490–1494

Wu HC, Du Toit ES, Reinhardt CF, Rimando AM, Van der Kooy F, Meyer JJM (2007) The phenolic, 3,4-dihydroxybenzoic acid, is an endogenous regulator of rooting in *Protea cynaroides*. Plant Growth Regul 52:207–215

Xie JT, Shao ZH, Hoek TLV, Chang WT, Li J, Mehendale S, Wang CZ, Hsu CW, Becker LB, Yin JJ, Yuan CS (2006) Antioxidant effects of ginsenoside Re in cardiomyocytes. Eur J Pharmacol 532:201–207

Yeh N, Chung JP (2009) High-brightness LEDs—energy efficient lighting sources and their potential in indoor plant cultivation. Renew Sust Energy Rev 13:2175–2180

Yorio NC, Goins GD, Kagie HR, Wheeler RM, Sager JC (2001) Improving spinach, radish, and lettuce growth under red light-emitting diodes (LEDs) with blue light supplementation. HortScience 36:380–383

Yu KW, Murthy HN, Hahn EJ, Paek KY (2005) Ginsenoside production by hairy root cultures of *Panax ginseng*: influence of temperature and light quality. Biochem Eng J 23:53–56

Zhong JJ, Seki T, Kinoshita SI, Yoshida T (1991) Effect of light irradiation on anthocyanin production by suspended culture of *Perilla frutescens*. Biotechnol Bioeng 38:653–658

# Chapter 12
# Influence of LED Lighting on In Vitro Plant Regeneration and Associated Cellular Redox Balance

**S. Dutta Gupta and A. Agarwal**

## 12.1 Introduction

Plant tissue culture, also referred to as in vitro or axenic culture, is an indispensable tool for research in plant biotechnology. It allows the propagation of genetically uniform and disease-free plant material possessing the desired traits. Commercial application of plant tissue culture has gained popularity in the past few decades as a means for rapid production of plant propagules in a large scale. It is also one of the most reliable techniques for ex situ germplasm conversation of rare and endangered plant species. The technology gives us the opportunity to control the development of plants by regulating numerous parameters that influence plant growth. Traditionally, the induction of changes in in vitro plant growth and development has been investigated by varying microenvironmental parameters such as medium composition, plant growth regulators, headspace temperature, $CO_2$ content and various chemical treatments (Kozai and Smith 1995; Kozai and Xiao 2008), while illuminating the cultures by gas discharge lamps (GDls), commonly fluorescent lamps (Fls). GDls emit white light with emission patterns from 400 to 700 nm at fixed intensities. Plant morphogenesis and its related aspects are mainly regulated by various photoreceptors which are activated by photons in the blue, red and far-red regions of the light spectrum. A significant portion of the spectral output emanated by GDls is not utilized by the plant cultures (Dutta Gupta and Jatothu 2013). Also, excess light irradiation causes photo inhibition and photooxidative damage in plants (Kasahara et al. 2004). Apart from the energy waste due to photons of undesirable wavelengths and excess light irradiation, GDls also dissipate

S. Dutta Gupta (✉) · A. Agarwal
Agricultural and Food Engineering Department, Indian Institute of Technology Kharagpur, Kharagpur 721302, India
e-mail: sdg@agfe.iitkgp.ernet.in



S. Dutta Gupta (ed.), *Light Emitting Diodes for Agriculture*,
DOI 10.1007/978-981-10-5807-3_12

a lot of energy as heat. Thus, GDIs may not be the ideal choice for illuminating the cultures maintained in vitro, and there is a need for an efficient light source to improve the rate of micropropagation with quality planting materials and reduce cost of propagation.

Light emitting diodes (LEDs) have been proposed recently as a versatile and energy-efficient light source for various applications of plant tissue culture. The narrow waveband emission and dynamic control of light intensity in LED-based illumination systems allows the customization of spectral quality to match the requirement of the plants (Dutta Gupta and Jatothu 2013). The precision in converting electrical energy to photons of specific wavelengths at the desired photosynthetic photon flux density (PPFD) with negligible heat loss makes LEDs more energy-efficient than all other available artificial lighting sources. Rapid advancements in the field of LED technology for reducing the manufacturing costs are expanding the scope for their application in commercial micropropagation.

The influence of light as an important signalling component in regulating plant growth and development has been well illustrated (Chen et al. 2004; Devlin et al. 2007; Samuolienė et al. 2010). In particular, spectral attributes along with their matching ability to different types of photoreceptors are the key factors that govern the morphogenesis of plants. Light-responsive signal transduction pathways operating within the plant system are mainly regulated by the red/far-red light-sensitive phytochromes and the blue light-sensitive cryptochromes and phototropins. Crosstalks or interactions between the red and blue photoreceptors help the plants in sensing and responding to the ambient light conditions (Franklin and Whitelam 2004). In this context, the success of in vitro plant regeneration largely relies on the spectral quality and photon efficiency of the artificial light sources. LEDs are the ideal source which can only provide required spectral attributes to stimulate organogenic as well as embryogenic responses. LED panels having emission spectra matching with the absorption spectra of plant photoreceptors may yield optimal in vitro productivity by influencing plant morphogenesis and metabolism (Massa et al. 2008). Numerous studies reported the successful applications of LEDs in promoting in vitro growth and morphogenesis from various plant species (Dutta Gupta and Jatothu 2013). Improvement in shoot organogenesis, ex vitro survival rate and biomass yield have been demonstrated under various LED treatments (Hahn et al. 2000; Nhut et al. 2003; Jao et al. 2005; Shin et al. 2008; Li et al. 2010; Dutta Gupta and Sahoo 2015). The impact of LED lighting on somatic embryogenesis has also been explored for a few plant species (Merkle et al. 2005; Park et al. 2010; Kim and Moon 2014; Botero Giraldo et al. 2015; Chen et al. 2016; Mai et al. 2016). LED lighting has also been found to have positive effects on the in vitro shoot and root development resulting in improved adaptability and growth of the plantlets after transfer to soil (Nhut et al. 2003; Moon et al. 2006; Daud et al. 2013; Hung et al. 2015; Ferreira et al. 2017). It is evident from these studies that the ideal light environment for each plant species is unique; i.e., the spectral composition and PPFD influencing the in vitro response of one plant species may not yield similar results for another plant species.

The first part of this review highlights the stimulatory responses of in vitro morphogenesis under LED lighting. However, it is noteworthy that the morphogenetic changes are the manifestation of numerous metabolic events occurring within the plant. Regulation of the reactive oxygen species (ROS) along with their crosstalk with antioxidative system is one such mechanism which significantly impacts plant growth and morphogenesis (Benson 2000; Gechev and Hille 2005; Dutta Gupta 2010). The ROS regulatory mechanisms affect in vitro growth and morphogenesis by alleviating the oxidative stress developed due to various metabolic activities. Plant cells host a plethora of ROS-producing and ROS-scavenging genes. Among them, antioxidant genes are key players in controlling free radical-mediated stress and cellular redox balance. ROS-mediated stress during in vitro culture and cellular antioxidant levels have been shown to affect regeneration responses in a number of plant species (Benson 2000; Dutta Gupta and Datta 2003/4; Batkova et al. 2008; Dutta Gupta 2010). The possible involvements of ROS-scavenging enzymes and non-enzymatic antioxidants during in vitro morphogenesis have recently been reviewed (Dutta Gupta 2010). These studies illustrated the role of ROS and concomitant oxidative protection system during regeneration of plants under fluorescence-based illumination system. In recent years, there has been a growing interest in the application of LED technology for the improvement of plant regeneration response. Apart from the changes in morphological and anatomical attributes, significant alterations in plant antioxidant status have also been reported in response to varying lighting conditions by means of specific LED treatments under in vitro conditions (Baque et al. 2010; Mengxi et al. 2011; Manivannan et al. 2015; Dutta Gupta and Sahoo 2015). There were also significant changes in the activity of antioxidative enzymes during ex vitro acclimatization (Faisal and Anis 2009; Ahmed and Anis 2014).Variations in regeneration response and ex vitro survival could be implicated to the changes in ROS metabolism induced by spectral quality and photon efficiency. Regulation of ROS network and antioxidants by means of different LED lighting regimes provides the scope for enhancing in vitro plant growth and development.

This chapter describes the effect of LEDs on in vitro morphogenic processes following shoot organogenesis and somatic embryogenesis and enumerates the role of spectral specificity using LED technology on changes in the steady-state level of ROS and their scavenging potential during plant regeneration and ex vitro survival.

## 12.2 Impact of LEDs on In Vitro Plant Regeneration

Asexual multiplication of plants by in vitro regeneration is essential for the propagation of plants, while maintaining their genetic fidelity. The two distinct developmental pathways involved in plant regeneration are shoot organogenesis and somatic embryogenesis. Regenerated plantlets thus obtained have to be acclimatized and finally transferred to ex vitro conditions for further growth. Influence of LED lighting on in vitro shoot organogenesis as well as somatic embryogenesis has been studied in a variety of plant species. Since plant photoreceptors are known to be primarily and most significantly stimulated by red and blue regions of the light spectrum, most of the studies have focused on evaluating the impact of monochromatic and mixed blue (440–480 nm) and red (630–665 nm) LED treatments. The effects of white, far-red, green, yellow and orange LEDs have also been reported in a few instances.

### *12.2.1 Effects of LED Lighting on Shoot Organogenesis and In Vitro Plantlet Development*

Light quality plays a significant role in inducing the differentiation of plant cells and regeneration of organs from cultured plant tissues (Hughes 1981; Lercari et al. 1986; Kozai et al. 1992). Shoot organogenesis under different LED lighting regimes has been evaluated for a wide variety of species with diverse parameters such as number of regenerated shoots per explant, elongation of regenerated shoots, formation of leaves, biomass accumulation and chlorophyll content. The major findings of these studies have been summarized in Table 12.1. Rooting, an essential component of in vitro plantlet formation, has also been assessed along with shoot organogenesis in many reports.

Spectral quality of LEDs significantly influenced the shoot regeneration response. However, it is difficult to draw a conclusion with regard to the specific nature of spectral effect of LEDs, as the response varied with the type of explants and plant species used in the culture system. A large number of studies demonstrated the promoting role of red plus blue LEDs in various combinations on shoot regeneration and subsequent growth of the regenerated plants (Nhut et al. 2000; Lian et al. 2002; Kim et al. 2004; Moon et al. 2006; Shin et al. 2008; Edesi et al. 2014; Hung et al. 2015; Al-Mayahi 2016). On the other hand, a considerable number of studies indicated the stimulatory effect of monochromatic red or blue LED treatments on shoot organogenesis (Jao et al. 2005; Poudel et al. 2008; Budiarto 2010; Lin et al. 2011; Wu and Lin 2012). A more or less equal response in in vitro shoot development under both monochromatic and mixed LED treatments has also been reported in certain plant species (Dewir et al. 2006; Chung et al. 2010; Lee et al. 2014; Ramírez-Mosqueda et al. 2016; Hung et al. 2016a, b). The use of far-red LEDs in combination with red and blue LEDs also improved the regenerative response in various species (Kim et al. 2004; Kurilčik et al. 2008; Chung et al. 2010; Park and Kim 2010). The differential response thus obtained may be attributed to the variable nature of synergistic interactions of the different light-harvesting photoreceptors in tune with genetic make-up of the plant species (Kim et al 2004).

Compared to conventional Fls, LED treatments significantly improved shoot organogenesis in various plant species. Mixed red and blue LED lighting regimes stimulated in vitro shoot proliferation in potato (Edesi et al. 2014), sugarcane

**Table 12.1** Effects of LED lighting on in vitro shoot organogenesis and plantlet growth

| Plant species and explant type | LED colour with wavelength (nm) | Shoot organogenesis and plantlet growth parameters assessed | Remarks | References |
|---|---|---|---|---|
| *Cymbidium* plantlets | BL (450), RL (660), RL + BL | Biomass, leaf development, chlorophylls | RL enhanced leaf growth, while BL + RL increased shoot and root biomass. Chlorophylls were highest under BL | Tanaka et al. (1998) |
| *Rehmannia glutinosa* single-node cuttings | BL (466), RL (665), RL + BL (1:1) | Shoot length, biomass, rate of photosynthesis | RL increased shoot length, whereas dry weight was higher under BL. Rate of photosynthesis was enhanced by BL + RL | Hahn et al. (2000) |
| *Musa* sp. (Banana), trifoliate shoot | BL, RL, RL + BL (9:1, 8:2, 7:3) | Biomass | RL + BL (8:2) yielded highest biomass accumulation | Nhut et al. (2000) |
| *Lilium* cv. 'Pesaro' (Lily), in vitro bulblets | BL (450), RL (660), RL + BL (1:1) | Bulblet enlargement and biomass accumulation, root induction | Bulblet size, biomass accumulation and root induction were enhanced by BL + RL | Lian et al. (2002) |
| *Fragaria* cv. 'Akihime' (Strawberry) trifoliate shoots | BL (450), RL (660), RL + BL (9:1, 8:2, 7:3) | Shoot length, leaf count, biomass, root growth, chlorophylls | Leaf count, shoot biomass, root length and chlorophyll content were highest under RL + BL (7:3). RL yielded longest shoots | Nhut et al. (2003) |
| *Solanum tuberosum* cv. 'Kennebec' (Potato) in vitro single-node cuttings | RL (645) + BL (460) continuous, intermittent and fluctuating light | Shoot height, growth rate, biomass | Growth was enhanced by intermittent LED lighting | Jao and Fang (2004a) |
| *Solanum tuberosum* cv. 'Kennebec' (Potato) in vitro single-node cuttings | Concurrent and alternating RL (645) and BL (460) lighting | Biomass | Highest biomass was observed under concurrent RL + BL treatments | Jao and Fang (2004b) |

(continued)

**Table 12.1** (continued)

| Plant species and explant type | LED colour with wavelength (nm) | Shoot organogenesis and plantlet growth parameters assessed | Remarks | References |
|---|---|---|---|---|
| *Zantedeschia jucunda* cv. 'Black Magic' (Calla lily) plantlets | RL (645), RL + BL (460) (3:2), fluctuating RL | Stem length, biomass, chlorophylls | Continuous RL increased stem length, while fluctuating RL enhanced biomass. RL + BL lighting increased chlorophyll content | Jao et al. (2005) |
| *Euphorbia millii* Plantlets | BL (440), RL (650), RL + BL (1:1), FrL (720) + RL (1:1), FrL + BL (1:1) | Stem length, biomass, leaf count, chlorophylls | BL + RL increased leaf count, total biomass and chlorophyll content. Stem length was enhanced by RL | Dewir et al. (2006) |
| *Vitis berlandieri* × *Vitis riparia* cv. 'Teleki 5BB' (Grape), Rootstock | BL (450), RL (660), RL + BL (1:1) | Shoot length, biomass | Shoot length and biomass were highest under RL | Heo et al. (2006) |
| *Tripterospermum japonicum* (Tsuru-rindo), apical shoots | BL (440), RL (650), RL + BL (1:1, 7:3) | Shoot length, biomass, rooting, chlorophylls | Longest shoots and highest rooting were observed under RL. RL + BL enhanced biomass and chlorophyll content | Moon et al. (2006) |
| *Vitis vinifera* cv. 'Gailiūne' (Grape) stem with axillary bud | RL (640, 660), BL (440) + RL (7:3), FrL (735) + RL (1:9), BL + RL + FrL (3:7:1, 4:7:3) | Shoot length, leaf count, root growth, chlorophylls | RL and FrL + RL increased shoot length. Treatments with BL increased root length and leaf count. Chlorophyll content was highest under BL + RL | Kurilčik et al. (2007) |
| *Chrysanthemum morifolium* cv. 'Ellen' in vitro plantlets | RL (640, 660), BL (440) + RL (7:3), FrL (735) + RL (1:9), BL + RL + FrL (3:7:1, 4:7:3) | Shoot length, biomass, leaf count, rooting, chlorophylls | RL and FrL + RL enhanced shoot length and root development. Treatments with BL increased chlorophyll content | Kurilčik et al. (2008) |

(continued)

**Table 12.1** (continued)

| Plant species and explant type | LED colour with wavelength (nm) | Shoot organogenesis and plantlet growth parameters assessed | Remarks | References |
|---|---|---|---|---|
| *Vitis* spp. cv. 'Hybrid Franc', var. 'Ryuukyuuganebu', cv. 'Kadainou R-1' (Grape) nodal explants | BL (480), RL (660) | Plant height, shoot number, leaf count, root development, chlorophylls | RL increased plant height and internode length. BL enhanced chlorophyll content. Root growth under LEDs improved in two genotypes | Poudel et al. (2008) |
| *Doritaenopsis* hort regenerated plantlets | BL (450), RL (660), RL + BL (1:1) | Leaf biomass, length and area, chlorophylls | Leaf biomass and area as well as chlorophyll content were highest under RL + BL | Shin et al. (2008) |
| *Phalaenopsis* hybrid in vitro flower-stalk node | RL, RL + BL (9:1, 8:2), RL + WL (1:1) | Shoot induction, biomass, shoot and leaf length | Highest shoot proliferation was recorded under RL + BL (9:1). Plantlet growth was improved by LEDs | Wongnok et al. (2008) |
| *Morinda citrifolia* leaf explants | BL, RL, RL + BL (1:1), FrL | Adventitious root formation | RL stimulated growth of adventitious roots | Baque et al. (2010) |
| *Anthurium andreanum* L. cv. 'Violeta', cv. 'Pink Lady' leaf cuttings | BL (450), RL (660), RL + BL (3:1, 1:1, 1:3) | Shoot induction and proliferation | Higher proportion of BL stimulated shoot induction and proliferation | Budiarto (2010) |
| *Oncidium* cv. 'Gower Ramsey' PLBs | BL (455), RL (660), BL + RL, FrL (730) + BL, FrL + RL, FrL + BL + RL | Shoot length, biomass, leaf length and count, number of roots, chlorophylls | Leaf count and length as well total biomass and root growth were enhanced by FrL + RL + BL and FrL + RL. Shoot length was highest under RL | Chung et al. (2010) |
| *Gossypium hirsutum* L. (Upland cotton) shoot bud apex cuttings | BL (460), RL (660), RL + BL (3:1, 1:1, 1:3) | Shoot length and diameter, leaf morphology, total biomass, chlorophylls | BL and BL + RL treatments improved growth and development as well as increased chlorophyll content | Li et al. (2010) |

(continued)

**Table 12.1** (continued)

| Plant species and explant type | LED colour with wavelength (nm) | Shoot organogenesis and plantlet growth parameters assessed | Remarks | References |
|---|---|---|---|---|
| *Castanea crenata* S. et Z. (Chestnut) in vitro plantlets | BL (440), RL (650), RL + BL (1:1), FrL (730) + RL (1:1) | Stem length, leaf area and number, root development | Stem and leaves exhibited enhanced growth under FrL + RL. Root growth was promoted by RL | Park and Kim (2010) |
| *Oncidium* cv. 'Gower Ramsey' shoot tips | BL (460), RL (660), RL + BL (9:1, 8:2, 7:3), FrL (715) + RL + BL (1:8:1), GL (530) + RL + BL (1:8:1) | Plant height, biomass, root growth, chlorophylls | RL and RL + BL treatments enhanced shoot elongation and biomass accumulation. Root growth was enhanced by RL + BL. Chlorophyll content increased under BL and RL +BL | Mengxi et al. (2011) |
| *Bletilla ochracea* Schltr. seeds | BL (470), GL (525), OL (590), RL (625), WL | Seedling leaf expansion, pseudobulb formation, rhizoid formation | Leaf width and pseudobulb thickness were highest under BL and WL. Rhizoid formation was promoted by RL and OL | Godo et al. (2011) |
| *Dendrobium officinale* PLBs | BL (450), RL (660), RL + BL (2:1, 1:1, 1:2) | PLB to shoot conversion, shoot biomass, chlorophylls | BL resulted in higher PLB conversion and RL increased shoot biomass. RL + BL yielded highest chlorophyll contents | Lin et al. (2011) |
| *Protea cynaroides* L. in vitro plantlets | BL (460), RL (630), RL + BL (1:1) | Biomass, leaf count | Number of leaves was highest under RL. Leaf dry weight was highest under RL + BL | Wu and Lin (2012) |
| *Jatropha curcas* L. regenerated shoots | BL (450), RL (660), RL + BL (1:1), WL | Root induction | RL promoted root induction | Daud et al. (2013) |
| *Saccharum officinarum* L. cv. CTC-07 (Sugarcane) seedlings | RL, BL, RL + BL (7:3, 3:7) | Shoot proliferation, shoot length, biomass, chlorophylls | Treatments with RL and RL + BL enhanced shoot development. Chlorophyll biosynthesis was highest under BL | Maluta et al. (2013) |

(continued)

**Table 12.1** (continued)

| Plant species and explant type | LED colour with wavelength (nm) | Shoot organogenesis and plantlet growth parameters assessed | Remarks | References |
|---|---|---|---|---|
| *Solanum tuberosum* L. cv. 'Agrie Dzeltenie', cv. 'Maret', cv. 'Bintje', cv. 'Anti', cv. 'Désirée' (Potato), cryopreserved shoot tips | BL, RL, RL + BL (9:1), RL + BL + FrL (7:1:2) | Survival and shoot regeneration | Survival and shoot regeneration percentage was highest under RL + BL in all cultivars | Edesi et al. (2014) |
| *Abeliophyllum distichum* Nakai., apical and axillary buds | BL (450), RL (660), RL + BL (1:1) | Shoot multiplication, shoot length, root formation | Shoot proliferation and root growth were promoted by BL and RL + BL, while RL increased shoot length | Lee et al. (2014) |
| *Saccharum officinarum* L. var. 'RB92579' (Sugarcane) in vitro plantlets | BL (460) + RL (630) (7:3, 1:1, 2:3, 3:7), WL | Shoot length, biomass, leaf number, tiller number, chlorophylls | High proportion of BL increased fresh weight and number of tillers. Shoot length was enhanced by higher proportion of RL. Chlorophyll content was highest under WL | Silva et al. (2014) |
| *Musa* spp. cv. 'Grande naine' AAA (Banana) in vitro shoots | WL | Shoot proliferation, shoot length, biomass, root growth | Shoot proliferation and root growth were enhanced by LEDs | Wilken et al. (2014) |
| *Achillea millefoilum* L. (Yarrow) in vitro shoot segments | BL, GL, RL, WL | Biomass, shoot length, root growth, chlorophylls | BL enhanced shoot biomass accumulation and root growth, whereas both BL and GL increased shoot length | Alvarenga et al. (2015) |
| *Curculigo orchioides* Gaertn. leaf explants | BL (470), RL (630), RL + BL (1:1) | Shoot organogenesis | Percentage of shoot organogenesis and shoot buds per explant were highest under BL | Dutta Gupta and Sahoo (2015) |
| *Fragaria* × *ananassa* Duch. cv. 'Camarosa' (Strawberry) encapsulated shoot tips | BL (460) + RL (660) (7:3, 1:1, 3:7, 1:9) | Shoot proliferation, shoot length, biomass, leaf count | Higher proportion of RL promoted shoot elongation, biomass accumulation, leaf growth, root | Hung et al. (2015) |

(continued)

**Table 12.1** (continued)

| Plant species and explant type | LED colour with wavelength (nm) | Shoot organogenesis and plantlet growth parameters assessed | Remarks | References |
|---|---|---|---|---|
| | | and area, root growth, chlorophylls | development and chlorophyll biosynthesis | |
| *Populus euramericana* (Poplar) regenerated shoots | BL (440), RL (650), RL + BL (7:3, 1:1), RL + GL (510) + BL (7:1:2) | Shoot regeneration, shoot growth and morphology, root development, biomass | RL + BL stimulated shoot regeneration along with shoot and leaf growth. RL impeded root growth | Kwon et al. (2015) |
| *Solanum tuberosum* L. cv. 'Shepody' (Potato) in vitro stem segments | BL (465) + RL (630, 660), BL + GL (520) + RL | Stem height and diameter, leaf area, biomass | Stem and leaf growth, biomass and chlorophyll content were enhanced by BL + GL + RL | Ma et al. (2015) |
| *Rehmannia glutinosa* Libosch. (Chinese foxglove) shoot tips | BL (450), RL (650) | Shoot length, leaf number and length, biomass, root growth, chlorophylls | RL stimulated shoot elongation and chlorophyll biosynthesis. Biomass accumulation, leaf development and root growth were enhanced by BL | Manivannan et al. (2015) |
| *Musa acuminata*cv. 'Nanicão Corupá' (Banana) in vitro apical buds | WL, RL + WL | Chlorophyll content | Both LED treatments enhanced chlorophyll biosynthesis | Vieira et al. (2015) |
| *Phoenix dactylifera* L. cv. 'Alshakr' (Date palm) in vitro shoot buds | RL + BL (9:1) | Shoot formation | LED treatment increased shoot proliferation | Al-Mayahi (2016) |
| *Lippia alba* (Mill.) N. E. Brown. In vitro hypocotyl segments | RL + BL, WL | Biomass, chlorophylls | Biomass accumulation and chlorophyll biosynthesis were highest under RL + BL | Batista et al. (2016) |
| *Vanilla planifolia* Andrews. Axillary buds | BL (460), RL (660), RL + BL (1:1), WL | Shoot proliferation, shoot length, | WL and RL + BL enhanced shoot proliferation, shoot length, biomass | Bello-Bello et al. (2016) |

(continued)

**Table 12.1** (continued)

| Plant species and explant type | LED colour with wavelength (nm) | Shoot organogenesis and plantlet growth parameters assessed | Remarks | References |
|---|---|---|---|---|
| | | biomass, leaf count, chlorophylls | accumulation and leaf formation. WL also increased the chlorophyll content | |
| *Saccharum* spp. var. RB98710 (Sugarcane), shoot segments | RL + BL (9:2) | Shoot multiplication, rooting | Shoot multiplication and root development were enhanced by LEDs | Ferreira et al. (2017) |
| *Vaccinium ashei* Reade cv. 'Titan' (Rabbiteye blueberry) in vitro shoot microcuttings | BL (460), RL (660), RL + BL (4:1, 1:1) | Shoot length, biomass, number of shoots and leaves, leaf area, chlorophylls | RL and RL + BL (4:1) were more suitable for shoot growth and development. Highest chlorophyll content was observed under BL | Hung et al. (2016a) |
| *Vaccinium corymbosum* L. cv. 'Huron' (Highbush blueberry) in vitro shoot microcuttings | BL (460), RL (660), RL + BL (4:1, 1:1) | Shoot length, biomass, number of shoots and leaves, leaf area, chlorophylls | Optimal shoot growth was reported for RL and RL + BL (1:1). Chlorophyll content was enhanced by BL | Hung et al. (2016b) |
| *Bacopa monnieri* (L.) Pennell (Water hyssop) leaf cuttings | RL + BL (4:1, 3:1, 2:1, 1:1), WL | Shoot induction and proliferation, shoot length | WL enhanced shoot regeneration. Shoot length was increased by WL and RL + BL | Karataş et al. (2016) |
| *Saccharum* spp. var. RB867515 (Sugarcane) plantlets | RL + BL (7:3, 1:1, 3:7), WL | Stem length, biomass, number of tillers and leaves | Shoot growth and development was enhanced under RL + BL (1:1) | Silva et al. (2016) |
| *Stevia rebaudiana* Bertoni var. 'Morita II' in vitro plantlets | BL (460), RL (660), RL + BL (1:1), WL | Shoot proliferation, shoot length, leaf count, root growth | Number of shoots was highest under RL. Shoot elongation and leaf growth were enhanced by RL + BL and BL. Chlorophyll content was highest under RL + BL. Root growth was reduced under LEDs | Ramírez-Mosqueda et al. (2016) |

*BL* blue LED, *RL* red LED, *FrL* far-red LED, *WL* white LED, *GL* green LED, *YL* yellow LED, *OL* orange LED, *PLB* protocorm-like body

(Maluta et al. 2013; Silva et al. 2016), vanilla (Bello-Bello et al. 2016) and date palm (Al-Mayahi 2016). The explants of *Lilium* cultured under red and blue mixed LED lighting exhibited enhanced bulblet regeneration as compared to monochromatic LED treatments (Lian et al. 2002). Monochromatic blue LED treatment promoted in vitro shoot induction and proliferation in *Anthurium* (Budiarto 2010), *Dendrobium* (Lin et al. 2011) and *C. orchioides* (Dutta Gupta and Sahoo 2015). Conversely, red LEDs enhanced shoot regeneration response in *S. rebaudiana* (Ramírez-Mosqueda et al. 2016). Higher shoot regeneration was observed in *A. distichum* under both blue and red mixed LED treatments (Lee et al. 2014), whereas exposure of cultures with white LEDs improved the shoot organogenesis in banana (Wilken et al. 2014), bacopa (Karataş et al. 2016) and vanilla (Bello-Bello et al. 2016).

The length of the regenerated shoots was found to be significantly influenced by the type of LED wavelength. High percentage of red PPFD was found to accelerate the elongation of regenerated shoots in various plant species including *R. glutinosa* (Hahn et al. 2000), chrysanthemum (Kim et al. 2004), grapes (Heo et al. 2006; Poudel et al. 2008), *Oncidium* (Chung et al. 2010), chestnut (Park and Kim 2010), *A. distichum* (Lee et al. 2014) and blueberry (Hung et al. 2016a, b). However, increase in the shoot length due to internode elongation under red LEDs made the regenerated shoots very weak (Kim et al. (2004). The elongation of regenerated shoots under blue LEDs has been observed in *A. millefolium* (Alvarenga et al. 2015), whereas sugarcane (Silva et al. 2016) and *S. rebaudiana* (Ramírez-Mosqueda et al. 2016) cultures exhibited increase in shoot length under various combinations of red and blue LEDs.

Proper leaf growth from regenerated shoots is indicative of a stable and sustainable in vitro developmental response. Leaf area, count and biomass were found to be significantly affected by specific LED treatments in a number of plant species. Red and blue mixed LED treatments in various combinations improved leaf growth in in vitro cultures of strawberry (Nhut et al. 2000), grapes (Kurilčik et al. 2007), *Doritaenopsis* (Shin et al. 2008), *P. euramericana* (Kwon et al. 2015), potato (Ma et al. 2015), *S. rebaudiana* (Ramírez-Mosqueda et al. 2016) and sugarcane (Silva et al. 2016). Monochromatic blue LED lighting induced leaf formation in cultures of cotton (Li et al. 2010), whereas regenerated shoots of *P. cynaroides* responded to develop leaves under red LEDs (Wu and Lin 2012). The application of far-red LEDs in combination with other LEDs resulted in improved leaf growth in *Oncidium* (Chung et al. 2010) and chestnut (Park and Kim 2010) shoot cultures.

In several instances, an increase in fresh and dry weights was observed in cultures maintained under LEDs as compared to the cultures raised under FLs. In vitro cultures of *Cymbidium* (Tanaka et al. 1998), banana (Nhut et al. 2000), *Lilium* (Lian et al. 2002), *Chrysanthemum* (Kim et al. 2004), *T. japonicum* (Moon et al. 2006), *Oncidium* (Mengxi et al. 2011), sugarcane (Maluta et al. 2013; Silva et al. 2016), *L. alba* (Batista et al. 2016) and blueberry (Hung et al. 2016b) irradiated under different combinations of red and blue LEDs were found to be associated with enhanced biomass production. Conversely, increased biomass production in cultures of *Dendrobium* (Lin et al. 2011) and *A. millefolium* (Alvarenga et al. 2015) was noted under monochromatic red and blue LEDs. The biomass of regenerated euphorbia (Dewir et al. 2006) and cotton (Li et al. 2010) was increased under monochromatic as well as mixed LED treatments.

The amount of chlorophyll in the cultured cells and/or tissues has been considered as one of the essential parameters to evaluate the effect of LEDs in most of the studies. Chlorophyll content is a reliable indicator of plant health. High chlorophyll content implies optimal photosynthetic activity and also indicates plant nutrient status (Liu et al. 2006). Cultures of *Euphorbia* (Dewir et al. 2006), *T. japonicum* (Moon et al. 2006), *Doritaenopsis* (Shin et al. 2008), grapes (Poudel et al. 2008), cotton (Li et al. 2010), *Oncidium* (Mengxi et al. 2011), sugarcane (Maluta et al. 2013), *L. alba* (Batista et al. 2016), *S. rebaudiana* (Ramírez-Mosqueda et al. 2016) and blueberry (Hung et al. 2016a, b) exhibited higher chlorophyll content under monochromatic blue LEDs or combinations of red and blue LEDs than cultures exposed to monochromatic red LED treatments. Conversely, red LED-induced increase in chlorophyll content was observed in the shoot cultures of banana (Vieira et al. 2015) and *R. glutinosa* (Manivannan et al. 2015). Enhanced chlorophyll biosynthesis was also noted with white LEDs in vanilla (Bello-Bello et al. 2016) and sugarcane (Silva et al. 2016). The LED treatments yielding higher chlorophyll contents are generally associated with improved shoot growth (Dewir et al. 2006; Moon et al. 2006; Shin et al. 2008; Li et al. 2010; Ma et al. 2015; Manivannan et al. 2015; Batista et al. 2016; Bello-Bello et al. 2016).

Besides shoot development, root induction and proliferation are also essential for in vitro plantlet formation. A promoting role of LED treatments on in vitro root induction and development has been reported for various plant species including *Lilium* (Lian et al. 2002), strawberry (Nhut et al. 2003; Hung et al. 2015) *T. japonicum* (Moon et al. 2006), grapes (Kurilčik et al. 2007; Poudel et al. 2008), chrysanthemum (Kurilčik et al. 2008), *Oncidium* (Chung et al. 2010; Mengxi et al. 2011), chestnut (Park and Kim 2010), *B. ochracea* (Godo et al. 2011), *A. distichum* (Lee et al. 2014), banana (Wilken et al. 2014), *A. millefolium* (Alvarenga et al. 2015), *P. euramericana* (Kwon et al. 2015) and *R. glutinosa* (Manivannan et al. 2015). On the contrary, LED lighting was found to deter the formation of roots in *S. rebaudiana* shoot cultures when compared to FLs (Ramírez-Mosqueda et al. 2016). It is worthy to note that the simultaneous improvement in root and shoot development under the same LED treatment was rare and occurred only in a few species such as chrysanthemum (Kurilčik et al. 2008), *A. millefolium* (Alvarenga et al. 2015), *P. euramericana* (Kwon et al. 2015) and sugarcane (Ferreira et al. 2017). Enhanced shoot growth of in vitro raised plantlets of *A. distichum* (Lee et al. 2014) and *R. glutinosa* (Manivannan et al. 2015) was obtained with the irradiation of monochromatic red LEDs, while root development was promoted under monochromatic blue LED treatments. The growth of adventitious roots in *M. citrifolia* (Baque et al. 2010) and *J. curcas* (Daud et al. 2013) cultures was reported to be highest under red LED treatments. The findings highlight the role of specific light wavelengths in triggering different developmental pathways in cultured cells and tissues.

### 12.2.2 Influence of LEDs on Somatic Embryogenesis

Among the modes of plant regeneration, somatic embryogenesis is the most widely adopted system for large-scale production of genetically uniform plants, in vitro induction of mutation and synthetic seeds. It is defined as a process in which a bipolar structure containing both shoot and root meristems develops from a somatic cell or cell cluster without any vascular connection to the parent tissue. The major factors that affect somatic embryogenesis include type of explants, media composition, plant growth regulators, temperature and light. In contrast to the effect of LEDs on shoot organogenesis, there exists little information on somatic embryo induction and development. The salient features of LED effects on somatic embryogenesis are summarized in Table 12.2.

**Table 12.2** Effect of LED lights on somatic embryo induction and germination in various plant species

| Plant species and explant type | LED colour with wavelength (nm) | Parameters assessed | Remarks | References |
|---|---|---|---|---|
| *Pinus taeda* L. (Loblolly pine), *Pinus elliottii* Engelm. (Slash pine), *Pinus palustris* Mill. (Longleaf pine) embryogenic cultures | RL, BL | SE germination and conversion | Germination and conversion of SEs was enhanced in the presence of RL | Merkle et al. (2005) |
| *Doritaenopsis* cv. 'Happy Valentine' leaf segments | RL (650), BL (440) + RL (1:1), FrL (730) + RL (1:1) | SE formation | FrL + RL promoted SE proliferation | Park et al. (2010) |
| *Pinus densiflora* (Japanese red pine) SEs | BL (450), RL (660), RL + BL (1:1), FrL (730) + RL (1:1) | Germination of SE | RL enhanced SE germination | Kim and Moon (2014) |
| *Panax vietnamensis* Ha *et* Grushv. embryogenic callus | BL, GL, YL, RL, WL, RL + BL (9:1, 8:2, 7:3, 6:4, 1:1, 4:6, 3:7, 2:8, 1:9) | Callus proliferation, plant, growth and development of plantlets, saponin content | The combination of RL (60%): BL (40%) was found to be most effective for plant formation. This condition was also the most efficient for plant growth and development | Nhut et al. (2015) |

(continued)

**Table 12.2** (continued)

| Plant species and explant type | LED colour with wavelength (nm) | Parameters assessed | Remarks | References |
|---|---|---|---|---|
| *Peucedanum japonicum* Thunb. Callus | BL (450), RL (660), FrL (730), RL + BL (8:1), RL + BL + FrL (1:1:1), RL + GL (525) + BL (7:1:1) | SE induction and conversion | SE formation was highest under RL + BL. Conversion of SE was highest under BL and RL | Chen et al. (2016) |
| *Saccharum* spp. var. RB98710 (Sugarcane) shoot segments | RL + BL (9:2) | SE induction | Formation of SEs was not enhanced by LEDs | Ferreira et al. (2017) |
| *Coffea canephora Explant* | RL + BL (4:1), RL + BL + WL (5:1:1, 3:1:1, 1:1:3, 2:1:2) | Induction and germination of SE | LEDs suppressed SE induction. RL + BL + WL enhanced SE germination | Mai et al. (2016) |

*BL* blue LED, *RL* red LED, *FrL* far-red LED, *WL* white LED, *GL* green LED, *SE* somatic embryo

A differential response of somatic embryos to different wavelengths of LED light was found in three southern pine species (Merkle et al. 2005). In all the three species, cultures irradiated with red LEDs resulted in higher frequencies of somatic embryo germination and conversion than the blue LEDs and conventional FLs. Among the various species, longleaf pine (*Pinus palustris* Mill.) was more responsive to LED treatments.

Spectral quality following the combination of red and far-red wavelengths provided by LEDs stimulates somatic embryogenesis in *Doritaenopsis* (Park et al. 2010). Interestingly, the embryogenic event was associated with a low degree of endoreduplication. This study for the first time suggested the stimulatory role of far-red light on somatic embryo regeneration.

A strong genotype-dependent influence of LED lighting was observed during germination of somatic embryos in Japanese red pine (*Pinus densiflora*). Germination was significantly inhibited by FLs followed by red plus blue LED treatments in the embryogenic suspensor mass (ESM) line 05–3, while lines 05–12, 05–29 and 05–37 responded to show positive effects under red LEDs (Kim and Moon 2014). The findings closely paralleled to those reported by Merkle et al. (2005). Chen et al. (2016) studied the influence of LED spectra on somatic embryogenesis of *Peucedanum japonicum* Thunb., a medicinal herb. The cultures irradiated with mixed (8:1) light spectra of red (660 nm) and blue (450 nm) produced the maximum number of globular and torpedo stages somatic embryos. The maximum percentage of vitrified somatic embryos was observed under warm white LEDs. In general, the light spectra blue, red and far-red (730 nm) alone failed to induce any somatic embryos. Beneficial effects of LED lighting on induction of somatic embryogenesis have also been reported in several orchids (see Chap. 13 for more details). The ability of LED lighting in establishing embryogenic cultures was also evaluated in *Coffea canefora* (Mai et al. 2016). Cultures exposed under LED R (41%):B (21%):W (21%) showed the maximum per cent regeneration of embryogenic cultures. Further LED light not only enhanced the germination rate of somatic embryos, but also shortened the germination time. Ferreira et al. (2017) investigated the effect of blue and red LEDs in mixed proportion of 18 and 82%, respectively, on somatic embryogenesis of sugarcane. SE induction was not favoured by LED treatment. The leaf explants exposed to LEDs formed callus and SE induced in the basal end of leaves that were exposed to FLs. However, the plants regenerated from SE and maintained under LEDs multiplied at a higher rate with more number of shoots than the Fls.

### 12.2.3 Effect of LED Irradiations on Ex Vitro Acclimatization

The final stage of micropropagation is the successful transfer of in vitro regenerated plants to field conditions following acclimatization. In fact, the success of plant regeneration is evaluated by its ability to produce plantlets that can survive outside the culture conditions. Proper rooting of the regenerated shoots is very crucial for ensuring the ex vitro survival of the plantlets as it allows uptake of nutrients and water from soil (Kozai et al. 1992). Well-rooted shoots are subjected to acclimatization or adaption before transfer to the field or greenhouse. Nevertheless, the stress imposed during transition from the in vitro to ex vitro environment is mainly responsible for low plant survival (Hazarika 2006; Kaur and Sandhu 2015). Along with a change in the ROS metabolism, improper development of photosynthetic machinery during in vitro culture may result in an array of morphological and physiological disorders to the regenerated plants. Photosynthetic competence of in vitro grown plantlets plays a vital role in ensuring its survival outside the culture vessel. In vitro light environment plays a significant role in developing the photosynthetic ability of the plantlets (Seon et al. 2000; see Chap. 6 for more details). The use of LEDs as the light source during in vitro plantlet development and subsequent acclimatization steps has been found to increase the rate of plant survival. An overview of the studies carried out to evaluate the spectral ability of LEDs on ex vitro acclimatization of in vitro regenerated plants is presented in Table 12.3.

In general, in vitro plantlets raised under LEDs performed well upon transfer to ex vitro environment compared to FLs. Successful acclimatization and subsequent growth of plants have been reported in various plant species including banana (Nhut et al. 2000), strawberry (Nhut et al. 2003; Hung et al. 2015), euphorbia (Dewir et al. 2006), *T. japonicum* (Moon et al. 2006), *J. curcas* (Daud et al. 2013), banana (Wilken et al. 2014; Vieira et al. 2015), vanilla (Bello-Bello et al. 2016), *S. rebaudiana* (Ramírez-Mosqueda et al. 2016), sugarcane (Ferreira et al. 2017) and

**Table 12.3** Effect of LED lights on the growth of in vitro raised plantlets during acclimatization

| Plant species | Light source | | Acclimatization response | References |
|---|---|---|---|---|
| | In vitro | Ex vitro | | |
| *Musa* sp. (Banana) | BL, RL, RL + BL (9:1, 8:2, 7:3), FL | Not mentioned | RL + BL (8:2) yielded best growth during acclimatization | Nhut et al. (2000) |
| *Fragaria* cv. 'Akihime' (Strawberry) | RL + BL (7:3), FL | Metal halide lamp | Plant height, leaf count and biomass accumulation were enhanced by RL + BL | Nhut et al. (2003) |
| *Euphorbia millii* | BL (440), RL (650), RL + BL (1:1), FrL (720) + RL (1:1), FrL + BL (1:1), FL | Halide lamps, high-pressure sodium lamps | All plantlets from LED and FL treatments were acclimatized successfully | Dewir et al. (2006) |
| *Tripterospermum japonicum* (Tsuru-rindo) | RL + BL (7:3), FL | FL | Plantlets raised under LEDs exhibited slightly better growth | Moon et al. (2006) |
| *Jatropha curcas* L. | BL (450), RL (660), RL + BL (1:1), WL, FL | Natural daylight conditions | RL yielded highest survival percentage | Daud et al. (2013) |
| *Musa* spp. cv. 'Grande naine' AAA (Banana) | WL, FL | Greenhouse conditions, no supplemental LED treatment | Survival rates did not differ significantly | Wilken et al. (2014) |
| Grapevine | WL, FL | WL, FL | LED lighting improved acclimatization response | Bleser et al. (2015) |
| *Fragaria* × *ananassa* Duch. cv. 'Camarosa' (Strawberry) | BL (460) + RL (660) (7:3, 1:1, 3:7, 1:9), FL | BL (460) + RL (660) (7:3, 1:1, 3:7, 1:9), FL | Plantlet survival, biomass accumulation, leaf growth and root development were enhanced under LEDs | Hung et al. (2015) |
| *Musa acuminata*cv. 'Nanicão Corupá' (Banana) | WL, RL + WL, FL | Greenhouse conditions, no supplemental LED treatment | All plantlets from LED as well as FL irradiations survived | Vieira et al. (2015) |

(continued)

**Table 12.3** (continued)

| Plant species | Light source | | Acclimatization response | References |
|---|---|---|---|---|
| | In vitro | Ex vitro | | |
| *Vanilla planifolia* Andrews | BL (460), RL (660), RL + BL (1:1), WL, FL | Greenhouse conditions, no supplemental LED treatment | Plantlet survival was high for all lighting conditions | Bello-Bello et al. (2016) |
| *Saccharum* spp. var. RB98710 (Sugarcane) | RL + BL (9:2), FL | Greenhouse conditions, no supplemental LED treatment | LED treatment influenced the acclimatization process with increased survival rate of plants | Ferreira et al. (2017) |
| *Vaccinium ashei* Reade cv. 'Titan' (Rabbiteye blueberry) | BL (460), RL (660), RL + BL (4:1, 1:1), FL | BL (460), RL (660), RL + BL (4:1, 1:1), FL | Shoot length and leaf area were increased under RL + BL. Chlorophyll content was increased by BL | Hung et al. (2016a) |
| *Stevia rebaudiana* Bertoni var. 'Morita II' | BL (460), RL (660), RL + BL (1:1), WL | Greenhouse conditions, no supplemental LED treatment | The highest percentage of acclimation was found with RL + BL | Ramírez-Mosqueda et al. (2016) |

*BL* blue LED, *RL* red LED, *FrL* far-red LED, *WL* white LED, *FL* fluorescent lamp

blueberry (Hung et al. 2016a). However, the abilities of LEDs to improve ex vitro performance with reference to their spectral quality varied from species to species. In vitro regenerated plantlets of *Coffea canefora* raised under LEDs adapted well compared to those cultured under FLs (Mai et al. 2016). The findings suggest that LED lighting not only enhances in vitro growth and development, but also improves ex vitro adaptability of the regenerated plants.

## 12.3 Effect of LEDs on ROS Regulatory Mechanisms During In Vitro Plant Morphogenesis and Ex Vitro Acclimatization

Physiological processes involving redox reactions release highly reactive, unstable free radicals as well as non-radical derivatives of oxygen, collectively termed as ROS. The oxygen radicals include the singlet oxygen ($^1O_2$), superoxide anion ($O_2^-$) and hydroxyl radical ($OH^{\cdot}$), whereas the non-radical derivatives are the peroxides, hypochlorites and ozone (Halliwell 1990). Generation of ROS mainly occurs in organelles such as chloroplast, mitochondria and peroxisomes due to high electron flux and highly oxidative metabolic reactions (Gill and Tuteja 2010). Uncontrolled production of ROS due to the saturation of antioxidative protection systems can cause cellular damage directly or indirectly through the synthesis of secondary toxic substances (Benson 2000; Bhattacharjee 2010). DNA and RNA damage due to oxidation by ROS may even lead to cell death (de Carvalho 2008). Cellular ROS generation is enhanced during conditions of biotic and abiotic stress for signalling and protection (Gill and Tuteja 2010). Various antioxidant molecules quench the ROS to maintain cellular oxidative stress at tolerable levels, ensuring the survival of the individual cells and the plant as a whole (Larson 1988). The antioxidant mechanisms are either enzymatic or non-enzymatic in nature. The enzymatic antioxidants mainly comprise of peroxidases (PODs), superoxide dismutases (SODs), catalase (CAT) and glutathione reductase (GR), while the non-enzymatic antioxidative role is performed by various secondary metabolites such as ascorbic acid, α-tocopherol, anthocyanins and carotenoids. However, ROS can also play an important role as signal molecules in regulating plant growth and morphogenesis (Pavlovic et al. 2002; Kwak et al. 2003; Gechev and Hille 2005; Kovalchuk 2010).

Plant tissue culture is generally perceived to occur under stress environment. Hormonal imbalances, suboptimal concentrations of macro- and micronutrients, accumulation of ethylene in the culture vessel and mechanical injury during inoculation and subculturing may induce various conditions of stress. Free radical-mediated stress along with concomitant changes in the cellular antioxidant levels has shown to affect in vitro morphogenesis in a number of plants (Benson 2000; Batkova et al. 2008; Dutta Gupta 2010). Oxidative stress may lead to browning of explant regeneration of hyperhydric plants and in vitro recalcitrance. Apart from the aforementioned stress-inducing factors, high photon flux with a wide range of wavelengths from the conventional artificial lighting systems can also induce photooxidative damage to the photosynthetic machinery (Demmig-Adams and Adams III 1992). On the contrary, light signal plays an important role in combating oxidative stress following the biosynthesis and activity of various phytohormones and antioxidants. Genes responsible for the biosynthesis of auxins, gibberellins and abscisic acid (ABA) are found to be regulated by various light stimuli (Hornitschek et al. 2012; Lee et al. 2012). The regulatory role of light in various ethylene-mediated signalling pathways and developmental processes has also been proposed (Zhong et al. 2012). Light-induced expression of antioxidative gene expression such as APX mRNA and CAT mRNA has been well demonstrated in *Arabidopsis* seedlings (Kubo et al. 1995).

Intensity and spectral quality of light exerts significant effects on cellular differentiation, growth and secondary metabolism. In recent years, several studies have demonstrated the potential of LEDs as an alternative photosynthetic radiation source for in vitro plant morphogenesis as well as for improved nutritional quality of horticultural plants (Dutta Gupta and Jatothu 2013).

Impact of LEDs on cellular redox balance has been scarcely studied and restricted to a few species. The salient features of studies carried out to evaluate the effect of different LED lighting treatments on the accumulation of ROS and antioxidant status during in vitro plant regeneration and ex vitro acclimatization are summarized in Table 12.4.

**Table 12.4** Effect of LEDs on ROS and antioxidant status during in vitro morphogenesis and acclimatization

| Plant species | LED colour and peak wavelength (nm) | Antioxidant/antioxidant enzyme/antioxidant capacity studied | Remarks | References |
|---|---|---|---|---|
| *Eleutherococcus senticosus* | BL (470), RL (660), BL + FrL (1:1) | Phenolics, flavonoids, SOD, CAT, MDHAR and GPX | High oxidative stress under RL resulted in poor biomass yield | Shohael et al. (2006) |
| *Morinda citrifolia* | BL, RL, RL + BL (1:1), FrL | Phenolic and flavonoid contents, SOD, CAT, G-POD and APX | Maximum numbers of adventitious roots were observed under RL, whereas RL + BL favoured callusing. CAT and G-POD activities were highest under RL + BL treatment | Baque et al. (2010) |
| *Oncidium* Gower Ramsey | BL (470 nm), GL (530 nm), YL (590 nm), RL (660 nm), FrL (715 nm), RL + BL (9:1, 8:2, 7:3), FrL + RL + BL (1:8:1), GL + RL + BL (1:8:1) | SOD, POD and CAT | PLB differentiation was associated with increase in antioxidant enzyme activities under BL | Mengxi et al. (2011) |
| *Cymbidium* Waltz 'Idol' | BL (450), GL (510), RL (640), BL + GL, RL + GL | SOD | Of all the light treatments, highest SOD activity was obtained with BL + GL treatment, whereas RL induced the lowest activity | Kaewjampa and Shimasaki (2012) |
| *Oryza sativa* cv. Ilmi | BL (450), GL (530), RL (660), RL + GL + BL (1:1:1) | Polyphenol and flavonoid contents, non-enzymatic antioxidant activities (ABTS, DPPH and FRAP) | BL, GL and RL + GL + BL treatment enhanced polyphenol and flavonoid accumulation. Antioxidant activities in rice leaves correlated with the amount of secondary metabolites | Jung et al. (2013) |
| *Curculigo orchioides* Gaertn. | BL (470), RL (630), BL + RL (1:1) | SOD, POD, CAT, APX and GR | Reduced oxidative stress due to high CAT activity under BL resulted in improved shoot organogenesis | Dutta Gupta and Sahoo (2015) |

(continued)

**Table 12.4** (continued)

| Plant species | LED colour and peak wavelength (nm) | Antioxidant/antioxidant enzyme/antioxidant capacity studied | Remarks | References |
|---|---|---|---|---|
| *Rehmannia glutinosa* Libosch. (Chinese foxglove) | BL (450), RL (650) | Phenolic and flavonoid contents, DPPH activity, reducing power, total antioxidant capacity, SOD, CAT, APX and GPX | LED treatments increased the total phenol and flavonoid contents. Improved growth under BL was associated with high enzymatic and non-enzymatic antioxidant activities | Manivannan et al. (2015) |
| *Phoenix dactylifera* L. cv. 'Alshakr' (Date palm) | RL + BL (9:1) | POD | RL + BL treatment increased shoot proliferation and POD activity | Al-Mayahi (2016) |
| *Thevetia peruviana* | RL (586-596), GL (525), BL (465), YL (590), WL | Total phenolic content, FRAP and ABTS assay | A higher phenolic content and antioxidant capacity were noted in cultures grown under darkness when compared to the different light conditions | Arias et al. (2016) |
| *Saccharum* spp. var. RB98710 (Sugarcane) | RL (82%) + BL (18%) | $H_2O_2$, MDA, SOD, APX and CAT | $H_2O_2$ content increased under LED. The plants exposed to LED showed lower SOD and CAT activity than the plants exposed to FL. APX activity was increased in plants under LED. There was no change in SOD and CAT activities during acclimatization | Ferreira et al. (2017) |
| *Stevia rebaudiana* Bertoni | BL (445), RL (638), RL + WL (1:1) | Phenolic content, SOD, POD and CAT | CAT and POD activities enhanced by BL | Simlat et al. (2016) |

*BL* blue LED, *RL* red LED, *FrL* far-red LED, *WL* white LED, *GL* green LED *YL* yellow LED

The occurrence of oxidative burst has been suggested in LED-treated somatic embryos of *Eleutherococcus senticosus* with a variation in secondary metabolite accumulation (Shohael et al. 2006). The study indicates that red light stimulated the production of eleutheroside E and E1 but with a reduction in somatic embryo dry weight. In contrast, the synthesis of phenol, total flavonoid and chlorogenic acid was enhanced by fluorescent light. Antioxidant enzymes, SOD, CAT, GST and MDHAR, were induced in RL-irradiated somatic embryos, and the findings reflect the differential sensitivity of the enzymes to the different light treatments.

Adventitious root induction from leaf explants of *Morinda citrifolia* was studied under different light sources generated from LEDs, including fluorescent light as control (Baque et al. 2010). Red light significantly induced adventitious roots. A differential response of adventitious root induction in relation to the light-induced ROS generation and antioxidant protection was observed. CAT and guaiacol peroxidase (G-POD) activities were highest under red. Higher APX activities were observed under fluorescent light, followed by blue light. However, there was no significant change in SOD activities under different light sources. The study suggests that higher APX activity in conjunction with CAT and G-POD mitigates the toxic effects of $H_2O_2$ under fluorescent light.

In *Oncidium*, the rate of differentiation of protocorm-like bodies (PLBs) was higher under BL than RL treatment. The POD, SOD and CAT activities were higher under BL than those under RL (Mengxi et al. 2011). The SOD activity in the leaves of *Oncidium* plantlets was found to be low under monochromatic RL treatment. In contrast, BL irradiation significantly enhanced the enzyme activities and accumulation of proteins in the leaves. However, there were no significant differences in the MDA content of the leaves between the light treatments. The study suggests the stimulatory role of BL in activating different antioxidant protection systems to scavenge excessive amounts of ROS.

The effect of composite spectra on in vitro PLB formation and shoot regeneration of hybrid *Cymbidium* was investigated by Kaewjampa and Shimasaki (2012). Combinational light treatments of RL + GL and BL + GL promote PLB multiplication and shoot formation with accompanying increase in SOD activity. Interval lighting using GL increased the SOD activity in PLBs most efficiently and induces organogenesis in PLB cultures of *Cymbidium* Waltz 'idol'.

Rice seedlings grown in vitro under monochromatic BL and GL and RL + GL + BL mixed LEDs exhibited higher polyphenol accumulation and enhanced non-enzymatic antioxidant activity than the seedlings raised under RL illumination (Jung et al. 2013). High levels of flavonoid glycosides were observed under BL. It has already been proposed that the blue light condition regulates the expression of flavonoid biosynthesis genes (Christie and Jenkins 1996). Metabolic profiles of rice leaves were significantly different between BL- and RL-irradiated samples. Thus, BL and RL showed dissimilar effects on plant growth and development. A difference in antioxidant activity was also noted under different lighting conditions, and the antioxidant activity followed the order BL = WL (white LED) = GL > $R < S$ (shade condition).

Incorporation of BL and/or RL light sources during in vitro culture of *Rehmannia glutinosa* was found to be beneficial for increasing the medicinal values of the plant via changes in the antioxidant enzyme activities (Manivannan et al. 2015). Activities of POD, SOD, CAT and G-POD were found to be highest under BL as compared to red RL and FL. The increase in antioxidant activity under both BL and RL treatments was correlated with the enhancement of phytochemicals. Modulation of spectral quality especially by the BL induced the antioxidant defence system.

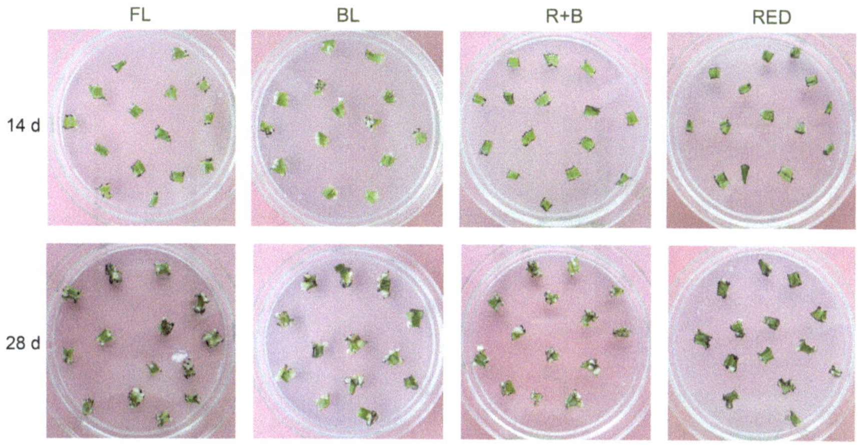


**Fig. 12.1** Response of *C. orchioides* leaf explants cultured on Murashige and Skoog's medium (Murashige and Skoog 1962) supplemented 4 mg $l^{-1}$ 6-benzylaminopurine (BAP) and 3% (w/v) sucrose under different LED treatments (*FL* fluorescent lamp; *BL* blue LEDs; *R* + *B* red and blue (1:1)-mixed LEDs; *RED* red LEDs; Adapted from Sahoo 2013)

LED irradiation-induced changes in the steady-state level of ROS and subsequent alterations in the antioxidative defence system were found to be closely associated with the shoot regeneration potential of *Curculigo orchioides* (Dutta Gupta and Sahoo 2015). Shoot regeneration response both in terms of per cent regeneration and mean number of shoot buds per responding explants was significantly improved under the irradiation of BL compared to FL and other light treatments (Fig. 12.1, Sahoo 2013, unpublished). The study illustrates the effect of spectral quality provided by LED source on the relationship between cellular redox balance and shoot organogenesis highlighting the role of LED in ROS modulation which ultimately orchestrates shoot regeneration. A significant increase in SOD activity was observed at 14-day cultures under RL treatment, while the activity decreased at 28 days. It is worthy to note that the differentiation and proliferation of shoot buds occurred during the period of 2–4 weeks of culture. On the contrary, SOD activity decreased gradually from 0 to 28 days under FL (Fig. 12.2a). Irradiation of cultures with RL + BL and RL decreased the CAT activity, whereas a substantial increase in the CAT activity was observed in cultures exposed to BL (Fig. 12.2b). Compared to the activity present in the leaf explants at the time of culture establishment, POX activity gradually decreased during the stages of shoot bud emergence and proliferation (Fig. 12.2c). Among the treated cultures, highest POX activity was observed under BL at 14-d-old cultures, whereas the trend was changed and 28-d-old cultures under RL resulted in the highest activity than that of FL, BL and RL + BL treatments. APX activity increased under BL and RL + BL treatments from 0 to 14 days and then decreased at 28 days. In contrast, a gradual decrease in the APX activity was observed in cultures treated with RL (Fig. 12.2d). The findings suggest that LED lighting was capable of creating a differential

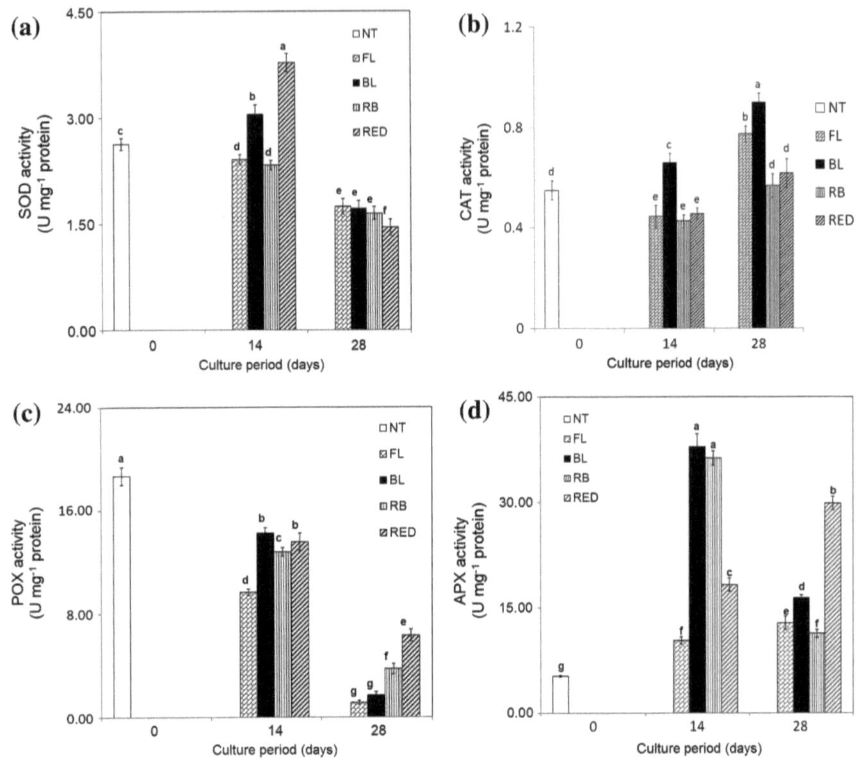


**Fig. 12.2** Changes in SOD (**a**), CAT (**b**), POX (**c**) and APX (**d**) activities at various culture intervals (0, 14 and 28 d) during shoot organogenesis of *C. orchioides* under different LED illuminations (*NT* untreated samples at 0 d; *FL* fluorescent lamp; *BL* blue LEDs; *R* + *B* red and blue (1:1) mixed LEDs; *RED* red LEDs). Values are represented as mean ± SD. Data indicated with different alphabets differ significantly according to LSD test at $p < 0.05$ (adapted from Dutta Gupta and Sahoo 2015)

response of ROS regulations with concomitant changes in the efficiency of shoot regeneration. Acquisition of organogenic competence and the emergence of shoot buds under BL appeared to be associated with the increased level of SOD activity. The study proposed a complex interplay between the spectral quality of LEDs, light signalling and redox metabolism.

The impact of LED spectra on seed germination and seedling growth of *S. rebaudiana* in vitro was studied by Simlat et al. (2016). Increased seed germination and the development of largest number of leaves were observed under BL with enhanced activity of CAT and POX. RL, however, significantly increased the shoot and root lengths and also enhanced the SOD activity. An opposite effect was exerted by BL. However, a significant change in the relative activities of the enzymes was observed upon varying the temperature, indicating the complexity of antioxidative response to various external stimuli (Simlat et al. 2016).

The effect of LED light source on shoot multiplication, phytochemicals and changes in the antioxidant enzyme activities has also been studied in date palm (Al-Mayahi 2016). In vitro shoot cultures of *P. dactylifera* exhibited elevated POD activity along with improved growth under red + blue (9:1) mixed LED treatment as compared to cultures exposed to fluorescent lamps (Al-Mayahi 2016), suggesting a close relationship between the light environment, antioxidative metabolism and morphogenesis.

Different LED lighting conditions had no significant impact on the total phenol content in suspension cultures of *T. peruviana* (Arias et al. 2016). Rather, production of phenols was highest in cultures maintained in the dark conditions. FRAP and ABTS assays of the LED-treated suspension cultures also exhibited lower antioxidant capacity than the cultures kept in dark conditions. The findings indicate that the antioxidant capacity of the cultures was effectively conditioned by the presence of phenolic compounds. Exposure to light had deterred the production of phenolic compounds and reduced the antioxidant activity of the *T. peruviana* suspension cultures.

Cultures of sugarcane exhibited increase in $H_2O_2$ content along with reduced SOD and CAT activities under a mixture of red and blue LEDs (Ferreira et al. 2017). However, the LED-treated plants had a high APX activity. The ability of APX to eliminate $H_2O_2$ may account for the better performance of plants regenerated under LED illumination during ex vitro acclimatization. The SOD and CAT activities of the in vitro grown plants exposed to LEDs did not differ significantly from the control during the acclimatization process.

From the reported examples, it appears that monochromatic LEDs and their combinations are able to modulate the cellular redox balance and have profound influence on shoot regeneration potential. However, response of antioxidant metabolism to light environment varies from species to species and is significantly influenced by the in vitro culture conditions. Further investigations are required to understand the influence of different LED wavebands on ROS and antioxidant status of in vitro plant cultures. The knowledge may prove to be crucial for enhancing in vitro growth and productivity of many economically important plants. Further, in vitro germplasm conservation techniques may also be improved by understanding the role of various LEDs in regulating the ROS network and oxidative stress response.

## 12.4 Conclusions

The present chapter demonstrates the potential of LEDs as a photosynthetically efficient and versatile lighting system for in vitro plant morphogenesis. The efficacy of various LED lighting regimes in promoting plant regeneration responses has been illustrated in a variety of plant species. The responses to spectral quality provided by LEDs vary according to plant species. Improved in vitro shoot proliferation, leaf development and root growth under LED illumination yielded healthy plantlets that were successfully acclimatized and transferred to field with a high rate of ex vitro survival. Induction and differentiation of somatic embryos into plantlets under LED irradiations have also been reported for a number of species. Further, the regulatory role of LEDs in modulating the ROS metabolism with a possible link to the regeneration potential has been discussed. The overall glimpse implies that the spectral energy distribution of LEDs could satisfy the requirements of various aspects of plant tissue culture including commercial micropropagation. Thus, LEDs may be used to develop micropropagation as well as transplant production systems that are more energy-efficient and have higher quantitative and qualitative yield as compared to the conventional GDL-based units.

## References

Ahmed MR, Anis M (2014) Changes in activity of antioxidant enzymes and photosynthetic machinery during acclimatization of micropropagated *Cassia alata* L. plantlets. In Vitro Cell Dev Biol—Plant 50:601–609

Al-Mayahi AMW (2016) Effect of red and blue light emitting diodes "CRB-LED" on in vitro organogenesis of date palm (*Phoenix dactylifera* L.) cv. Alshakr. World J Microbiol Biotechnol 32:160

Alvarenga ICA, Pacheco FV, Silva ST, Bertolucci SKV, Pinto JEBP (2015) In vitro culture of *Achillea millefolium* L.: quality and intensity of light on growth and production of volatiles. Plant Cell Tiss Organ Cult 122:299–308

Arias JP, Zapata K, Rojano B, Arias M (2016) Effect of light wavelength on cell growth, content of phenolic compounds and antioxidant activity in cell suspension cultures of *Thevetia peruviana*. J Photochem Photobiol B Biol 163:87–91

Baque MA, Hahn EJ, Paek KY (2010) Induction mechanism of adventitious root from leaf explants of *Morinda citrifolia* as affected by auxin and light quality. In Vitro Cell Dev Biol—Plant 46:71–80

Batista DS, de Castro KM, da Silva AR, Teixeira ML, Sales TA, Soares LI, Cardoso MDG, Santos MDO, Viccini LF, Otoni WC (2016) Light quality affects in vitro growth and essential oil profile in *Lippia alba* (Verbenaceae). In Vitro Cell Dev Biol—Plant 52:276–282

Baťková P, Pospíšilová J, Synková H (2008) Production of reactive oxygen species and development of antioxidative systems during in vitro growth and ex vitro transfer. Biol Plant 52:413–422

Bello-Bello JJ, Martínez-Estrada E, Caamal-Velázquez JH, Morales-Ramos V (2016) Effect of LED light quality on in vitro shoot proliferation and growth of vanilla (*Vanilla planifolia* Andrews). Afr J Biotechnol 15(8):272–277

Benson EE (2000) Do free radicals have a role in plant tissue culture recalcitrance? In Vitro Cell Dev Biol—Plant 36:163–170

Bhattacharjee S (2010) Sites of generation and physicochemical basis of formation of reactive oxygen species in plant cell. In: Dutta Gupta S (ed) Reactive oxygen species and antioxidants in higher plants. CRC Press, Florida, pp 1–30

Bleser E, Tittmann S, Rühl EH (2015) Effects of LED-illumination and light intensity on the acclimatization of in vitro plantlets to ex vitro conditions. Acta Hort 1082:131–139

Botero Giraldo C, Urrea Trujillo AI, Naranjo Gómez EJ (2015) Regeneration potential of *Psychotria ipecacuanha* (Rubiaceae) from thin cell layers. Acta Biol Colomb 20(3):181–192

Budiarto K (2010) Spectral quality affects morphogenesis on Anthurium plantlet during in vitro culture. Agrivita 32(3):234–240

Chen M, Chory J, Fankhauser C (2004) Light signal transduction in higher plants. Annu Rev Genet 38:87–117

Chen CC, Agrawal DC, Lee MR, Lee RJ, Kuo CL, Wu CR, Tsay HS, Chang HC (2016) Influence of LED light spectra on in vitro somatic embryogenesis and LC–MS analysis of chlorogenic acid and rutin in *Peucedanum japonicum* Thunb.: a medicinal herb. Bot Stud 57(9):1–9

Christie JM, Jenkins GI (1996) Distinct UV-B and UV-A/blue light signal transduction pathways induce chalcone synthase gene expression in *Arabidopsis* cells. Plant Cell 8(9):1555–1567

Chung JP, Huang CY, Dai TE (2010) Spectral effects on embryogenesis and plantlet growth of *Oncidium* 'Gower Ramsey'. Sci Hortic 124:511–516

Daud N, Faizal A, Geelen D (2013) Adventitious rooting of *Jatropha curcas* L. is stimulated by phloroglucinol and by red LED light. In Vitro Cell Dev Biol—Plant 49:183–190

de Carvalho MHC (2008) Drought stress and reactive oxygen species. Plant Signal Behav 3(3):156–165

Demmig-Adams B, Adams WW III (1992) Photoprotection and other responses of plants to high light stress. Annu Rev Plant Biol 43:599–626

Devlin PF, Christie JM, Terry MJ (2007) Many hands make light work. J Exp Bot 58:3071–3077

Dewir YH, Chakrabarty D, Hahn EJ, Paek KY (2006) The effects of paclobutrazol, light emitting diodes (LEDs) and sucrose on flowering of *Euphorbia millii* plantlets in vitro. Eur J Hortic Sci 71(6):240–244

Dutta Gupta S (2010) Role of free radicals and antioxidants in in vitro morphogenesis. In: Dutta Gupta S (ed) Reactive oxygen species and antioxidants in higher plants. CRC Press, Florida, pp 229–247

Dutta Gupta S, Datta S (2003/4) Antioxidant enzyme activities during in vitro morphogenesis of gladiolus and the effect of application of antioxidants on plant regeneration. Biol Plant 47: 79–183

Dutta Gupta S, Jatothu B (2013) Fundamentals and applications of light-emitting diodes (LEDs) in in vitro plant growth and morphogenesis. Plant Biotechnol Rep 7:211–220

Dutta Gupta S, Sahoo TK (2015) Light emitting diode (LED)-induced alteration of oxidative events during in vitro shoot organogenesis of *Curculigo orchioides* Gaertn. Acta Physiol Plant 37:233

Edesi J, Kotkas K, Pirttilä AM, Häggman H (2014) Does light spectral quality affect survival and regeneration of potato (*Solanum tuberosum* L.) shoot tips after cryopreservation? Plant Cell Tiss Organ Cult 119:599–607

Faisal M, Anis M (2009) Changes in photosynthetic activity, pigment composition, electrolyte leakage, lipid peroxidation, and antioxidant enzymes during ex vitro establishment of micropropagated *Rauvolfia tetraphylla* plantlets. Plant Cell Tiss Organ Cult 99:125–132

Ferreira LT, de Araújo Silva MM, Ulisses C, Camara TR, Willadino L (2017) Using LED lighting in somatic embryogenesis and micropropagation of an elite sugarcane variety and its effect on redox metabolism during acclimatization. Plant Cell Tiss Organ Cult 128:211–221

Franklin KA, Whitelam GC (2004) Light signals, phytochromes and cross-talk with other environmental cues. J Exp Biol 55(395):271–276

Gechev TS, Hille J (2005) Hydrogen peroxide as a signal controlling plant programmed cell death. J Cell Biol 168:17–20

Gill SS, Tuteja N (2010) Reactive oxygen species and antioxidant machinery in abiotic stress tolerance in crop plants. Plant Physiol Biochem 48:909–930

Godo T, Fujiwara K, Guan K, Miyoshi K (2011) Effects of wavelength of LED-light on in vitro asymbiotic germination and seedling growth of *Bletilla ochracea* Schltr. (Orchidaceae). Plant Biotechnol 28:397–400

Hahn EJ, Kozai T, Paek KY (2000) Blue and red light-emitting diodes with or without sucrose and ventilation affect in vitro growth of *Rehmannia glutinosa* plantlets. J Plant Biol 43(4):247–250

Halliwell B (1990) How to characterize a biological antioxidant. Free Radic Res Commun 9:1–32

Hazarika BN (2006) Morpho-physiological disorders in in vitro culture of plants. Sci Hortic 108:105–120

Heo JW, Shin KS, Kim SK, Paek KY (2006) Light quality affects in vitro growth of grape 'Teleki 5BB'. J Plant Biol 49(4):276–280

Hornitschek P, Kohnen MV, Lorrain S, Rougemont J, Ljung K, López-Vidriero I, Franco-Zorrilla JM, Solano R, Trevisan M, Pradervand S, Xenarios I, Fankhauser C (2012) Phytochrome interacting factors 4 and 5 control seedling growth in changing light conditions by directly controlling auxin signaling. Plant J 71:699–711

Hughes KW (1981) In vitro ecology: exogenous factors affecting growth and morphogenesis in plant culture systems. Environ Exp Bot 21:281–288

Hung CD, Hong CH, Jung HB, Kim SK, Van Ket N, Nam MW, Choi DH, Lee HI (2015) Growth and morphogenesis of encapsulated strawberry shoot tips under mixed LEDs. Sci Hortic 194:194–200

Hung CD, Hong CH, Kim SK, Lee KH, Park JY, Dung CD, Nam MW, Choi DH, Lee HI (2016a) In vitro proliferation and ex vitro rooting of microshoots of commercially important rabbiteye blueberry (*Vaccinium ashei* Reade) using spectral lights. Sci Hortic 211:248–254

Hung CD, Hong CH, Kim SK, Lee KH, Park JY, Nam MW, Choi DH, Lee HI (2016b) LED light for in vitro and ex vitro efficient growth of economically important highbush blueberry (*Vaccinium corymbosum* L.). Acta Physiol Plant 38:152

Jao RC, Fang W (2004a) Effects of frequency and duty ratio on the growth of potato plantlets in vitro using light-emitting diodes. HortScience 39(2):375–379

Jao RC, Fang W (2004b) Growth of potato plantlets in vitro is different when provided concurrent versus alternating blue and red light photoperiods. HortScience 39(2):380–382

Jao RC, Lai CC, Fang W, Chang SF (2005) Effects of red light on the growth of *Zantedeschia* plantlets in vitro and tuber formation using light-emitting diodes. HortScience 40(2):436–438

Jung ES, Lee S, Lim SH et al (2013) Metabolite profiling of the short-term responses of rice leaves (*Oryza sativa* cv. Ilmi) cultivated under different LED lights and its correlations with antioxidant activities. Plant Sci 210:61–69

Kaewjampa N, Shimasaki K (2012) Effects of green LED lighting on organogenesis and superoxide dismutase (SOD) activities in protocorm-like bodies (PLBs) of *Cymbidium* cultured in vitro. Environ Control Biol 50(3):247–254

Karataş M, Aasim M, Dazkirli M (2016) Influence of light-emitting diodes and benzylaminopurin on adventitious shoot regeneration of water hyssop (*Bacopa monnieri* (L.) Pennell) in vitro. Arch Biol Sci 68(3):501–508

Kasahara M, Kagawa T, Sato Y, Kiyosue T, Wada M (2004) Phototropins mediate blue and red light-induced chloroplast movements in *Physcomitrella patens*. Plant Physiol 135:1388–1397

Kaur A, Sandhu JS (2015) High throughput in vitro micropropagation of sugarcane (*Saccharum offiinarum* L.) from spindle leaf roll segments: cost analysis for agri-business industry. Plant Cell Tiss Organ Cult 120:339–350

Kim YW, Moon HK (2014) Enhancement of somatic embryogenesis and plant regeneration in Japanese red pine (*Pinus densiflora*). Plant Biotechnol Rep 8:259–266

Kim SJ, Hahn EJ, Heo JW, Paek KY (2004) Effects of LEDs on net photosynthetic rate, growth and leaf stomata of chrysanthemum plantlets in vitro. Sci Hortic 101:143–151

Kovalchuk I (2010) Multiple roles of radicals in plants. In: Dutta Gupta S (ed) Reactive oxygen species and antioxidants in higher plants. CRC Press, Florida, pp 31–44

Kozai T, Smith MAL (1995) Environmental control in plant tissue culture—general introduction and overview. In: Aitken-Christie J, Kozai T, Smith MAL (eds) Automation and environmental control in plant tissue culture. Kluwer Academic Publishers, The Netherlands, pp 301–318

Kozai T, Xiao Y (2008) A commercialized photoautotrophic micropropagation system. In: Dutta Gupta S, Ibaraki Y (eds) Plant tissue culture engineering. Springer, The Netherlands, pp 355–371

Kozai T, Fujiwara K, Hayashi M, Aitken-Christie J (1992) Thein vitro environment and its control in micropropagation. In: Kurata K, Kozai T (eds) Transplant production systems. Kluwer Academic Publishers, The Netherlands, pp 247–282

Kubo A, Saji H, Tanaka K, Kondo N (1995) Expression of *Arabidopsis* cytosolic ascorbate peroxidase gene in response to ozone orsulfur dioxide. Plant Mol Biol 29:479–489

Kurilčik A, Miklušytė-Čanova R, Žilinskaitė S, Dapkūnienė S, Duchovskis P, Kurilčik G, Tamulaitis G, Žukauskas A (2007) In vitro cultivation of grape culture under solid-state lighting. Sodinink Daržinink 26(3):235–245

Kurilčik A, Miklušytė-Čanova R, Dapkūnienė S, Žilinskaitė S, Kurilčik G, Tamulaitis G, Duchovskis P, Žukauskas A (2008) In vitro culture of *Chrysanthemum* plantlets using light-emitting diodes. Cent Eur J Biol 3(2):161–167

Kwak JM, Mori IC, Pei ZM, Leonhardt N, Torres MA, Dangl JL, Bloom RE, Bodde S, Jones JDG, Shroeder JI (2003) NADPH oxidases *AtrbohD* and *AtrbohF* genes function in ROS-dependent ABA signaling in *Arabidopsis*. Eur Mol Biol Organ J 22(11):2623–2633

Kwon AR, Cui HY, Lee H, Shin H, Kang KS, Park SY (2015) Light quality affects shoot regeneration, cell division, and wood formation in elite clones of *Populus euramericana*. Acta Physiol Plant 37:65

Larson RA (1988) The antioxidants of higher plants. Phytochem 27:969–978

Lee KP, Piskurewicz U, Turečková V, Carat S, Chappuis R, Strnad M, Fankhauser C, Lopez-Molina L (2012) Spatially and genetically distinct control of seed germination by phytochromes A and B. Genes Dev 26:1984–1996

Lee NN, Choi YE, Moon HK (2014) Effect of LEDs on shoot multiplication and rooting of rare plant *Abeliophyllum distichum* Nakai. J Plant Biotechnol 41:94–99

Lercari B, Tognoni F, Anselmo G, Chapel D (1986) Photocontrol of in vitro bud differentiation in *Saintpaulia ionantha* leaves and *Lycopersicon esculentum* cotyledons. Physiol Plant 67 (3):340–344

Li H, Xu Z, Tang C (2010) Effect of light-emitting diodes on growth and morphogenesis of upland cotton (*Gossypium hirsutum* L.) plantlets in vitro. Plant Cell Tiss Organ Cult 103:155–163

Lian ML, Murthy HN, Paek KY (2002) Effects of light emitting diodes (LEDs) on the in vitro induction and growth of bulblets of *Lilium* oriental hybrid 'Pesaro'. Sci Hortic 94:365–370

Lin Y, Li J, Li B, He T, Chun Z (2011) Effects of light quality on growth and development of protocorm-like bodies of *Dendrobium officinale* in vitro. Plant Cell Tiss Organ Cult 105:329–335

Liu Y, Tong Y, Zhu Y, Ding H, Smith EA (2006) Leaf chlorophyll readings as an indicator for spinach yield and nutritional quality with different nitrogen fertilizer applications. J Plant Nutr 29:1207–1217

Ma X, Wang Y, Liu M, Xu J, Xu Z (2015) Effects of green and red lights on the growth and morphogenesis of potato (*Solanum tuberosum* L.) plantlets in vitro. Sci Hortic 190:104–109

Mai NT, Binh PT, Gam DT, Khoi PH, Hung NK, Ngoc PB, Ha CH, Thanh Binh HT (2016) Effects of light emitting diodes—LED on regeneration ability of *Coffea canephora* mediated via somatic embryogenesis. Tap Chi Sinh Hoc 38(2):228–235

Maluta FA, Bordignon SR, Rossi ML, Ambrosano GMB, Rodrigues PHV (2013) In vitro culture of sugarcane exposed to different light sources. Pesqui Agropecu Bras 48(9):1303–1307

Manivannan A, Soundararajan P, Halimah N, Ko CH, Jeong BR (2015) Blue LED light enhances growth, phytochemical contents, and antioxidant enzyme activities of *Rehmannia glutinosa* cultured in vitro. Hortic Environ Biotechnol 56(1):105–113

Massa GD, Kim HH, Wheeler RM, Mitchell CA (2008) Plant productivity in response to LED lighting. HortScience 43(7):1951–1956

Mengxi L, Zhigang X, Yang Y, Yijie F (2011) Effects of different spectral lights on *Oncidium* PLBs induction, proliferation, and plant regeneration. Plant Cell Tiss Organ Cult 106:1–10

Merkle SA, Montello PM, Xia X, Upchurch BL, Smith DR (2005) Light quality treatments enhance somatic seedling production in three southern pine species. Tree Physiol 26:187–194

Moon HK, Park SY, Kim YW, Kim CS (2006) Growth of Tsuru-rindo (*Tripterospermum japonicum*) cultured in vitro under various sources of light-emitting diode (LED) irradiation. J Plant Biol 49(2):174–179

Murashige T, Skoog F (1962) A revised medium for rapid growth and bio assays with tobacco tissue cultures. Physiol Plant 15:473–497

Nhut DT, Hong LTA, Watanabe H, Goi M, Tanaka M (2000) Growth of banana plantlets cultured in vitro under red and blue light-emitting diode (LED) irradiation source. Acta Hortic 575:117–124

Nhut DT, Takamura T, Watanabe H, Okamoto K, Tanaka M (2003) Responses of strawberry plantlets cultured in vitro under superbright red and blue light-emitting diodes (LEDs). Plant Cell Tiss Organ Cult 73:43–52

Nhut DT, Huy NP, Tai NT, Nam NB, Luan VQ, Hien VT, Tung HT, Vinh BT, Luan TC (2015) Light-emitting diodes and their potential in callus growth, plantlet development and saponin accumulation during somatic embryogenesis of *Panax vietnamensis* Ha *et* Grushv. Biotechnol Biotechnol Equip 29(2):299–308

Park SY, Kim MJ (2010) Development of zygotic embryos and seedlings is affected by radiation spectral compositions from light emitting diode (LED) system in Chestnut (*Castanea crenata* S. et Z.). J Korean For Soc 99(5):750–754

Park SY, Yeung EC, Paek KY (2010) Endoreduplication in *Phalaenopsis* is affected by light quality from light-emitting diodes during somatic embryogenesis. Plant Biotechnol Rep 4(4):303–309

Pavlović D, Đorđević V, Kocić G (2002) A "cross-talk" between oxidative stress and redox cell signaling. Med Biol 9(2):131–137

Poudel PR, Kataoka I, Mochioka R (2008) Effect of red-and blue-light-emitting diodes on growth and morphogenesis of grapes. Plant Cell Tiss Organ Cult 92:147–153

Ramírez-Mosqueda MA, Iglesias-Andreu LG, Bautista-Aguilar JR (2016) The effect of light quality on growth and development of in vitro plantlet of *Stevia rebaudiana* Bertoni. Sugar Tech. doi:10.1007/s12355-016-0459-5

Sahoo TK (2013) Effects of light emitting diodes (LEDs) on changes in antioxidative enzyme activities and compounds during in vitro shoot organogenesis of *Curculigo orchioides* Gaetrn. M. Tech thesis, Indian Institute of Technology Kharagpur, India

Samuolienė G, Brazaitytė A, Urbonavičiūtė A, Šabajevienė G, Duchovskis P (2010) The effect of red and blue light component on the growth and development of frigo strawberries. Zemdirb Agric 97:99–104

Seon JH, Cui YY, Kozai T, Paek KY (2000) Influence of in vitro growth conditions on photosynthetic competence and survival rate of *Rehmannia glutinosa* plantlets during acclimatization period. Plant Cell Tiss Organ Cult 61:135–142

Shin KS, Murthy HN, Heo JW, Hahn EJ, Paek KY (2008) The effect of light quality on the growth and development of in vitro cultured *Doritaenopsis* plants. Acta Physiol Plant 30:339–343

Shohael AM, Ali MB, Yu KW, Hahn EJ, Islam R, Paek KY (2006) Effect of light on oxidative stress, secondary metabolites and induction of antioxidant enzymes in *Eleutherococcus senticosus* somatic embryos in bioreactor. Process Biochem 41:1179–1185

Silva MMA, de Oliveira ALB, Oliveira-Filho RA, Gouveia-Neto AS, Camara TJR, Willadino LG (2014) Effect of blue/red LED light combination on growth and morphogenesis of *Saccharum officinarum* plantlets in vitro. SPIE 8947.doi:10.1117/12.2036200

Silva MMA, de Oliveira ALB, Oliveira-Filho RA, Camara T, Willadino L, Gouveia-Neto A (2016) The effect of spectral light quality on in vitro culture of sugarcane. Acta Sci Biol Sci 38 (2):157–161

Simlat M, Ślęzak P, Moś M, Warchol M, Skrzypek E, Ptak A (2016) The effect of light quality on seed germination, seedling growth and selected biochemical properties of *Stevia rebaudiana* Bertoni. Sci Hortic 211:295–304

Tanaka M, Takamura T, Watanabe H, Endo M, Yanagi T, Okamoto K (1998) In vitro growth of *Cymbidium* plantlets cultured under superbright red and blue light-emitting diodes (LEDs). J Hort Sci Biotechnol 73:39–44

Vieira LDN, de Freitas Fraga HP, dos Anjos KG, Puttkammer CC, Scherer RF, da Silva DA, Guerra MP (2015) Light-emitting diodes (LED) increase the stomata formation and chlorophyll content in *Musa acuminata* (AAA) 'Nanicão Corupá' in vitro plantlets. Theor Exp Plant Physiol 27(2):91–98

Wilken D, Gonzalez EJ, Gerth A, Gómez-Kosky R, Schumann A, Claus D (2014) Effect of immersion systems, lighting, and TIS designs on biomass increase in micropropagating banana (*Musa* spp. cv. 'Grande naine' AAA). In Vitro Cell Dev Biol—Plant 50:582–589

Wongnok A, Piluek C, Techasilpitak T, Tantivivat S (2008) Effects of light emitting diodes on micropropagation of *Phalaenopsis* orchids. Acta Hortic 788:149–156

Wu HC, Lin CC (2012) Red light-emitting diode light irradiation improves root and leaf formation in difficult-to-propagate *Protea cynaroides* L. plantlets in vitro. HortScience 47(10):1490–1494

Zhong S, Shi H, Xue C, Wang L, Xi Y, Li J, Quail PH, Deng XW, Guo H (2012) A molecular framework of light-controlled phytohormone action in *Arabidopsis*. Curr Biol 22:1530–1535

# Chapter 13
# Impact of Light-Emitting Diodes (LEDs) on Propagation of Orchids in Tissue Culture

**E. Hanus-Fajerska and R. Wojciechowska**

## 13.1 Introduction

Orchidaceae Juss. is one of the largest and most fascinating families of angiosperms which comprises more than 800 genera, with over 25,000 species, widely distributed around the world (Steward and Griffiths 1995; Freudenstein and Chase 2015). In the wild, orchids' growth is extremely slow. It takes several years to obtain a flowering specimen. In natural conditions, the majority of flowers are not pollinated, thus their ovules cannot be fertilized, and in consequence capsules with viable seeds are rarely formed. Furthermore, the growth and development of orchids are markedly influenced by the climatic conditions, as well as by the protective canopy of vegetation existing in local habitats, like in case of the numerous known epiphytic species (Zotz 2013; Zotz and Winkler 2013; de la Rosa-Manzano et al. 2014, 2015; Zang et al. 2015). This is why orchids are listed in the Red Data Book, which has been prepared by the International Union of Conservation of Nature and Natural Resources (IUCN). Thus, there is an urgent need to adopt appropriate methods and ways to tackle this issue. The entire Orchidaceae family is currently also included in the second Appendix of Convention on the International Trade in Endangered Species of Wild Fauna and Flora (CITIES). In order to conserve the orchid germplasms, the biosphere reserves, national parks, or national reserves are being established in the orchid-rich regions

E. Hanus-Fajerska (✉) · R. Wojciechowska
Unit of Botany and Plant Physiology, Institute of Plant Biology and Biotechnology, Faculty of Biotechnology and Horticulture, University of Agriculture, Al. 29-Listopada 54, 31-425 Kraków, Poland
e-mail: e.hanus@ogr.ur.krakow.pl



S. Dutta Gupta (ed.), *Light Emitting Diodes for Agriculture*,
DOI 10.1007/978-981-10-5807-3_13

of the world. The large number of regional research programmes have already given benefits, such as studies on the orchid flora carried out in Dirbu-Saikhowa National Park and Biosphere Reserve in upper Assam (Gogoi 2005; Gogoi et al. 2010) or in National Nature Reserves of China (Zang et al. 2015). Nonetheless, in some complex study areas the greatest challenge for both researchers and workers of state administration is to draw precise conservation recommendations for future conservation topics from existing partial biodiversity data. Therefore, the research should be continued, especially in underexplored countries characterized by high biodiversity level of orchid flora. An excellent example could be the Andean rain forest of Bolivia, located between the Madidi National Park on the Peruvian border and Amboró National Park near Santa-Cruz de la Sierra (Müller et al. 2003) or Torres del Paine Biosphere Reserve in Chile (Vidal et al. 2012). On the contrary, in some European counties there are still some problems with compliance of existing legislation regarding protecting areas (Křenová and Kidlmann 2015). On the other hand, the description of the orchid flora of Britain and Ireland with precise distribution of species has been worked out with details (Bateman 2011; Fay 2015). Additionally, in the majority of other carefully monitored European stands located in the wild, local orchid species are frequently being threatened or endangered (Pindel and Pindel 2004; Nordström and Hedrén 2009; Jakubowska-Gabara et al. 2012; Rewicz et al. 2015; Seaton et al 2015; Kul et al. 2016). Unfortunately, in situ conservation, that is suffering from fragmentation of habitats or different other kinds of negative effects caused by human beings to most of biomes, is an insufficient option (Palomo et al. 2013; Sheshukova et al. 2014; Ballantyne and Pickering 2015). Furthermore, wild plants with extremely small populations are particularly difficult to protect. For this reason, a special conservation programme has been introduced in China (Chen et al. 2015), and in other countries around the world. In order to overcome the limitations of in situ conservation of orchids, there is a need to optimize existing propagation protocols under in vitro conditions to produce quality planting material of threatened, endangered, and rare species, and preserve them using available biotechnological tools. Efficient protocols have been developed for propagation of numerous species, such as endangered *Renantheraim schootiana* Rolfe (Seeni and Latha 1992; Wu et al. 2014), naturally overexploited *Dendrobium candidum* Wall *ex* Lindl (Zhao et al. 2008), rare *Dendrobuim wangliangii* Hu, Long and Jin, (Zhao et al. 2013), rare species of both medicinal and ornamental importance such as *Colelogyne cristata* Lindl., (Naing et al. 2011), *Eulophia nuda* Lindl. (Panwar et al. 2012), threatened species such as *Cymbidium aloifolium* (L.) Sw. (Prahdan et al. 2014), and *Dendrobium thyrsiflorum* Rchb. ex André (Battacharyya et al. 2015). Although considerable progress has been achieved recently in this important field of orchid protection and conservation approach, there is still a need to develop new methods for dealing with the problem, and improve known techniques of in vitro propagation.

## 13.2 The Modes of Propagation Under In Vitro Conditions Using Representatives of Orchidaceae

The plant cell and tissue culture has found to be applicable to some representatives of the Orchidaceae family since the development of a non-symbiotic seed germination method by Knudson (1946). Germination of tiny orchid seeds with an undifferentiated embryo consisting of a few hundred cells is quite a peculiar phenomenon in the plant kingdom. Because of the complexity of the long-lasting germination process, more attention is presently paid to the orchid's mycoheterotrophy than to any other plant group (Whigham et al. 2006; Sathiyadash et al. 2014; Khamchatra et al. 2016). Besides, attempts are being made to develop additional high technology tool for the establishment of successful conservation programmes of endangered species (Zettler and Hofer 1998; Swangmaneecharern et al. 2012; Ercole et al. 2013; Hossain et al. 2013; Fracchia et al. 2014).

The techniques that are currently in use should be optimized for each particular species or even subspecies under study. For instance, it should be brought to notice that during initiation of in vitro culture, rapid browning of explanted tissue, just after excision from the mother plant, is a frequent phenomenon. The oxidation of phenolic compounds, resulting from mechanical damage of tissue, accounts for the continuous browning of the most of orchid explants. Thus, inhibition of growth and decrease in regeneration ability of initiated cultures may be a side effect of oxidation of phenolic compounds (Tsai et al. 2004). This effect can be notably important in commercial production for horticultural purposes. Orchids are grown as valuable long-life pot plants or cut flowers. The stable interest of producers is mainly caused by elegant beauty of their flowers (Rudall et al. 2012). Plants are chiefly appreciated for their blossoms, and consequently the possibility to incite particular species to flowering phase under in vitro conditions has received a great deal of attention (Zhao et al. 2013; da Silva et al. 2014), hence, the process will require additional optimization.

As far as market production is concerned, the great progress was achieved in orchid production after it has been demonstrated that during a comparatively short time, producers can obtain millions of plantlets regenerated from meristem culture (or callus tissue) by repeatedly sectioning and subculturing so-called protocorm-like bodies (PLBs) (Chia et al. 1998; Begum et al. 1994; Saiprasad et al. 2004; Martin and Madsssery 2006; Zhao et al. 2008; Naing et al. 2011; Ng and Saleh 2011). Moreover, as PLBs can be successfully cryopreserved, it should also be emphasized that this approach has got results in preservation of existing orchid biodiversity (Ramah et al. 2015a, b).

Next question involving some doubt that might be important for market production is that orchids are usually out-breeders, and as a result of generative propagation, heterozygous plants are often obtained. That is why, in the case of *Cattleya, Cymbidium, Dendrobium, Oncidium, Phalenopsis, Paphiopedillum,* and many other genera, it is essential to elaborate effective in vitro protocols in order to obtain numerous regenerated plantlets (clones). Using this potent technique of in vitro propagation, which can be applied during the whole year, the above-cited genera with numerous species and cultivars have been regenerated from different types of explants, such as meristems, shoot tips, axiliary buds, leaves, inflorescence part, aerial roots, pseudobulbs, and even so-called thin cross sections (TCS) (Shimasaki and Uemoto 1990; Tokuhara and Mii 1993; Begum et al. 1994; Nayak et al. 1997, 1998, 2002; Bhadra and Hossain 2003; Martin and Madsssery 2006; Chugh et al. 2009; Mohanthy et al. 2012; Chen et al. 2015; da Silva et al. 2014; Bhattacharyya and Van Staden 2016; Khamchatra et al. 2016). This area has been routinely aided by scientific activity, and this is why the tissue culture technology is currently available for large-scale propagation of a great number of orchid species and their hybrids (Arditti 2008; Mohanthy et al. 2012; Bhattacharyya et al. 2016). Thus, numerous efficient protocols have been worked out to reinforce the economic potential of this important international market of valuable ornamentals.

Similarly, a specific mode of orchid propagation via protocorms has resulted in quite different requirements during artificial seed production technology, than it is usually observed for other ornamental plants (Prahdan et al. 2014). However, we should also be aware of the challenges that are being faced today in that domain, such as a need for extensive investigation of numerous factors controlling in vitro flowering or an elaboration of technology of artificial seed production from different source materials like protocorms or protocorm-like bodies (Prahdan et al. 2014; Lemay et al. 2015). Furthermore, we still need to make orchid propagation methods much more efficient, but at the same time, acceptable from ecological point of view. Possible solution could be the use of economically justified lighting sources. LED technology can be unquestionably useful in this respect.

## 13.3 The Role of Light in the In Vitro Propagation of Orchids

Vascular plants, as typical sessile organisms, are especially influenced by their environment. Light is one of the most significant factors that affect each photosynthetic plant, regulating its growth and development. Plants can sense irradiation intensity (quantity of light), its wavelength (quality), and light duration (photoperiod) (Batschauer 1999). These different aspects of light are as well of great importance during plant propagation under in vitro conditions. The photoperiod, photosynthetic flux density (PPFD), and spectral composition connected with the source of light (fluorescent, incandescent, metal halide, LED lighting) are enumerated as those vitally important among different factors regulating growth and morphogenetic response of plant tissue culture. In orchid tissue culture, light affects photosynthesis, morphogenetic responses, and biochemical composition of plantlets (Cybularz-Urban and Hanus-Fajerska 2008; Cybularz-Urban et al. 2007a, b, 2015; Swiderski et al. 2007). Orchids are frequently illuminated during most phases of in vitro production. However, there are also instances in which cultures must be kept under darkness. The transfer from dark to light treatment should be gradual. The selection of the appropriate artificial light treatment to orchid illumination is very interesting area of studies (Dutta Gupta and Jatothu 2013). Among various lighting sources, the most popular in orchid cultures are white fluorescence lamps (FL) which have a wide range of wavelengths (350–750 nm). However, maxima of FL emission spectrum are not matching the light requirements of plants. These needs are designated by the light absorption spectra specific for plant photoreceptors such as chlorophylls and carotenoids. In this instance, important parameters for the intensity of photosynthesis are the level of photosynthetic pigments and the density of applied photons flux (or irradiation intensity) in the optimal wavelength ranges (430–450 and 640–660 nm). Therefore, the most promising source of light for plants seems to be light-emitting diodes (LEDs) especially red and blue ones. It is worth mentioning that one of the first applications of LED in horticulture was connected with small plant cultivations such as in vitro cultures (Kim et al. 2004). Currently, LED lamps are increasingly applied to manipulate organogenesis and morphogenesis of various species propagated in tissue culture (Kurilčik et al. 2008; Dutta Gupta and Jatothu 2013; Li et al. 2013; Habiba et al. 2014a).

## 13.4 Application of LED Light in Improving Tissue Culture of Orchids

One of the important attributes of LED is the light emission in a very narrow spectrum of wavelengths (Morrow 2008). Modern LED systems for plants enable a dynamic control of spectral composition of emitted light. In some of them, it is possible to mix a number of wavebands, from violet, blue, green, orange, red, to far-red light (Yano and Fujiwara 2012). Both spectrum and photon flux density can be adjusted to physiological and morphological needs of plants, therefore the LED light may be used in a more efficient manner in comparison with fluorescent or sodium lamps. Moreover, the unique feature of diode is lack of the generation of heat. This allows locating plant organs close to the LED light without the risk of damage caused by heat stress. These are the main reasons why light-emitting diodes are successfully used in tissue culture of many plant species, including orchids (Dutta Gupta and Jatothu 2013). Due to the advancement of LED technology and increasing knowledge about the impact of light quality on plants, LED systems specific for orchids' tissue culture were also used (Pan et al. 2015).

Many reports have shown that plant's response to definite spectral composition of light is species specific and may vary depending on the cultivation method and other growth factors (Singh et al. 2015; Ouzounis et al. 2015). When study is conducted in the greenhouse environment, plants receive the amount of natural light which is supplemented with artificial light most commonly up to 16 h per day at intensity from 100 to 200 $\mu$mol $m^{-2}$ $s^{-1}$. Knowing that weather conditions are not repeatable, the results of such experiments in successive years may vary

(Wojciechowska et al. 2015). However, what is specific for in vitro cultures is a strictly controlled environment. In such conditions, the only radiation sources are lamps emitting selected type of light (most frequently at photoperiod 16 h per day and intensity about 50 μmol $m^{-2}$ $s^{-1}$). For this reason, the results dependent on light might to be more predictable than those obtained in greenhouse. Many satisfactory effects were obtained just with the use of LEDs solely. The aim of a number of studies on orchid propagation was to investigate the effects of monochromatic and/or mixed LED light on morphogenesis features such as protocorm-like bodies production, axillary shoot formation, shoot elongation, rhizogenesis, and also chemical composition of plantlets (Table 13.1). The main conclusion of reports cited in Table 13.1 reveals that LED lamps might successfully replace fluorescent lamps that are usually exploited in orchid propagation due to the

**Table 13.1** Current status on the in vitro propagation of orchids under LED lighting

| Species/cultivar | LED treatments | Lighting conditions (PPFD, photoperiod); control treatment | Remarks | References |
|---|---|---|---|---|
| *Bletilla ochracea* | (a) Blue (470 nm); (b) green (525 nm); (c) orange (590 nm); (d) red (625 nm); (e) white (460 nm and 560 nm) | 40 μmol $m^{-2}$ $s^{-1}$, 24 h light/0 h dark (continuous light); darkness | Green and orange light yielded the highest frequencies of seed germination. Orange and red light effectively stimulated rhizogenesis. Leave width was bigger under white and blue LED light | Godo et al. (2011) |
| *Calanthe* (two hybrids) | (a) Red (660 nm) and blue (450 nm); 0.7:1 (b) red (660 nm) and far-red (730 nm); 1:1.1 | 20 μmol $m^{-2}$ $s^{-1}$, 16 h light/8 h dark; FL (fluorescent white) | Red + blue LEDs enhanced in vitro growth of plantlets, while red + far-red had an inhibitory effect | Baque et al. (2011) |
| *Cymbidium* '*Golden Bird*' | Red (660 nm) and blue (450 nm) in ratio: (a) 1:0; (b) 0:1; (c) 1:1 | 45 μmol $m^{-2}$ $s^{-1}$, 16 h light/8 h dark; FL (40 μmol $m^{-2}$ $s^{-1}$ PPFD) | Red light (1:0) stimulated leaves' growth (especially the length of leaves), and blue (0:1) promoted chlorophylls content | Tanaka et al. (1998) |

(continued)

**Table 13.1** (continued)

| Species/cultivar | LED treatments | Lighting conditions (PPFD, photoperiod); control treatment | Remarks | References |
|---|---|---|---|---|
| *Cymbidium* 'Idol' | (a) Red (640 nm); (b) blue (450 nm), (c) green (510 nm); (d) FL and green; (e) red and green; (f) blue and green; (d–f): one day under green light in seven day cycle) | 50 μmol $m^{-2}$ $s^{-1}$, 16 h light/8 h dark; FL | Red with green LEDs promoted the PLBs formation, whereas fluorescent light with green LEDs stimulated the rate of shoot formation (93.3%) | Kaewjampa and Shimasaki (2012) |
| *Cymbidium* (nine hybrid cultivars) | Red (660 nm) and blue (450 nm) in ratio: (a) 1:0; (b) 1.5:1; (c) 1:1; (d)1:1.5; (e) 0:1 | 45 μmol $m^{-2}$ $s^{-1}$, 16 h light/8 h dark; FL | Monochromatic red LEDs or blue ones had negative effect on PLBs proliferation, but mixed red with blue in the ratio 1:1.5 favored PLBs proliferation more than FL (PLBs were of poor quality) | (da Silva 2014) |
| *Cymbidium dayanum* and *Cymbidium finlayso-nianum* | (a) Red (640 nm); (*b*) blue (450 nm), (c) green (510 nm) | 50 μmol $m^{-2}$ $s^{-1}$, 16 h light/8 h dark; FL | Green and blue light enhanced PLBs production | Nahar et al. (2016) |
| *Dendrobium officinale* | Red (660 nm) and blue (450 nm) in ratio: (a) 1:0; (b) 0:1; (c) 1:1; (d) 2:1; (e) 1:2 | 70 μmol $m^{-2}$ $s^{-1}$, 16 h light/8 h dark; FL and darkness | Monochromatic blue and mixed red to blue in the ratio 1:2 promoted PLBs formation and increased dry matter content of PLBs | Lin et al. (2011) |
| *Dendrobium kingianum* | (a) Red (640 nm); (b) blue (450 nm); (c) green (510 nm); (d) white (broad spectrum); (e) red and blue; 1:1 | 50 μmol $m^{-2}$ $s^{-1}$, 16 h light/8 h dark; FL | Monochromatic blue light, in comparison with red, increased chlorophyll content compared to red. The rate of shoot formation was promoted by red LEDs | Habiba et al. (2014b) |

(continued)

**Table 13.1** (continued)

| Species/cultivar | LED treatments | Lighting conditions (PPFD, photoperiod); control treatment | Remarks | References |
|---|---|---|---|---|
| *Doritaenopsis* | (a) Red (660 nm); (b) blue (450 nm); (c) red and blue; 1:1 | 70 μmol $m^{-2}$ $s^{-1}$, 16 h light/8 h dark; FL | Red light mixed with blue affected leaf and root fresh/dry weight, as well as leaf photosynthetic pigments and carbohydrate contents | Shin et al. (2008) |
| *Oncidium 'Gower Ramsey'* | Red—R (660 nm), blue—B (455 nm), and far-red-Fr (730 nm) mixed or monochromatic: (a) RBFr; (b) RFr; (c) BFr; (d) RB; (e) R; (f) B | 50 μmol $m^{-2}$ $s^{-1}$, 16 h light/8 h dark; FL | Far-red light in mixture with blue and red or with red LEDs significantly increased leaf expansion, the number of leaves and roots, chlorophyll content and fresh/dry weight of plantlets | Chung et al. (2010) |
| *Oncidium* | Blue (460 nm), red (660 nm), yellow (590 nm), green (530 nm), and far-red (715 nm): monochromatic used in PLBs induction and mixed in rooting | Monochromatic light (11 μmol $m^{-2}$ $s^{-1}$ for PLB induction, proliferation and differentiation) with mixed LED light (50 μmol $m^{-2}$ $s^{-1}$ for study of plantlets rooting), 16 h light/8 h dark | Red light enhanced induction and proliferation of PLBs, and lengths of regenerated plantlets. Blue spectrum promoted differentiation and increased antioxidant enzyme activities in PLBs | Mengxi et al. (2011) |
| *Oreorchis patens* | (a) Red (660 nm) (b) blue (450 nm) | PPFD—no data, 16 h light/8 h dark; FL | Red light promoted swelling of embryo, and protocorm formation | Bae et al. (2014) |
| *Paphiopedilum delenatii* Guillaumin | Red (660 nm) and blue (450 nm) in ratios: (a) 1:0; | PPFD 30 μmol $m^{-2}$ $s^{-1}$, | Blue LED light (100%) affected the highest shoot | Luan et al. (2015) |

(continued)

**Table 13.1** (continued)

| Species/cultivar | LED treatments | Lighting conditions (PPFD, photoperiod); control treatment | Remarks | References |
|---|---|---|---|---|
| | (b) 0:1; (c) 9:1; (d) 1:1 | 24 h photoperiod; darkness | elongation and number of nodes after 4 months of culture | |
| *Phalenopsis 'Cassandra Rose'* | Red (640–700 nm) and blue (450–480 nm) in ratios: (a) 1:0; (b) 9:1; (c) 4:1; (d) red with white (blue + yellow 585–600 nm) in ratio 1:1 (red: white) | PPFD and photoperiod—no data; FL | The best protocorm development was under red and blue LEDs in the ratio 4:1 and 9:1. The ratios 1:0; 9:1 and red + white LEDs were better than FL in stimulating growth of the plantlets | Wongnok et al. (2008) |
| *Phalenopsis* | (a) Red; (b) blue; (c) red and blue; 1:1; (d) white: used in study of (1) vegetative growth (in growth chamber) and (2) flower initiation (in the greenhouse) | PPFD—no data; 8/16 h (light/dark); natural light | Blue, red + blue (1:1) and white LEDs enhanced seedlings growth and increased chlorophyll content. During 3 months of greenhouse vegetation, young orchids' flowering was not forced | Dewi et al. (2014) |

proper regulation of emitted spectrum. The effects highlighted are strictly connected with the species/cultivar of orchids, its development phase, and spectral characteristics of tested LED light.

Red light is mainly involved in vegetative growth. It has been observed that the use of monochromatic red LED light enhanced fresh and dry matter of *Phalenopsis* plantlets (Wongnok et al. 2008; Hsu and Chen 2010), the rate of shoot formation of *Dendrobium kingianum* (Habiba et al. 2014b), and the length of *Cymbidium* leaves (Tanaka et al. 1998). However, such effects were usually accompanied by a decrease in chlorophylls content which, in turn, was stimulated by blue LED light (Wongnok et al. 2008; Habiba et al. 2014b). Other authors have obtained the best results with the use of mixed red with blue light, for example, in equal proportions in *Doritaneopsis* propagation (Shin et al. 2008). It is interesting that low intensity of blue LED light resulted in higher shoot elongation in ex vitro potted *Paphiopedilum delenatii* when compared to red or red + blue LED light treatments (Luan et al. 2015).

As shown in Table 13.1, attempts have been made to find such spectrum of light that would be the most suitable for micropropagation of various cultivars of orchids. In cited works, the authors try to discuss light-induced responses of tested plants but the mechanisms of these reactions are not fully recognized. It is worth emphasizing that the knowledge about plant photoreceptors which detect the quality, quantity, and direction of light is constantly expanding. Blue and UV-A wavelength ranges are absorbed by identified three forms of cryptochromes, two phototropins, and ZTL/ADO family (associated with circadian clock and flowering) (Devlin et al. 2007). Plants are able to detect red and far-red light via specific forms of phytochromes, but the mechanism of its action is still not fully recognized (Batschauer 1999). Green light (500–550 nm) appears necessary in regulatory or signaling function (Folta and Marunich 2007). According to recent studies, orchid micropropagation might be significantly enhanced by the use of green LED light. For instance, it stimulates seed germination of *Bletilla ochracea* (Godo et al. 2011) and PLBs production or organogenesis of two *Cymbidium* cultivars (Nahar et al. 2016). Moreover, green LED light in an interesting combination with red LEDs or with fluorescent light may significantly improve regeneration of *Cymbidium* Waltz 'Idol' (Kaewjampa and Shimasaki 2012). On the other hand, monochromatic red or blue LED light had detrimental effect on PLBs production of nine hybrid *Cymbidium* cultivars (da Silva 2014). Fascinating was the use of far-red in the specific treatment with red light to stimulate somatic embryogenesis in *Doritaneopsis* (Park et al. 2010). Exceptional results were also obtained in *Oncidium* cultures, in which a mixture of two combinations, far-red with red + blue and far-red with red LED light, clearly increased the quality of plantlets (Chung et al. 2010). However, growth of *Calanthe* plantlets was inhibited under mixed far-red with red LED light (Baque et al. 2011). It has been observed that the response of orchids to far-red light seems to be dependent on the species and proportions of FR to other wavelengths.

It is interesting to know, whether the positive effects obtained under specific LED light during in vitro propagation of orchid are maintained in ex vitro conditions. After transferring to quite different environment, delicate plantlets were easily damaged and grew very slowly. Due to the changes in intensity and quality of light, humidity, $CO_2$ concentration etc., acclimatization of plantlets requires significant rebuilding of some anatomical features and adjustment of physiological processes to new conditions. An increase in the antioxidant enzymes activities was noted with increasing light intensity (60, 160, 300 μmol $m^{-2}$ $s^{-1}$, PHILIPS SON-T lamps) in the leaves of micropropagated *Phalenopsis* during acclimatization (Ali et al. 2005). However, the issue of which spectrum and intensity of LED light in ex vitro conditions might improve the quality of young orchids and accelerate the flowering is still not clarified. Recently, Dewi et al. (2014) reported that in comparison with natural light, blue, red-blue (1:1), and white LEDs enhanced vegetative growth of *Phalenopsis*. The authors suggest that in the case of inducing the first flowering of orchid, the most probable to act as a trigger for this process is white or blue LED light. We are witnessing an increasing number of intriguing LED light applications during in vitro propagation of orchids. However, further experiments are necessary with the dynamic control of LED lighting to obtain the best results connected with growth and first flowering of orchids.

## 13.5 Conclusions

In recent years, considerable advances have been made in orchid plant tissue culture when LEDs are applied as the sole source of lighting. Compared to fluorescent light, LED lighting improves the quality of orchid plantlets obtained via tissue culture technology. Nevertheless, it is crucial to adjust spectral composition of LEDs accurately as per with the cultured species and this issue is still open for further research. An emerging question is how to exploit LED technology during acclimatization and ex vitro transfer for better survival of regenerated plantlets. Moreover, the effectiveness of application of LED lighting to accelerate flowering in orchid plantlets is another major concern. This issue deserves special attention in the near future.

## References

Ali M, Hahn EJ, Peak HY (2005) Effects of light intensities on antioxidant enzymes and malondialdehyde content during short-term acclimatization on micropropagated *Phalenopsis* plantlet. Environ Exp Bot 54:109–120

Arditti J (2008) Micropropagation of orchids. Blackwell Publ, USA. ISBN-13:978-1-4058–6088-9

Bae KH, Oh KH, Kim SY (2014) Sodium hypochlorite treatment and light-emitting diode (LED) irradiation effect on in vitro germination of *Oreorchis patens* (Lindl.) Lindl. J Plant Biotech 41:44–49

Ballantyne M, Pickering CM (2015) The impacts of trail infrastructure on vegetation and soils: current literature and future directions. J Environ Man 164:53–64

Baque AM, Shin YK, Elshmari T, Lee EJ, Paek KY (2011) Effect of light quality, sucrose and coconut water concentration on the microporpagation of *Calanthe* hybrids ('Bukduseong' × 'Hyesung' and 'Chunkwang' × 'Hyesung'). Aust J Crop Sci 5(10):1247–1254

Bateman RM (2011) Two steps forward, one step back: deciphering British and irish marsch-orchids. Watsonia 14:347–376

Batschauer A (1999) Light perception in higher plants. Cell Mol Life Sci 55(2):153–166

Battacharyya P, Kumaria S, Job N, Tandon P (2015) Phyto-molecular profiling and assessment of antioxidant activity within micropropagated plants of *Dendrobium thyrsiflorum*: a threatened, medicinal orchid. Plant Cell Tissue Organ Cult 122:535–550

Begum AA, Tamaki M, Kako S (1994) Formation of protocorm-like bodies (PLBs) and shoot development through in vitro culture of outer tissue of *Cymbidium* PLB. J. Jpn Hortic Sci 63(3):663–637

Bhadra SK, Hossain MM (2003) In vitro germination and micropropagation of *Geodorum densiflorum* (Lam.) Schltr., an endangered orchid species. Plant Tissue Cult 13(2):165–171

Bhattacharyya P, Van Staden J (2016) *Ansellia africana* (Leopard orchid): a medicinal orchid species with untapped reserves of important biomolecules—a mini review. S Afr J Bot 106:181–185

Bhattacharyya P, Kumaria S, Tandon P (2016) High frequency regeneration protocol for *Dendrobium nobile*: a model tissue culture approach for propagation of medicinally important orchid species. S Afr J Bot 104:232–243
Chen Y, Goodale UM, Fan XL, Gao JY (2015) Asymbiotic seed germination and in vitro seedling development of *Paphiopedilum spicerianum*: an orchid with an extremely small population in China. Global Ecol Conserv 3:367–378
Chia TF, Hew CS, Loh CS, Lee YK (1998) Carbon/nitrogen ratio and greening and protocorm formation in orchid callus tissue. HortScience 13:599–601
Chugh S, Satyakam G, Rao IU (2009) Micropropagation of orchids: a review on the potential of different explants. Sci Hortic 122:507–520
Chung JP, Huang CY, Dai TE (2010) Spectral effects on embryogenesis and plantlet growth of *Oncidium* 'Gower Ramsey'. Sci Hortic 124:511–516
Cybularz-Urban T, Hanus-Fajerska E (2008) The morphogenetic capability and the viability of regenerants in micropropagated orchid hybrids infected with viral pathogens. Folia Hortic 20 (2):93–102
Cybularz-Urban T, Hanus-Fajerska E, Swiderski A (2007a) Effect of wavelength on in vitro organogenesis of *Cattleya* hybrid. Act Biol Cracoc, Series Bot 49(1):113–118
Cybularz-Urban T, Hanus-Fajerska E, Swiderski A (2007b) Preliminary morphological, anatomical and biochemical characteristic of micropropagated *Cattleya* under UV-A and white light illumination. Zesz Probl Post Nauk Roln 523:59–67
Cybularz-Urban T, Hanus-Fajerska E, Bach A (2015) Callus induction and organogenesis in vitro of *Cattleya* protocorm-like bodies (PLBs) under different light conditions. Act Sci Pol, Seria Hort Cult Hortic 14(6):19–28
da Silva JAT (2014) The response of protocorm-like bodies of nine hybrid *Cymbidium* cultivars to light-emitting diodes. Environ Exp Biol 12:155–159
da Silva JAT, Zeng S, Cardoso J, Dobránszki J, Kerbauy GB (2014) In vitro flowering of *Dendrobium*. Plant Cell Tissue Organ Cult 119:447–456
de la Rosa-Manzano E, Andarade JL, Zotz G, Reyes-García C (2014) Epiphytic orchids in tropical dry forests in Yucatan, Mexico–species occurrence, abundance and correlations with host tree characteristics and environmental conditions. Flora 209:100–019
de la Rosa-Manzano E, Andarade JL, García-Mendosa E, Zotz G, Reyes-García C (2015) Photoprotection related to xantophyll cycle pigments in epiphytic orchids acclimated at different microevironments in two tropical dry forests of the Yucatan Penisula, Mexico. Planta 242:1425–1438
Devlin PF, Christie JM, Terry MJ (2007) Many hands make light work. J Exp Bot 58(12):3071–3077
Dewi K, Purwestri YA, Astuti YTM, Natasaputra L, Parmi (2014) Effects of light quality on vegetative growth and flower initiation in *Phalaenopsis*. Indonesian J Biotech 19(1):33–42
Dutta Gupta S, Jatothu B (2013) Fundamentals and applications of light-emitting diodes (LEDs) in in vitro plant growth and morphogenesis. Plant Biotech Rep 7:211–220
Ercole E, Rodda M, Molinatti M, Voyron S, Perotto S, Girlanda M (2013) Cryopreservation of orchid mycorrhizal fungi: a tool for conservation of endangered species. J Micr Meth 93:134–137
Fay MF (2015) British and Irish orchids in a changing world. Curtis's Bot Mag 2(1):3–23
Folta KM, Maruhnich SA (2007) Green light: a signal to slow down or stop. J Exp Bot 58:3099–3111
Fracchia S, Aranda-Rickert A, Flachsland E, Terada G, Sede S (2014) Mycorrhizal compatibility and symbiotic reproduction of *Gaviela australis*, an endangered terrestial orchid from south Patagonia. Mycorrhiza 24:627–634
Freudenstein JV, Chase MW (2015) Phylogenetic relationships in Epidendroideae (Orchideaceae), one of the great flowering plant radiations: progressive specialization and diversification. Ann Bot 115:665–681
Godo T, Fujiwara K, Guan K, Miyoshi K (2011) Effects of wavelength of LED-light on in vitro asymbiotic germination and seedling growth of *Bletilla ochracea* Schltr. (Orchidaceae). Plant Biotech 28:397–400

Gogoi K (2005) The genus *Dendrobium* in Dibru-Saikhowa National Park and Biosphere Reserve. J Orchid Soc India 19(1/2):17–25

Gogoi K, Borah RL, Shrama GC (2010) Orchid flora of Dibru-Saikhowa National Park and Biosphere Reserve, Assam India. Pleione 4(1):124–134

Habiba SU, Shimasaki K, Ahasan MM, Alam MM (2014a) Effect of 6-benzylaminopurine (BA) and hyaluronic acid (HA) under white light emitting diode (LED) on organogenesis in protocorm-likebodies (PLBs) of *Dendrobium kingianum*. American-Eurasian J Agric Environ Sci 14(7):605–609

Habiba SU, Shimasaki K, Ahasan MM, Alam MM (2014b) Effects of different light quality on growth and development of protocorm-like bodies (PLBs) in *Dendrobium kingianum* cultured in vitro. Bangladesh Res Public J 10(2):223–227

Hossain MM, Rahi P, Gulati A, Shrama M (2013) Improved ex vitro survival of asymbiotically raised seedlings of *Cymbidium* using mycorrhizal fungi isolated from distant orchid taxa. Sci Hortic 159:109–112

Hsu HC, Chen C (2010) The effect of light spectrum on the growth characteristics of in vitro cultures of *Phalaenopsis*. Prop Ornam Plants 10(1):3–8

Jakubowska-Gabara J, Kurzac M, Kiedrzyński M, Kopeć D, Kucharski L, Kołodziejek J, Niedźwiedzki P, Popkiewicz P, Witosławski P, Zielińska K (2012) New stations of rare, protected and threatened species of vascular plants in Central Poland. Part II Fragm Florist Geobot 19(2):349–359

Kaewjampa N, Shimasaki K (2012) Effects of green LED lighting on organogenesis and superoxide dismutase (SOD) activities in protocorm-like bodies (PLBs) of *Cymbidium* cultured in vitro. Environ Control Biol 50(3):247–254

Khamchatra N, Dixon KW, Taniwiwat S, Piapukiew J (2016) Symbiotic seed germination of an endangered epiphytic slipper orchid, *Paphiopedillum villosum* (Lindl.) Stein. from Thailand. S Afr J Bot 104:76–81

Kim SJ, Hahn EJ, Heo JW, Paek KY (2004) Effects of LEDs on net photosynthetic rate, growth and leaf stomata of chrysanthemum plantlets in vitro. Sci Hortic 101:143–151

Knudson L (1946) A new nutrient medium for germination of orchid seeds. Am Orchid Soc Bull 39:214–217

Křenová Z, Kidlmann P (2015) Natura 2000—solution for Eastern Europe or just a good start? The Šumava National Park as a test case. Biol Conserv 186:287–275

Kull T, Selgis U, Peciῆa MV, Metsare M, Ilves A, Tali K, Sepp K, Kull K, Shefferson RP (2016) Factors influencing threat levels to orchids across Europe on the basis of national red lists. Ecol Evol 6(17):6245–6265

Kurilčik A, Dapkūniene S, Kurličik G, Žilinskaite S, Žukauskas A, Duchovskis P (2008) Effect of photoperiod duration on the growth of *Chrysanthemum* plantlets in vitro. Sci Works Lith Inst Hort Lith Univ Agric Sodininkyste Ir Darzyninkyste 27(2):39–43

Lemay MA, De Vriendt L, Pellerin S, Poulin M (2015) Ex situ germination as a method for seed viability assessment in a peatland orchid *Palanthera blephariglottis*. Am J Bot 102(3):390–395

Li H, Tang C, Xu Z (2013) The effects of different qualities on rapeseed (*Brassica napus* L.) plantlet growth and morphogenesis in vitro. Sci Hortic 150:117–124

Lin Y, Li J, Li B, He T, Chun Z (2011) Effects of light quality on growth and development of protocorm-like bodies of *Dendrobium officinale* in vitro. Plant Cell Tissue Organ Cult 105:329–335

Luan YQ, Huy NP, Nam NB, Huong TT, Hien FT, Hien NTT, Hai NT, Thinh DK, Nhut DT (2015) Ex vitro and in vitro *Paphiopedilum delenatii* Guillaumin stem elongation under light-emitting diodes and shoot regeneration via stem node culture. Acta Physiol Plant 37. doi:10.1007/s11738-015-1886-8

Martin KP, Madsssery J (2006) Rapid in vitro propagation of *Dendrobium* hybrids through direct shoot formation from foliar explants, and protocorm-like bodies. Sci Hortic 108:95–99

Mengxi L, Zhigang X, Yang Y, Yijie F (2011) Effects of different spectral lights on *Oncidium* PLBs induction, proliferation, and plant regeneration. Plant Cell Tiss Organ Cult 106:1–10
Mohanthy P, Paul S, Das MC, Kumaria S, Tandon P (2012) A simple and efficient protocol for the mass propagation of *Cymbidium maestrii*: an ornamental orchid of Northeast India. AoB Plants pls023. doi:10.1093/aobpl/pls023
Morrow RC (2008) LED lighting in horticulture. HortScience 43(7):1947–1950
Müller R, Nowicki C, Barthlott W, Ibish PL (2003) Biodiversity and endemism maping as a tool for regional conservation planning—case study of the Pleurothallidinae (Orchidaceae) of the Andean rain forests in Bolivia. Biodiv Conserv 12:2005–20024
Nahar SJ, Haque SM, Kazuhiko S (2016) Application of chondroitin sulfate on organogenesis of two *Cymbidium* spp. under different sources of lights. Not Sci Biol 8(2):156–160
Naing AH, Chung JD, Park IS, Lim KB (2011) Efficient plant regeneration of the endangered medicinal orchid, *Colelogyne cristata* using protocorm-like bodies. Acta Physiol Plant 33:659–666
Nayak NR, Rath SP, Patnaik S (1997) In vitro propagation of three epiphytic orchids, *Cymbidium aloifolium* (L.) Sw., *Dendrobium aphyllum* (Roxb.) Fisch. and *Dendrobium moschatum* (Buch-Ham) Sw. through thidiazuron-induced high frequency shoot proliferation. Sci Hortic 71:243–250
Nayak NR, Chand PK, Rath SP, Patnaik SN (1998) Influence of some plant growth regulators on the growth and organogenesis of *Cymbidium aloifolium* (L.) Sw. seed-derived rhizomes in vitro. In Vitro Cell Dev Biol Plant 34:185–188
Nayak NR, Sahoo S, Patnaik S, Rath SP (2002) Establisment of thin cross section culture (TCS) culture method for rapid micropropagation of *Cymbidium aloifolium* (L.) Sw and *Dendrobium nobile* Lind. (Orchidaceae). Sci Hortic 94:107–116
Ng ChY, Saleh NM (2011) In vitro propagation of *Paphiopedilum* orchid through formation of protocorm-like bodies. Plant Cell Tissue Organ Cult 105:193–202
Nordström S, Hedrén M (2009) Evolution, phylogeography and taxonomy of allopolyploid *Dactylorhiza* (Orchidaceae) and its implications for conservation. Nordic J Bot 27:458–556
Ouzounis T, Rosenqvist E, Ottosen CO (2015) Spectral effects of artificial light on plant physiology and secondary metabolism: a review. HortScience 50(8):1128–1135
Palomo I, Martīn-López B, Potschin M, Hainez-Young H, Montes C (2013) National parks, buffer zones and surrounding lands: mapping ecosystem service flows. Ecosyst Serv 4:104–116
Pan X, Zhou H, Li X, Wang W, Huang H (2015) LED mixed lighting for tissue culture of orchids. United States Patent No.: US 8,944,631 B2, 3 Feb 2015
Panwar D, Ram K, Shekhwat NS (2012) In vitro propagation of *Eulophia nuda* Lindl., an endangered orchid. Sci Hortic 139:46–52
Park SY, Yeung EC, Paek KY (2010) Endoreduplication in *Phalaenopsis* is affected by light quality from light-emitting diodes during somatic embryogenesis. Plant Biotech Rep 4:303–309
Pindel A, Pindel Z (2004) Initiation of in vitro cultures of chosen endangered species of orchids. Folia Hortic 16(2):111–117
Prahdan S, Tiwura B, Subedee BR, Pant B (2014) In vitro germination and propagation of a threatened medicinal orchid, *Cymbidium aloifolium* (L.) Sw. through artificial seed. Asian Pac J Trop Biomed 4(12):971–976
Ramah S, Mubbarakh SA, Sinniah UR, Subramaniam S (2015a) Effect of droplet-vitrification on *Brassidium* Shooting Star's orchid protocorm-like bodies (PLBs). Sci Hortic 197:254–260
Ramah S, Mubbarakh SA, Ping KS, Subramaniam S (2015b) Effects of dropled-vitrification cryopreservation based on physiological and antioxidant enzyme activities of *Brassidium* shooting star orchid. Sci World J. doi:10.1155/2015/961793
Rewicz A, Zielińska K, Kiedrzyński M, Kucharski L (2015) Orchidaceae in the antropogenic landscape of central Poland: diversity, extinction and conservation perspectives. Arch Biol Sci 67(1):119–130
Rudall PJ, Perl CD, Bateman RM (2012) Organ homologies in orchid flowers re-interpreted using the Musk Orchid as a model. PeerJ 1:e26. doi:10.7717/peerj.26

Saiprasad GVS, Raghuveer P, Khetarpal S, Chandra R (2004) Effect of various polyamines on production of protocorm-like bodies in orchid—*Dendrobium* 'Sonia'. Sci Hort 100:161–168
Sathiyadash K, Muthukumar T, Murgan SB, Sathishkumar R, Pandey RR (2014) In vitro symbiotic seed germination of South Indian endemic orchid *Coelogyne nervosa*. Mycoscience 55:183–189
Seaton PT, Pritchard HW, Marks TR (2015) Aspects of orchid conservation: seed and polen storage and their value in re-introduction projects. Univ J Plant Sci 3(4):72–76
Seeni S, Latha PG (1992) Foliar regeneration of the endangered Red Vanda, *Renanthera imschootiana* Rolfe (Orchidaceae). Plant Cell Tissue Organ Cult 29:167–172
Sheshukova L, Klimenko E, Miryugina T, Olshetyn A, Vychuzhanina A (2014) Ecotourism in Western Siberia: issues and topical solutions. Middle-East J Sci Res 19(1):105–109
Shimasaki K, Uemoto S (1990) Micropropagation of terrestial *Cymbidium* species using rhizomes developed from seeds and pseudobulbs. Plant Cell Tissue Organ Cult 22:237–244
Shin KS, Murthy HN, Heo JW, Hahn EJ, Paek KY (2008) The effect of light quality on the growth and development of in vitro cultured *Doritaenopsis* plants. Acta Physiol Plant 30:339–343
Singh D, Basu C, Meinhardt-Wollweber M, Roth B (2015) LEDs for energy efficient greenhouse lighting. Renew Sustain Energ Rev 49:139–147
Steward J, Griffiths M (1995) Manual of orchids. Timber Press, New York
Swangmaneecharern P, Serivichyaswat P, Nontachaiyapoom S (2012) Promoting effect of orchid mycorrhizal fungi *Epulorhiza* isolates on seed germination of *Dendrobium* orchids. Sci Hortic 148:55–58
Swiderski A, Cybularz-Urban T, Hanus-Fajerska E (2007) The influence of UV radiation on the level of phenolic compounds in *Cattleya* leaves. Acta Physiol Plant 29(1 suppl.):115
Tanaka M, Takamura T, Watanabe H, Endo M, Yanagi T, Okamoto K (1998) In vitro growth of *Cymbidium* plantlets cultured under superbright red and blue light-emitting diodes (LEDs). J Hortic Sci Biotech 73:39–44
Tokuhara K, Mii M (1993) Micropropagation of *Phalenopsis* and *Doritaneopsis* by culturing shoot tips of flower stalk buds. Plant Cell Rep 13:7–11
Tsai CC, Peng CI, Huang SC, Pl Huang, Chou Ch (2004) Determination of genetic relationship of *Dendrobium* species (Orchidaceae) in Taiwan based on the sequence of the internal transcribed spacer of ribosomal DNA. Sci Hortic 101:315–325
Vidal OJ, san Martin C, Mardones S, Bauk V, Vidal CF (2012) The orchids of Torres del Paine biosphere reserve: the need for species monitoring and ecotourism planning for biodiversity conservation. Gayana Bot 69(1):136–146
Whigham DF, O'Neil JP, Rasmussen HN, Caldwell BA, McCormick MK (2006) Seed longevity in terrestial orchids—potential for persistent in situ seed banks. Biol Conserv 129:24–30
Wojciechowska R, Długosz-Grochowska O, Kołton A, Żupnik M (2015) The effects of LED supplemental lighting on yield and some quality parameters of lamb's lettuce in two winter cycles. Sci Hortic 187:80–86
Wongnok A, Piluek C, Techasilpitak T, Tantivivat S (2008) Effects of light emitting diodes on micropropagation of *Phalaenopsis* orchids. Acta Hortic 788:149–156
Wu K, Zeng S, Lin D, da Silva JAT, Bu Z, Zhang J, Duan J (2014) In vitro propagation and reintroduction of the endangered *Renanthera imschootiana* Rolfe. PLoS One 9(10):e110033. doi:10.1371/journal.pone0110033
Yano A, Fujiwara K (2012) Plant lighting system with five wavelength-band light-emitting diodes providing photon flux density and mixing ratio control. Plant Meth 8:46
Zang Z, Yan Y, Li J, He JS, Tang Z (2015) Distribution and conservation of orchid species richness in China. Biol Conserv 181:64–72
Zettler LW, Hofer CJ (1998) Propagation of the little club-spur orchid (*Palanthera clavellata*) by symbiotic seed germination and its ecological implications. Environ Exp Bot 39:189–195
Zhao P, Wu F, Feng FS, Wang WJ (2008) Protocorm-like body (PLB) formation and plant regeneration from the callus culture of *Dendrombium candidum* Wall *ex* Lindl. In Vitro Cell Dev Biol Plant 44:178185

Zhao D, Hu G, Chen Z, Shi Y, Zheng L, Tang A, Long C (2013) Micropropagation and in vitro flowering of *Dendrobuim wangliangii*: a critically endangered medicinal orchid. J Med Plant Res 7(28):2098–2110
Zotz G (2013) The systematic distribution of vascular epiphytes—a critical update. Bot J Linn Soc 171:453–481
Zotz G, Winkler U (2013) Aerial roots of epiphytic orchids: the velamen radicum and its role in water and nutrient uptake. Oecologia 171:733–741

# Chapter 14
# LEDs and Their Potential in Somatic Embryogenesis of *Panax vietnamensis* Ha *et* Grushv.

**Duong Tan Nhut, Nguyen Phuc Huy, Hoang Thanh Tung, Vu Quoc Luan and Nguyen Ba Nam**

## 14.1 Introduction

Ngoc Linh ginseng, which is known worldwide for its scientific name *Panax vietnamensis* Ha *et* Grushv., is an endemic species of Vietnam. It has high content of triterpenoid saponin (among the ginseng species that have the highest dammarane of MR2, Rg1 and Rb1, 12–15%), and low level of oleanolic acid. MR2, which accounts for almost 50% of total saponin content in Vietnamese ginseng, plays a major role in the antidepressant, antistress, and memory-improving effect of this plant (Dong et al. 2007). Research on the subject has been mostly restricted to saponin content analysis and pharmacology effects. Nhut et al. (2006) investigated the influence of different media on callus, shoot, and adventitious root multiplication and examined the saponin content of *P. vietnamensis*.

The effect of light irradiation on plant cell and tissue growth and secondary metabolite biosynthesis has been reported in many instances. Compared with traditional fluorescent lamp (FL), LED lamps showed to have more advantages and could be used as a new lighting source in micropropagation (Bula et al. 1991). However, little attention has been paid to *P. vietnamensis* cultures using LEDs, and no research has been conducted to examine the role of yellow (Y), green (G), and white (W) LEDs in *P. vietnamensis* cultures.

This chapter describes the effect of various types of LEDs (B—blue, G—green, Y—yellow, R—red, W—white, and R—in combination with B at different ratios) on callus formation and subsequent plantlet formation along with saponin accumulation in *P. vietnamensis* Ha *et* Grushv.—a high-value medicinal crop native to Vietnam.

---

D.T. Nhut (✉) · N.P. Huy · H.T. Tung · V.Q. Luan · N.B. Nam
Tay Nguyen Institute for Scientific Research, Vietnam Academy of Science and Technology, Dalat 670000, Vietnam
e-mail: duongtannhut@gmail.com



S. Dutta Gupta (ed.), *Light Emitting Diodes for Agriculture*,
DOI 10.1007/978-981-10-5807-3_14

## 14.2 Establishment of Callus Cultures and Growth of Callus

Leaves of 3-month-old in vitro plants were excised to 1 $cm^2$ and cultured on Murashige and Skoog's (1962) (MS) medium containing 1.0 mg/l 2, 4-dichlorophenoxyacetic acid (2.4-D) and 0.5 mg/l thidiazuron (TDZ) in the darkness for callus formation. Callus clusters (70 mg) were cultured on Schenk and Hildebrandt (SH) medium (1972) supplemented with 0.2 mg $l^{-1}$ TDZ, 1.0 mg $l^{-1}$ 2,4-D, 30 g $l^{-1}$ sucrose, and 9 g $l^{-1}$ agar for growth (Cuong et al. 2012). Sixteen lighting conditions [(1) compact fluorescent lamps (3U), (2) FL, (3) B, (4) G, (5) Y, (6) R, and (7) W-LEDs, and combinations of R and B-LEDs at different ratios—(8) 90:10, (9) 80:20, (10) 70:30, (11) 60:40, (12) 50:50, (13) 40:60, (14) 30:70, (15) 20:80, and (16) 10:90) were used with the light intensity of 20–25 μmol $m^{-2}$ $s^{-1}$ and darkness (D).

Jie et al. (2003) found that light intensity and spectral quality affect callus growth in *Cistanche deserticola* and biosynthesis of phenylethanoid glycosides. It is apparent from this study that yellow light (570–590 nm) was effective for callus growth of *P. vietnamensis*. The higher fresh and dry weights of callus were obtained with Y-light, compared with the cultures treated with FL and other light sources.

Soni and Swarnkar (1996) demonstrated that blue and yellow light-induced callus and shoot bud formation from leaf cultures of *Vigna aconitifolia*. Ouyang et al. (2003) also found that light intensity and the spectral quality influence the growth of callus culture in *C. deserticola* and the production of phenylethanoid glycosides. In this study, there were significant differences in callus formation and growth under various lighting conditions (Fig. 14.1). The highest fresh weight (1,197 mg) and dry weight (91.7 mg) of callus were scored under Y-LED light. This finding has important implications for improving callus growth using Y-, G-, and W-LED lights, and Y-LED light was identified to promote this process in *P. vietnamensis*. Following the treatment with Y-LED light, a considerable improvement in callus growth (Fig. 14.4a) was recorded when the callus clusters were maintained under 60:40 red to blue, compared to those cultured under fluorescent light. There was no significant difference in callus growth under 3U compact, FL, G- and W-LED lights, combinations of R- and B-LEDs with the ratios of 70:30 and 50:50, and the darkness. The proliferation of the callus was inhibited by R and B, and the combination of R and B at the ratios of 90:10, 80:20, 40:60, 30:70, 20:80, and 10:90. Under R-LED, results showed that callus fresh and dry weights were the lowest than those cultured under other lighting sources.

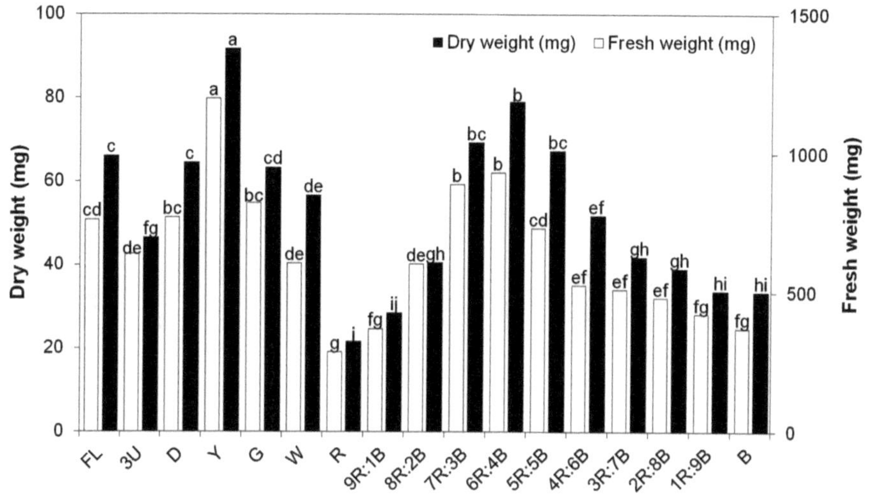


**Fig. 14.1** Influence of different lighting conditions on callus growth of *P. vietnamensis* after 12 weeks of culture

## 14.3 Development of Embryogenic Cultures

In order to develop embryogenic cultures, calli were transferred onto Murashige and Skoog (MS) medium (1962) containing 1 mg $l^{-1}$ 2,4-D, 0.2 mg $l^{-1}$ kinetin, 0.5 mg $l^{-1}$ α-naphthaleneacetic acid (NAA), 30 g $l^{-1}$ sucrose, and 8.5 g $l^{-1}$ agar.

Numerous studies have attempted to establish tissue culture of *Panax ginseng*. Among the sources of explants used, roots (Chang and Hsing 1980; Yang 1992), zygotic embryos (Lee et al. 1989; Arya et al. 1993; Zhong and Zhong 1992), cotyledons (Choi and Soh 1994; Jie et al. 2003), leaves (Cellárová et al. 1992), and flower buds (Shoyama et al. 1995) were found suitable for callus and embryo formation. In this work, in vitro leaf segments of *P. vietnamensis* were exploited to obtain calli, which were subsequently used as explants for the establishment of embryogenic cultures.

Synthetic auxins have been demonstrated to have an important role in regeneration of *Panax* species through somatic embryogenesis. Among the auxins evaluated, 2,4-D was demonstrated to be the best choice for callogenesis and somatic embryogenesis in *P. ginseng* (Chang and Hsing 1980; Zhong and Zhong 1992; Arya et al. 1993). Somatic embryogenesis could be further improved when kinetin was used in combinations with 2,4-D (Choi et al. 1984; Lee et al. 1989). Wang et al. (1999) investigated influences of auxins on somatic embryogenesis in *Panax quinquefolius* and demonstrated that 2,4-D in combinations with NAA gave a better result compared to sole 2,4-D. Another study on *P. vietnamensis* Ha *et* Grushv. found that somatic embryo formation could not be obtained when calli were cultured on media with sole 2,4-D or NAA (Nhut et al. 2009). In this study, calli derived from in vitro leaf segments were cut into segments (1.0 × 1.0 cm) and

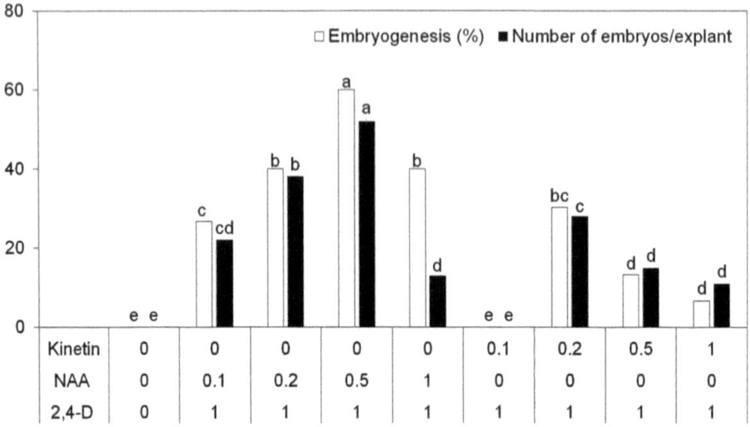


**Fig. 14.2** Effects of 2,4-D in combination with NAA or kinetin on somatic embryogenesis of *P. vietnamensis*

cultured on MS medium supplemented with 1.0 mg $l^{-1}$ 2,4-D in combinations with NAA and/or kinetin (0.1–1.0 mg $l^{-1}$). As shown in Figs. 14.2 and 14.3, 1.0 mg $l^{-1}$ 2,4-D plus 0.5 mg $l^{-1}$ NAA gave the highest rate of somatic embryogenesis.

Borkird et al. (1986) showed that low concentration of 2,4-D promotes cell division and dedifferentiation of plant tissues. In the present work, 2,4-D in combination with kinetin resulted in a lower percentage of somatic embryogenesis (Fig. 14.2). When adding 1.0 mg $l^{-1}$ 2,4-D together with NAA (0.2–1.0 mg $l^{-1}$), high frequency of somatic embryogenesis (40–60%) was recorded, and the highest number of embryos (52 embryos/explant) was obtained on MS medium containing 1.0 mg $l^{-1}$ 2,4-D and 0.5 mg $l^{-1}$ NAA. Zhou and Brown (2006) found similar results when studying the effects of auxins on somatic embryogenesis in *P. quinquefolius*. They demonstrated that 1.0 mg $l^{-1}$ 2,4-D in combination with 1.0 mg $l^{-1}$ NAA gave a higher percentage embryogenesis than media with sole 2,4-D. Wang et al. (1999) reported that optimal concentrations of 2,4-D and NAA were 1.1 mg $l^{-1}$ 2,4-D or 2.8 mg $l^{-1}$ NAA, respectively, for somatic embryogenesis in *P. quinquefolius*, and the effect of 2,4-D was independent of the concentrations of NAA when they were used together.

In this study, the effects of 2,4-D in combination with NAA and kinetin on somatic embryogenesis in *P. vietnamensis* were investigated. We used 1.0 mg $l^{-1}$ 2,4-D and 0.2 mg $l^{-1}$ kinetin in combinations with NAA at different concentrations. Among all the treatments, MS medium supplemented with 1.0 mg $l^{-1}$ 2,4-D, 0.2 mg $l^{-1}$ kinetin, and 0.5 mg $l^{-1}$ NAA showed to be the most effective for somatic embryogenesis (80%) with 117 embryos per explant (Figs. 14.3 and 14.4b, c).

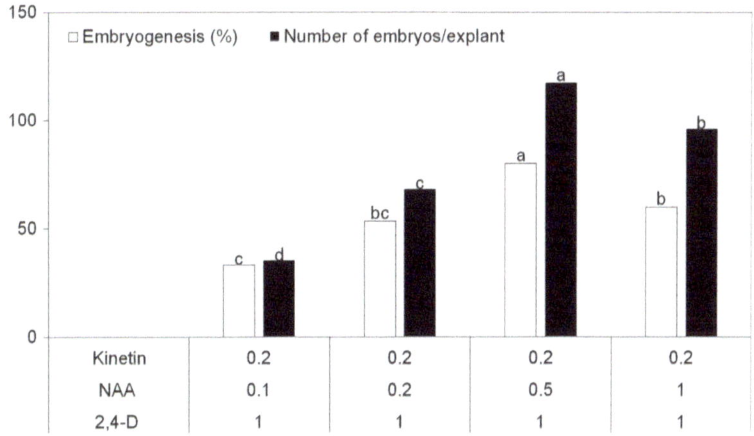


**Fig. 14.3** Effects of 2,4-D in combination with NAA and kinetin on somatic embryogenesis of *P. vietnamensis*

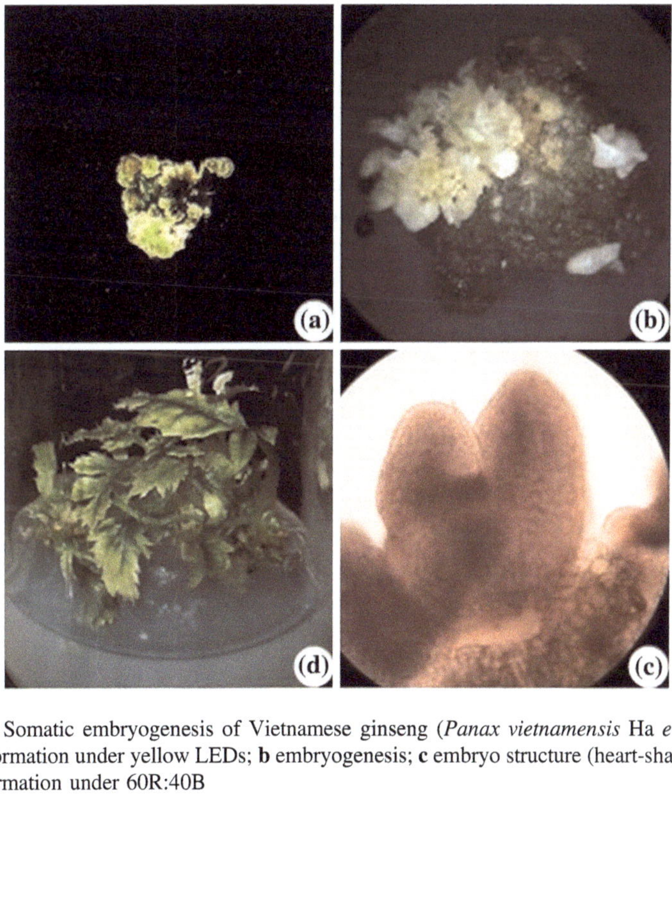


**Fig. 14.4** Somatic embryogenesis of Vietnamese ginseng (*Panax vietnamensis* Ha *et* Grushv.). **a** Callus formation under yellow LEDs; **b** embryogenesis; **c** embryo structure (heart-shaped stage); **d** plant formation under 60R:40B

## 14.4 Plantlet Development

Clusters of embryogenic callus (30 mg) were cultured on SH medium supplemented with 1 mg $l^{-1}$ BA, 0.5 mg $l^{-1}$ NAA, 30 g $l^{-1}$ sucrose, and 9 g $l^{-1}$ agar to attain plantlets (Cuong et al. 2012). SH medium supplemented with 0.5 mg $l^{-1}$ BA, 0.5 mg $l^{-1}$ NAA, 30 g $l^{-1}$ sucrose, and 9 g $l^{-1}$ agar (Nhut et al. 2010) was used to explore the further growth and development of plantlets (2 cm in height).

Different lighting conditions including compact fluorescent lamps, combinations of R- and B-LEDs with different ratios, and the darkness were used as the lighting sources.

Another interesting observation is that the type of lighting source affected the development of *P. vietnamensis* plantlets which were derived from embryogenic callus (Figs. 14.5 and 14.6). As shown in Figs. 14.5 and 14.6, 60% R-LED plus 40% B-LED resulted in the best plant development after 12 weeks of culture (fresh weight: 1,147 mg; dry weight: 127 mg; average plant height: 3.1 cm; and 11.21 plants per explant). The plant development under this lighting condition was much better when compared with that grown under fluorescent lamps. The fresh and dry weights of 505 and 49 mg, average plant height of 1.88 cm, and 5.83 plants per explant were obtained from plantlets cultured under FL (Fig. 14.4d). In addition, statistical analysis indicated that the mixtures of R- and B-LEDs with the ratios of 80:20, 70:30, and 50:50 were most effective for plant regeneration. On the other hand, no significant difference was found among explants maintained under other combinations of R- and B-LEDs (90:10, 40:60, 30:70, and 20:80). Dark condition was identified to be unsuitable for plant development from somatic embryos (Fig. 14.5). From these data, it is clear that 60% R-LED in combination with 40% B-LED was the optimal lighting condition for plant regeneration with highest values of fresh and dry weights (540 and 82 mg), average plant height (5.4 cm), leaf diameter (1.62 cm), and leaf length (2.90 cm).

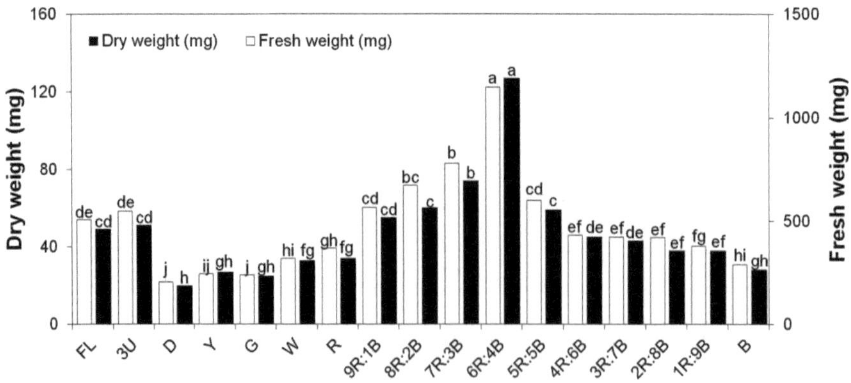


**Fig. 14.5** Influence of different lighting conditions on fresh and dry weights of regenerated plants of *P. vietnamensis* after 12 weeks of culture

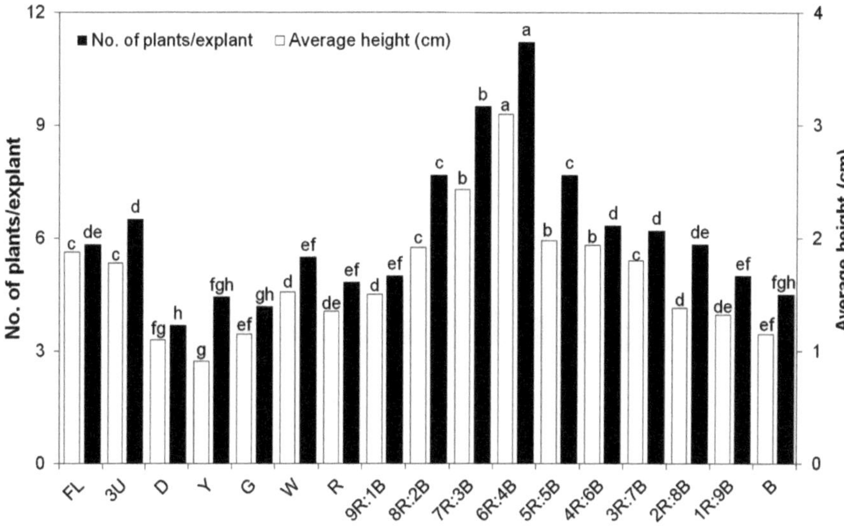


**Fig. 14.6** Influence of different lighting conditions on average height and number of plants per explant of *P. vietnamensis* after 12 weeks of culture

## 14.5 Influence of LEDs on Saponin Accumulation

Light is an essential factor in the biosynthesis of secondary metabolites. In 1973, Krewzaler and Hahlbrock demonstrated that light is a major factor in flavonoid glycoside synthesis in cell culture of *Petroselinum hortense*. Previous studies reported the influence of light on metabolite accumulation of *Perilla frutescens* and *Artemisia annua* (Zhong et al. 1991; Liu et al. 2002). Another study, which set out to determine the effect of light on the metabolic processes of ginseng (*P. ginseng* C.A. Mayer) adventitious roots, was carried out by Park et al. (2013). However, there has been little discussion about the biosynthesis of secondary metabolites in *P. vietnamensis* using different lighting systems. In the present study, LEDs were used to explore the correlation between lighting conditions and ginsenoside production. Thin-layer chromatography was able to detect the Rg1, Rb1, and MR2 bands in the plantlets cultured under different lighting conditions (Nhut et al. 2015). Moreover, bands of other ginsenosides detected in *P. vietnamensis* plants in the native habitat were also found in the in vitro samples. The results showed that there was no significant difference between the ginsenosides of in vitro *P. vietnamensis* plantlets compared to that of the native ones.

The influence of lighting conditions on saponin accumulation of in vitro *P. vietnamensis* plantlets is shown in Fig. 14.7. The highest content of Rg1 (0.41%) was recorded when plants maintained under FL, while the lowest one (0.23%) was scored under Y-LEDs. The highest content of MR2 (0.52%) was found under 20% R-LED combined with 80% B-LEDs, whereas the lowest one was observed under G-LEDs. Plants cultured under FL not only showed the highest Rg1 content but

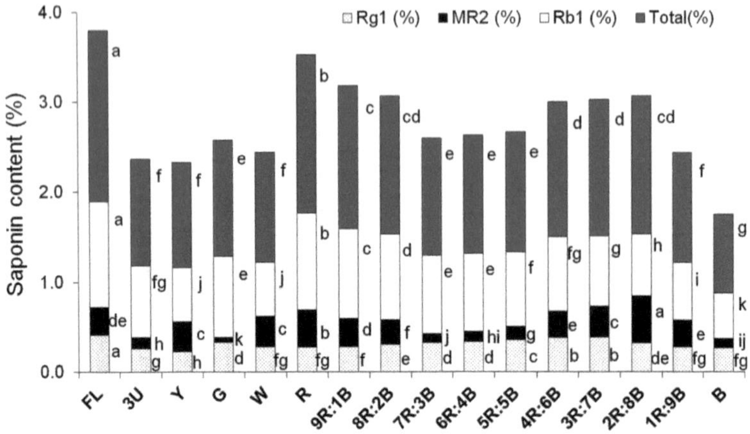


**Fig. 14.7** Influence of different lighting conditions on saponin content of *P. vietnamensis* after 12 weeks of culture

also Rb1 and total ginsenoside content (1.18%) compared with those cultured under other lighting sources (Fig. 14.7).

The finding from this research suggests that there is no correlation in saponin content and the growth and development of in vitro regenerated plantlets of *P. vietnamensis* cultured under different lighting conditions and those in the natural habitat.

## 14.6 Conclusion

It appears that appropriate light had regulatory role in callus growth, plantlet development, and saponin accumulation, and every developmental stage of *P. vietnamensis* in vitro requires specific lighting condition for callus growth and plantlet development. In commercial tissue culture laboratories, LED lighting system provides additional advantages including lower energy consumption, small size, durability, long operating lifetime, wavelength specificity, relatively cool emitting surfaces, and the user's ability to determine their spectral composition. This study suggests that the combination of embryogenic callus formation technique with proper lighting condition seems to favor micropropagation of *P. vietnamensis*.

## References


Arya S, Arya ID, Eriksson T (1993) Rapid multiplication of adventitious somatic embryos of *Panax ginseng*. Plant Cell Tiss Org Cult 34:157–162

Borkird C, Choi Jung H, Sung R (1986) Effect of 2,4-D on the expression of embryogenic program in carrot. Plant Physiol 81:1143–1146

Bula RJ, Morrow RC, Tibbitts TW, Ignatius RW, Martin TS, Barta DJ (1991) Light-emitting diodes as a radiation source of plants. Hortic Sci 26:203

Cellárová E, Rychlová M, Vranová E (1992) Histological characterization of in vitro regenerated structures of *Panax ginseng*. Plant Cell Tiss Org Cult 30:165–170

Cuong HV, Nam NB, Luan TC, Vinh BT, Nhut DT (2012) Effect of LED-light on growth and saponin accumulation of callus and plantlets of *Panax vietnamensis* Ha *et* Grushv. in vitro. J Sci Tech Vietnam 50(4):475–491

Chang WC, Hsing YI (1980) Plant regeneration through somatic embryogenesis in root-derived callus of ginseng (*Panax ginseng* C.A. Meyer). Theor Appl Genet 57:133–135

Choi KT, Yang DC, Kim NW, Ahn IO (1984) Redifferentiation from tissue culture and isolation of viable protoplasts in *Panax ginseng* C.A. Meyer. In: Proceedings of 4th International Ginseng Syrup, Korean Ginseng and Tobacco Res Inst, Korea, pp 1–11

Choi YE, Soh WY (1994) Origin of somatic embryo induced from cotyledons of zygotic embryos at various developmental stages of ginseng. J Plant Biol 37:365–370

Jie QY, Wang XD, Zhao B, Wang YC (2003) Light intensity and spectral quality influencing the callus growth of *Cistanche deserticola* and biosynthesis of phenylethanoid glycosides. Plant Sci 165:657–661

Lee HS, Yang SG, Jeon JH, Liu JR, Lee KW (1989) Plant regeneration through somatic embryogenesis from mature zygotic embryos of ginseng (*Panax ginseng* C.A. Meyer) and flowering of plantlets. Kor J Bot 32:145–150

Liu CZ, Guo C, Wang YC, Ouyang F (2002) Effect of light irradiation on hairy root growth and artemisinin biosynthesis of *Artemisia annua* L. Biochemistry 38:581–585

Dong NT, Luan TC, Huong NTT (2007) Ngoc Linh Ginseng and some medicinal plants belong to Ginseng family. Science and Technology Publishing House

Murashige T, Skoog F (1962) A revised medium for rapid growth and bioassay with tobacco tissue cultures. Physiol Plant 15:473–496

Nhut DT, Chien HX, Truc NB, Nam NB, Tinh TX, Luan VQ, Binh NV, Hien VT, Huong TT, Nhan NCT, Thuy LNM, Nga LTM, Hien TT, Hai NT (2010) Micropropagation of *Panax vietnamensis* Ha *et* Grushv. J Biotechnol Vietnam 8(3B):1211–1219

Nhut DT, Huy BN, Phong PT, Hai NT, Luan TC (2006) Primary study on multiplication of adventitious root of *Panax vietnamensis*-a valuable material source of saponin isolation. Biotech Agro Plant 118–121

Nhut DT, Luan VQ, Binh NV, Phong PT, Huy BN, Ha DTN, Tam PQ, Nam NB, Hien VT, Vinh BT, Hang LTM, Ngoc DTM, Thao LB, Luan TC (2009) Effect of some factors on in vitro biomass of Vietnamese ginseng (*Panax vietnamensis* Ha *et* Grushv.) and primary analysis of saponin content. J Biotechnol Viet 7(3):365–378

Nhut DT, Huy NP, Tai NT, Nam NB, Luan VQ, Hien VT, Tung HT, Vinh BT, Luan TC (2015) Light-emitting diodes and their potential in callus growth, plantlet development and saponin accumulation during somatic embryogenesis of *Panax vietnamensis* Ha *et* Grushv. Biotechnol Biotechnol Equip 29:299–308

Ouyang J, Wang X, Zhao B, Yuan X, Wang Y (2003) Effects of rare earth elements on the growth of *Cistanche deserticola* cells and the production of phenylethanoid glycosides. J Biotechnol 102(2):129–134

Park SY, Lee JG, Cho HS, Seong ES, Kim HY, Yu CY, Kim JK (2013) Metabolite profiling approach for assessing the effects of colored light-emitting diode lighting on the adventitious roots of ginseng (*Panax ginseng* C.A. Mayer). Plant Omics J 6:224–230

Schenk RU, Hildebrandt AC (1972) Medium and techniques for induction and growth of monocotyledonous and dicotyledonous plant cell cultures. Can J Bot 50:199–204

Shoyama Y, Matsushita H, Zhu XX, Kishira H (1995) Somatic embryogenesis in ginseng (*Panax* species). In: Bajaj YPS (ed) Biotechnology in agriculture and forestry. Somatic embryogenesis and synthetic seed II, vol 31. Springer, Berlin, pp 343–356

Soni J, Swarnkar PL (1996) Mophogenetics and biochemical variations under different spectral light in leaf cultures of *Vigna aconitifolia*. J Phytol Res 9(1):89–93

Wang X, Proctor JTA, KrishnaRaj S, Saxena PK, Sullivan JA (1999) Rapid somatic embryogenesis and plant regeneration in American ginseng: effects of auxins and explants. J Ginseng Res 23:148–163

Yang JS (1992) Plant regeneration from adventitious root-derived calli of ginseng (*Panax ginseng* C.A. Meyer). J Agric Assoc China 159:41–48

Zhong JJ, Seki T, Kinoshita SI, Yoshida T (1991) Effect of light irradiance on anthocyanin production by suspended culture of *Perilla frutescens*. Biotechnol Bioeng 38:653–658

Zhong SL, Zhong SG (1992) Morphological and uhrastructural characteristics of the embryogenic callus of American ginseng. Chin J Bot 4:92–98

Zhou S, Brown DCW (2006) High efficiency plant production of North American ginseng via somatic embryogenesis from cotyledon explants. Plant Cell Rep 25:166–173

# Index




S. Dutta Gupta (ed.), *Light Emitting Diodes for Agriculture*,
DOI 10.1007/978-981-10-5807-3

MIX
Papier aus verantwortungsvollen Quellen
Paper from responsible sources
FSC® C105338

If you have any concerns about our products,
you can contact us on
**ProductSafety@springernature.com**

In case Publisher is established outside the EU,
the EU authorized representative is:
**Springer Nature Customer Service Center GmbH**
**Europaplatz 3, 69115 Heidelberg, Germany**

Printed by Libri Plureos GmbH
in Hamburg, Germany